MODELING CROP PRODUCTION SYSTEMS
Principles and Application

Modeling Crop Production Systems

Principles and Application

PHOOL SINGH

Emeritus Professor of Plant Physiology
College of Basic Sciences and Humanities
CCS Haryana Agricultural University, Hisar
India

CRC Press
Taylor & Francis Group
Boca Raton   London   New York

CRC Press is an imprint of the
Taylor & Francis Group, an informa business

Science Publishers

Enfield (NH)   Jersey   Plymouth

SCIENCE PUBLISHERS
An Imprint of Edenbridge Ltd., British Isles.
Post Office Box 699
Enfield, New Hampshire 03784
United States of America

Website: *http://www.scipub.net*

sales@scipub.net (marketing department)
editor@scipub.net (editorial department)
info@scipub.net (for all other enquiries)

Library of Congress Cataloging-in-Publication Data

Singh, Phool.
Modeling crop production systems: principles and application / Phool Singh.
  p. cm.
Includes bibliographical references and index.
ISBN 1-57808-418-0
1. Food crops—Mathematical models. 2. Agricultural systems—Mathematical models. I. Title.

SB175.S56 2006
631.501'5118—dc22
                                                          2006050610

ISBN 978-1-57808-418-0

Published by Science Publishers, Inc., Enfield, NH, USA
An Imprint of Edenbridge Ltd.
Printed in India.

Dedicated to my professor
Dr. S.M. Welch

Preface

Lord Kelvin had once stated, 'When you cannot express it in numbers, your knowledge is a meagre and unsatisfactory kind ...'. In this new era of agricultural research and development, simulation partially substitutes for experiments. Simulation can be instrumental in determining recommendations for various agro-technology packages. Researchers and technologists are convinced with the degree of accuracy achieved in simulating crop and soil systems, thus enabling accurate prediction of the outcomes needed in agricultural decision making.

Until recently, emphasis was more on food and feed production by way of agronomic research and development. Now environmental concerns have added another dimension. Our aim must be to maintain adequate food production and minimize the level of environmental degradation. To achieve this objective, we need to include simulation of all or part of the soil-plant-atmosphere systems. Arriving at these goals through experimental research only will require huge resources. The use of simulation models is a necessity as also an aide to help with the decision making progress in sustainable agricultural systems. During these times of low funding for agricultural research and extension, experimental research is both time consuming and expensive. For the present, however, the evaluation of key simulated research must continue along with field experimentation.

This book is directed at undergraduate and post-graduate students in the disciplines of agronomy, plant breeding, agricultural meteorology, crop physiology, agriculture economics, entomology, plant pathology, soil science and ecology (environmental science). This book may also be useful for administrators in various agricultural universities in order to direct research, extension and teaching activities. Planners at national and state levels may also benefit from this book.

A number of texts exist that deal with the application of mathematics in the field of biology. A good book on modeling agricultural system should examine the issue in sufficient depth to enable the reader to understand the relation between an individual mathematical expression and its biological intuition. Readers are advised that the relation between a certain pair of variables obeys a particular equation. The reader has to decide why such an equation should be used, what physical and biological assumptions are implied by the use of the equation, and when the equation should not be used.

The reader should also try to answer the query that why do certain physical characteristics of a system lead to particular types of mathematical descriptions? For example, the exponential growth of an organism is because of the fact that the organism does not face any environmental resistance.

Crop production systems are among the most complex ones studied by modern agricultural science. Organizing the experimental knowledge of crop production system without the bookkeeping and deductive methods of mathematics is very difficult. Research literature of experimental agriculture is witness to the increasing use of mathematical methods. However, agricultural scientists and students often lack the training to make proper use of these methods, and agricultural scientists/students are often unable to grasp the biological relevance of mathematical models encountered in literature.

This present book does possess the minimal mathematical prerequisites necessary to familiarize oneself with the concepts of calculus, matrix algebra and statistics. Here, the intention is to teach the students the process by which the properties of the systems can be grasped in the framework of mathematical structure. The primary concern is not with the manipulation of equations, but the conceptual content of the relevant mathematical structure. A student whose formal mathematical background includes an elementary introduction to calculus, matrix algebra and statistics, will be able to comprehend the conceptual content. It is essential for the student/reader to have a sufficient grounding in those specific areas of crop production system to which the concepts are applied.

In this book, two type of examples have been given: (1) those which are not related to the agriculture; and (2) those which are related to the crop production system. The former relates to prototypes and the latter deals with the complexities.

It is advisable to the students/readers to select the exercises which are related to their field of specialization and apply the principles from their own field of biology.

Modeling facility can be taken care of the experience of developing one's own models and the critical study of models developed by others. If the reader or student takes the first step in modeling, i.e he or she defines the problem or system of his own interest, my objective is fulfilled. The first step will lead to his or her destination if he continues the efforts till the problem is solved.

The book is based on course 'Entom 891, Sys. Modl. Biol' (Modeling of Biological Systems, 3+1), taught to me by Professor S.M. Welch in 1980 at Kansas State University, Manhattan, Kansas, USA.

This book consists of six chapters, namely:

1. Philosophy, role and terminology of system science;
2. Development of model structure;
3. Specification of component behavior;
4. Computer implementation;
5. Model testing and validation; and
6. Biological application of models.

The author acknowledges the support of DST and USER Scheme, catalyzed and supported by the Department of Science and Technology under the utilization of scientific expertise of superannuated scientists' scheme to accomplishment of this project.

No man is an island into himself. This is particularly true in the field in crop modeling which weaves together skeins from many sources; each worker owes part of his intellectual development to many other peoples. I express my gratitude to E.S. Pine and Steinlitz-Hammacher Co. New Jersy; B.F.J. Manly and Chapman & Hall; P.C. Muchow & J.A. Bellamy and CAB International; J. Sack & J. Meadows and Science Research Associates, Inc.; R. Elmasri & S.B. Navathe and Addison Wesley; K.R. Baker, C.C. Carter & J.N. Sasser, and North Carolina State University Graphics; S.E. Jorgensen and Elsevier; C.H. Goulden and Asia Publishing House; A.E. Lewis and East West Press Pvt. Ltd; W.J. Conover and John Wiley & Sons; M.J. Kropff & H.H.V. Laar and CAB International; J. Hanks and J.T. Ritchie and ASA, Inc., CSSA, Inc., SSSA Inc; K.F.E. Watt and McGraw-Hill Book Co.; J.B. Dent and Blackie and Applied Science Publishers Ltd; R.W. Poole and Mc Graw-Hill Book Co.; W.L. Quirin and Harper & Row publishers; R.R. Sokal & F.J. Rohlf and W.H. Freeman & Co.; H.V. Keulen & J. Wolf and PUDOC; R.M. Peart & R.B. Curry and Marcel Dekker, Inc.; J. France & J.H.M. Thornley and Butterworths; C.R. Searle and John Wiley & Sons; J.D. Spain and Addison-Wesley Publishing Co.; and L.J. Fritschen & L.W. Gay and Springer-Verlag.

I am deeply thankful to Regent Professor E.T. Kanemasu, Professor, G.M. Paulsen and Professor S.M. Welch, who catalyzed and supported this project. I also express my gratitude to UNESCO for the Fellowship at K.S.U. Manhattan, without whose help, I could not write this book on modeling. I owe a great deal to Prof. Maharaj Singh, the then Director of Research, CCS HAU, Hissar. I express my gratitude to Dr D.P. Singh, the then Vice-Chancellor, JNKVV Jabalpur for the inspiration to complete this work.

I sincerely thank the distinguished Prof. J.W. Jones, Department of Agricultural and Bio-Engineering, University of Florida, USA, for his valuable comments to include the well thought out exercises and their solutions in this text.

I shall always be in debt to CCS HAU, Hisar, the Departments of Botany, Entomology, Plant Pathology, Soil Science, Agronomy Agril. Meteorology, Agril.

Economics and Directorate of Research. I also express my sincere thanks to ISRO, Bangalore; Centre for Advanced Studies in Botany, Madras University; NRSA, Hyderabad; IARI, New Delhi, SAC, Ahmedabad; CRRI, Cuttack for being instrumental for improving the quality of the book.

I am extremely thankful to Dharmesh for formating of the manuscript and Abhishek Saini for helping to review the literature.

I owe a continuing debt of gratitude to my family, particularly my wife, Uma, for her forebearance during the long gestation of this book.

In the last but not the least, I express my gratitude greater than ever to Valentina Chauhan, my daughter, for helping in correcting the proofs.

PHOOL SINGH, Ph.D.

Contents

4. COMPUTER IMPLEMENTATION — 205

1

Philosophy, Role and Terminology of System Science

1.1 HISTORY OF SYSTEM SCIENCE

In the post second world war scenario, system science remained confined to the design of electrical circuits in laboratories under the aegis of electrical engineering.

The concepts of potential and transfer apply directly to simple electrical circuits. Ohm's law relates the applied voltage, E, in a circuit to the product of the current, i and resistance R,

$$E = i\,R, \tag{1.1}$$

where E is the potential difference or the work done in moving one charge in the field of another charge. The unit of potential is volt, V, with one volt being equal to 1 joule per Coulomb. Generally the potential difference is expressed with respect to some reference level. Often, the level is called the ground, considered to be an infinite electrical sink.

The current represents the movement of electrons or charge. Since the charge of an electron is 0.1603×10^{-18} Coulomb, $A\,s$, current is thought of as the rate of movement of coulombs. Current is expressed in ampere, A, where one ampere is equal to one coulomb per second.

The resistance of a circuit is the restriction to the flow of current which is related to the electron field of the material used. Resistance here is expressed in Ohms, Ω. One Ohm is equal to one volt per ampere.

Power, P, is used to express the dissipation of energy. It is the product of volt and ampere, or joules per second. Other useful expressions are:

$$P = Ei = i^2\,R = \frac{E^2}{R} \tag{1.2}$$

Resistors may be connected either in parallel or in series, giving rise to parallel, series, and combination circuit. The series circuit is illustrated in Figure 1.1.

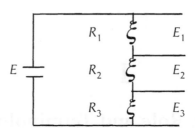

Fig. 1.1 The series circuit.

The same current flows through all of the resistors since they are connected in the series. However, the total voltage drop across the individual resistors is equal to the sum of the individual voltage drops.

$$E = E_1 + E_2 + E_3 \tag{1.3}$$

By substitution, $$E = iR_1 + iR_2 + iR_3 \tag{1.4}$$

or $$\frac{E}{i} = R_1 + R_2 + R_3 = \Sigma R \tag{1.5}$$

Assuming $R_1 = 20\ \Omega$, $R_2 = 50\ \Omega$, $R_3 = 30\ \Omega$, and $E = 5$ volts, we have

$$i = \frac{E}{\Sigma R} = \frac{5}{(20+50+30)} = 0.05\ \text{A} \tag{1.6}$$

$$E_1 = 0.05 \times 20 = 1\ \text{V} \tag{1.7}$$

$$E_2 = 0.05 \times 50 = 2.5\ \text{V} \tag{1.8}$$

$$E_3 = 0.05 \times 30 = 1.5\ \text{V} \tag{1.9}$$

In contrast to a series circuit, the voltage across all the legs of a parallel circuit is the same and the total current is the sum of the individual currents. Consider a parallel circuit in Figure 1.2.

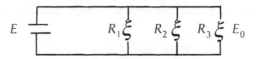

Fig. 1.2 The parallel circuit.

$$i = i_1 + i_2 + i_3 \tag{1.10}$$

or
$$i = \frac{E_1}{R_1} + \frac{E_2}{R_2} + \frac{E_3}{R_3} \qquad (1.11)$$

Since
$$E_1 = E_2 = E_3 = E$$

$$i = \frac{E}{R_p} = E\left(\frac{1}{R_1} + \frac{1}{R_2} + \frac{1}{R_3}\right) \qquad (1.12)$$

where R_p is the parallel resistance of the network. The equation for parallel resistance is

$$\frac{i}{R_p} = \frac{1}{R_1} + \frac{1}{R_2} + \frac{1}{R_3} \qquad (1.13)$$

When only two resistances are in parallel, the above equation becomes

$$R_p = \frac{R_1 \ R_2}{R_1 + R_2} = \frac{20 \times 50}{20 + 50} = 14.3 \ \Omega \qquad (1.14)$$

The parallel resistance in foregoing Figure 1.2 is

$$R_p = \frac{R_{p_1} \ R_3}{R_{p_1} + R_3} = \frac{14.3 \times 30}{14.3 + 30} = 9.7 \ \Omega \qquad (1.15)$$

The current flow is

$$i = \frac{E}{R_p} = \frac{5}{9.7} = 0.52 \ A \qquad (1.16)$$

and
$$i = \frac{E}{R_1} = \frac{5}{20} = 0.25 \ A$$

$$i_2 = \frac{5}{50} = 0.1 \ A$$

$$i_3 = \frac{5}{30} = 0.17 \ A$$

A combination series and parallel circuit, sometimes called a loaded voltage divider, is used to illustrate one of the most common errors in environmental measurements, 'the error of parallel resistors'. Consider a combination circuit

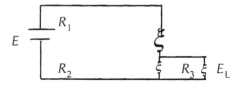

Fig. 1.3 A combination circuit.

If R_1 = 400 Ω, R_2 = 100 Ω, and R_3 is infinite, then unloaded output voltage E_0 is 1/5 of E or 10 mv if E = 50 mv. As the resistance of R_3 decreases, the parallel resistance of R_2 and R_3 decrease with a resulting decrease in the loaded output voltage E_L. The difference between E_L and E_0 can be thought of as an output error and represented as $100\,(E_L-E_0)/E_0$ for various values of R_3/R_2, which is shown in the following table indicating the error associated with a loaded voltage:

Table 1.1 The error associated with a loaded voltage divider

R_3 ohms	$\dfrac{R_3}{R_2}$	$\dfrac{E_L}{mv}$	$100\left(\dfrac{E_0 - E_L}{E_0}\right)$
1	0.010	0.123	98.765
10	0.100	1.111	88.889
100	1.000	5.556	44.444
1000	10.000	9.259	7.407
10000	100.000	9.921	0.794
100000	1000.000	9.952	0.080
1000000	10000.000	9.999	0.008

Source: Fritschen, L.J. and L.W. Gay, September 1979, Environmental Instrumentation. Springer-Verlag, New York, Heidelberg, Berlin, pp. 216.

The error decreases with a corresponding increase in the ratio R_3/R_2. The error is approximately 0.1% when R_3/R_2 is 1000 (Table 1.1). Either the impedance of a measuring device, e.g. R_3, must be at least 1000 times the resistance of the device being measured, or the resistance of a voltage divider must be atleast 1000 times the resistance of the transducer.

A system (circuit network) might be having many subsystems (components, resistors). A component has its own input and output. The output of the system will be decided by the inputs and outputs of individual components. The efficiency of the system will be controlled by (1) input, (2) output, and (3) component.

Thus,

(1) represent the system as a network of components;

(2) describes the relationship between the input and output of each component;

(3) on this basis, it predicts the system behavior; and

(4) incorporates model results into a larger system.

The second discipline, aided by system science, is economical in cost and reliability analyses.

Since the seventeenth century, biology has been using the principles of systems science to study the population growth vis-a-vis birth and death rates. In the beginning of the nineteenth century, demographers used the concepts of system science to predict the future population of human beings.

The prediction of the spread of malaria has been made since the beginning of twentieth century with the help of the following equations:

$$\frac{d\,Ih}{dt} = \frac{Ih}{t} - \frac{Rh}{t} \tag{1.17}$$

where $\dfrac{d\,Ih}{dt}$ is the rate of increase of infected human, $\dfrac{Ih}{t}$ is the number of new infection per unit time; and Rh is the number of recovery per unit time.

Similarly,

$$\frac{d\,Im}{dt} = \frac{Im}{t} - \frac{Dm}{t} \tag{1.18}$$

where $\dfrac{d\,Im}{dt}$ is the rate of increase of infected mosquitoes; $\dfrac{Im}{t}$ is the number of infection per unit time; and $\dfrac{Dm}{t}$ is the number of death of infected mosquitoes.

The discipline of crop science did not lag behind in using the principles of systems science. In 1971, the first crop model came into existence on cotton crop with the emphasis on interfacing photosynthesis with other plant physiological processes.

International Benchmark Sites Network for Agrotechnology Transfer (IBSNAT) project did a commendable job from September 1, 1982 to

August 31, 1993, on modeling on varieties of cereals: wheat, maize, barley, sorghum, millet and rice; grain legumes: soybean, peanut and dry bean; root crops: cassava, aroid and potato; and other crops: sunflower, sugarcane, pineapple, cotton (IBSNAT Decade, 1993).

After the completion of the IBSNAT project, the International Consortium for Agricultural Systems Applications (ICASA) came into existence. ICASA is an consortium of individuals and organizations involved in or interested in systems research and applications. A multidisciplinary effort is anticipated and expected in continuation of the IBSNAT project. (International Benchmark Sites Network for Agrotechnology Transfer, November, 1993, The IBSNAT decade, IBSNAT, Dept. of Agronomy and Soil Science, College of Tropical Agriculture and Human Resources, University of Hawaii, Honolulu, Hawaii).

Sirotenko (2001) presented a brief history of crop modeling activities in the former USSR. The author's view on the problems and perspective of further development is delineated here. The history of crop-modeling in the former Soviet Union is not tedious and monotonous. It started with the almost simultaneous development of two competitive versions (radiation and carbon dioxide) of the quantitative theory of plant canopy photosynthesis. The first approach had been developed by Budagovsky et al., (1964) and the second one was presented by Budyko (1964) and Budyko and Gandin (1964). While young scientists were delighted with these works, the leading specialists rejected them as a mere mathematical game. Still, an informal society called Weather-Yield Mathematics (WYM) was established in 1968 to develop these new ideas under the leadership of Professor Juhan Ross. In the following three decades, extensive and rather effective activities on developing crop simulation models in the former USSR have been carried out within the framework of WYM, resulting in more than 20 monographs and some 100 papers being published.

Due to the economic crisis generated by the disintegration of the USSR, the optimism of the participants of WYM society has changed to disappointment associated with the lack of progress in the application of mathematical models and computer in Agronomy. Activities in this field practically stopped for some years and only recently, there have been some signs of reanimation. What are the conclusions that can be drawn from an analysis of the experience gained in crop modeling in the former USSR?

Sinclair and Saligman (1996) provided the historical aspects of the crop modeling in describing the entire period of crop modeling into four phases: (1) infancy, (2) juvenility, (3) adolescence, and (4) maturity.

1.1.1 Infancy

Infancy refers to the period following World War II (1963-70), when system analysis and computer science provided convenient and relatively friendly techniques to emulate the interaction of components in a complex system stimulated by the Cold War and space exploration. The earliest models were developed to estimate light interception and photosynthesis in crop canopies (Loomis and Williams, 1963; de Wit, 1965; Duncan et al., 1967; Sinclair and Seligman, 1996). These models calculated the light profile in a canopy and made it possible to assess the sensitivity of crop photosynthetic rates to solar angles, leaf angle distribution, and the latitudinal position of the crop. These models were relatively simple, but they opened the way to quantitative, mechanistic estimates of the maximum attainable growth rates. Crop growth and potential yield became quantitatively and demonstrably linked via biochemical and biophysical mechanisms to the amount of solar energy available for the accumulation of chemical energy and biomass by plants.

1.1.2 Juvenile Phase

In the juvenile phase, a further stimulus to crop modeling followed by tremendous advances in equipments for field experimentation provided entirely new sets of data to use in models. These advances in equipments included photocells to measure canopy light level, improved anemometers to monitor the wind speed in and above the crop, and data loggers with magnetic data storage. The new experimental data encouraged a physico-chemical view of crop growth based on a detailed description of the crop microclimate and the response of the plants to this environment (Lemon et al., 1971; as quoted in Sinclair and Seligman, 1996).

Such models offered the promise of experimentation in the evaluation of improved genetic material and new management techniques in the context of a wide range of cropping environments (Bowen et al., 1973; as quoted in Sinclair and Seligman, 1996).

Important advances in describing the various subcomponents of carbon assimilation were made during this period and the significance of stomatal conductance in regulating leaf gas exchange was quantitatively described (Cowan, 1977; as quoted in Sinclair and Seligman, 1996). The fate of photoassimilates in respiratory pathways was carefully analyzed (Penning de Vries, 1975; as quoted in Sinclair and Seligman, 1996).

Developmental processes of plants became an important consideration as the time frame of models was lengthened to include the entire growing season. Expressions for the partitioning of the assimilate among various tissues, particularly to the grain, were important. The addition of these various components led to a number of models of complexity such as GOSSYM (Whisler et al., 1986; as quoted in Sinclair and Seligman, 1996), CERES (Ritchie et al., 1985; as quoted in Sinclair and Seligman, 1996), and SOYGRO (Wilkerson et al., 1985; as quoted in Sinclair and Saligman, 1996).

1.1.3 Adolescence

The processes that determine the manner in which materials are partitioned within the plant are not well understood. In order to describe these processes within the plant, model builders have defined hypothetical pools of compounds that responded to supply and demand. Such reductionism, when inappropriately applied, can be misleading. When a high level of plant organization is being modeled, its use may well give a more distorted representation of organ growth than the use of conservative allometric relationships (Sinclair and Seligman, 1996). A simple water balance model was found to be superior to COTTAM and GOSSYM in approximating crop water stress and field water balance (Asare et al., 1992; as quoted in Sinclair and Seligman, 1996). An empirical equation was found to be superior to CERES in predicting the annual potential wheat yields in Maxico (Bell and Fischer, 1994; as quoted in Sinclair and Seligman, 1996). In simulating the water runoff from various agricultural watersheds, Loague and Freeze (1985; as quoted in Sinclair and Seligman, 1996) found a regression model superior to a quasi-physically-based model.

Increasing reductionism in models did not result in less variability in predictions among the complex models (Sinclair and Seligman, 1996). The practical consequence is that it is impossible to create universal crop model (Spitters, 1990; as quoted in Sinclair and Seligman, 1996). Not surprisingly, it has been found that each new season or a new location brings new challenges that were not foreseen in the original model, and the expectation of universality fails (Sinclair and Seligman, 1996). Attempts made to use the existing crop models developed for higher latitudes failed an experiment to simulate crops in the semiarid tropics of Australia (Carberry and Abrecht, 1991; as quoted in Sinclair and Seligman, 1996). Important deficiencies were found in each of three complex wheat models even after they had been calibrated for a new set of conditions in New Zealand (Porter et al., 1993, as quoted in Sinclair and Seligman, 1996).

This is a belief that crop model must be verified or validated. All models are basically a collection of hypotheses, so they inherently cannot be validated (Pease and Bull, 1992; Oreskes et al., 1994; as quoted in Sinclair and Seligman, 1996). Not only can other collections of hypotheses approximate the experimental results equally well, but the validation data themselves are flawed by substantial experimental and observational error (Sinclair and Seligman, 1996).

Despite all these shortcomings, crop models can be used effectively to study the possible implications of various assumptions about a crop or an environment. This viewpoint has been explicitly adopted by the American Society of Agronomy towards publication of agronomic models in the Agronomy Journal. Modeling papers are desired that 'deal with both concepts and integration of agronomic information into model' and model 'Validation' is not to be considered as a major factor in the acceptance of a paper (Hatfield, 1993, as quoted in Sinclair and Seligman, 1996).

Overall, three of the original basic ideas of crop models have been discredited: models are not necessarily improved by extensive reduc-tionism; universal crop models can not be constructed; and models cannot be validated. All of these changes have necessitated a new perspective on the construction and benefits of crop models (Sinclair and Seligman, 1996).

1.1.4 Maturity

One can visualize crop models in teaching, research and applied models as powerful aids where in reasoning about the performance of a crop or about the relative benefits of alternative management strategies. The crop models allow us to set our knowledge and assumptions about the behavior of a crop in an organized, logical and dynamic framework. After studying or using the models, faulty assumptions can be usually identified and a more structural insight to the importance of specific feedback effects acquired (Sinclair and Seligman, 1996).

The heuristic benefit of crop models in teaching is clear. Crop models were introduced into the classroom more than 20 years ago, with continual upgradation (e.g. Waldren, 1984; Hart and Hanson, 1990; Wullschleger et al., 1992; as quoted in Sinclair and Seligman, 1996). Crop modeling exercises are perceived by students as an effective tool for teaching factors that influence crop production (Meisner et al., 1991; as quoted in Sinclair and Seligman, 1996). Relatively simple, transparent models allow students to explore the major factors that influence crop production under various circumstances. Learning is likely to be facili-tated by using a model that is simple and transparent enough in its structure to allow students to dissect it and to understand the logic underlying behavior (Sinclair and Seligman, 1996).

Research on crop systems or subsystems can use models to organize concepts and information that reflect the current understanding of the system as also to determine their adequacy in explaining relevant phenomena. Shortcomings of the model can highlight important but poorly understood aspects of the crop. The model needs to be structured in such a manner that variables become physically or physiologically meaningful and can be investigated either experimentally or by observing the system behavior. Crop models can then prove to be quite useful in analyzing experimental results by virtue of their ability to substantiate possible causes of differences, thus providing a level of interpretation beyond the bounds of statistical significance that currently guide the analyses of crop experiments (Sinclair and Seligman, 1996).

Even the use of crop models in farm management has succeeded more in an heuristic role rather than as an on-line decision aid. Examples can be given of the SIRATAC model for cotton pest management in Australia (Ives and Hearn, 1987; as quoted in Sinclair and Seligman, 1996) and the EPIPRE model for wheat pest management in the Netherlands (Rabbinge and Rijsdijk, 1983, as quoted in Sinclair and Seligman, 1996). Each model requires the growers to pay for membership and supply field observations to a central processing centre. At the central processing centre, model simulations provided the growers with updated pest management recommendations. In each case, there was an initial steady increase in grower membership, which resulted in a general improvement in pest management. However, both systems suffered a loss of membership after the initial successes. The decline in participation has been ascribed not to dissatisfaction with the model results but to the fact that the growers believed they had learned the lessons of the models and could now manage on their own (Weiss, 1994, as quoted in Sinclair and Seligman, 1996). The models were a success in the sense that they taught the growers improved pest management by helping them interpret their own field observations more effectively (Sinclair and Seligman, 1996).

While crop models cannot produce all the answers to crop production problems, when reasonably constructed, they can prove to be important heuristic tools in teaching, research and in management and administrative applications. They can also be used to produce hypotheses and knowledge, thereby allowing the user to reason more consistently and transparently about factors or conditions that deserve thought by students, additional experimental study by researchers, or more attention from growers. Intelligent, consistent, transparent reasoning—as well as observations, experimentations and experience—cannot be replaced by crop models, but they can be well supported by them. Because of the large number of situations when the heuristics functions of crop models can prove to be crucial, if not an indispensible tool, it is believed that crop modeling can be expected to have a long and production maturity (Sinclair and Seligman, 1996).

1.2. GENERAL TOPOLOGY AND TERMINOLOGY OF SYSTEMS

1.2.1 Variable

As is obvious, a variable will vary with times. A useful thumb rule is to takc the variables to be the inputs and outputs of the individual components. At the same time, when we define the variables, it is convenient to assign them symbols so that we may refer to them later without unduly overworking ourselves.

At the time, when the variables are listed, it is usually desirable to specify their dimensions. The dimensions of a variable essentially reveal what kind of physical entity the variable represents.

The overall input along with the overall output represent the interaction of the system with its environment. ln the investigation of a natural system, the overall input includes the experimental conditions and external stimuli applied to the system. The overall output includes all the observable responses. An investigator often modifies the system so as to obtain an output not normally yielded by the system or to render it abnormally sensitive to a particular type of input. When this is done, the investigator must carefully consider the relationship between the original system he or she wishes to learn about and the modified system under study.

Even though most of the discussion is in terms of the rate of change with respect to time, the same consideration applies in discussing the rate of change with respect to any other variables taken as an independent variable. Time is itself an independent variable; so is space may. Mind, matter, energy may be classified as dependent variables.

Certain measurements increase by vanishingly small amounts, the smallness being limited by one's ability to discriminate between correspondingly fine differences. For example, if we were to take 1000 people, all of them weighing between 150 and 151 pounds, we could, if the scales were sufficiently sensitive, arrange them in order of increasing weight. Obviously, in order to do this, we would need a scale that can discriminate differences smaller than 1/1000 of a pound. Theoretically, with an unlimited population to draw from, we could take 1000 people weighing people between 150.002 and 152.003 pounds and also arrange them in order of weight if the scales were sensitive to less than 1/1000000 of a pound. Needless to say, in real life, there would never be any occasion to carry these measurements to such hair-splitting accuracy. Nevertheless, a mean value of, say, 150.01 pounds has a conceptual reality. Such variables are known as continuous variables.

On the other hand, if a statistician states that the average family has 2.3 children, we hinder at the image of three tenths of a child. We do not for a moment deny the utility of this mean for certain economic uses, but we immediately realize that another class of values is involved. Such measures are called discontinous variables. They can be arrived at by counting rather than by measuring against a scale of some kind. Generally, we enumerate progeny with distinct characteristics; in studying epidemics, we count cases; in bacteriology, we count organisms. In all of these examples the units are indivisible, the count moves up, discontinues and stops instead of rolling up a continuous slope.

1.2.2 Parameter

The distinction between parameters and overall inputs is not always clear. Practically, quantities that display a great deal of variability are usually treated as inputs, whereas those that are more nearly constant are treated as parameters. So, for example, if we have a growing bacterial population, the rate of growth as determined by temperature, nutrient supply, acidity, and so on. The 'rate of growth', which is often treated as a parameter, includes the dependence of the bacterial system upon its environment.

A distinction may be made between inputs, outputs, parameters, and state variables on the basis of their mutual dependences, as accounted for by the equations of the models. Inputs and parameters depend upon none of the other quantities. Outputs depend upon inputs, parameters, and state variables, while state variables depend upon inputs, parameters, and each other.

The overall output depends on the instantaneous values of the variables, that are to be taken as defining the instantaneous condition or the state of the system and may be referred as state variables. If we know the form of equations, the value of the input, and the values of the parameters, then the model tells us how the state variables change.

Parameters may change with the age. For example, the permeability of the cell membrane may change with time. The parameter value may also depend upon the time of the day. For example, in bee foraging, the system parameters may depend upon time of the day.

The Gaussian or normal curve is described by the equation

$$Y = \frac{1}{\sqrt[q]{2\pi}} \ e^{-(x-\mu)^2/2\sigma^2}$$

For non-mathematics students, this equation will certainly appear incomprehensible, but it is usually not used directly in biological statistics. These are a few significant features of this equation, which can be appreciated by even those who are relatively not comfortable with figures. In this equation, Y represents the relative frequency of some variable quantity, *x*. The values for π and e are constant. π is the familiar ratio of the circumference to the diameter of any circle, 3.1416; and e is the base of the Neparian or natural logarithms. For those new to such terminology, e may be taken as a constant in the same way that the value of π is accepted.

The important features to be noted are the two parameters (i.e., numerical characteristics) μ and σ, where μ is the arithmatic mean, and, σ the standard deviation. The standard deviation is a measure of the spread of the data about the mean. Since the value for π and e are fixed forever and constant, the entire curve is completely defined or characterized by the two parameters, the mean and standard deviation. Thus, parameters are the fixed values varying from one population to another. These are not fixed forever and, hence, are not constant.

In systems science, the topic of discussion might be the form of the relationship between the variables and the manner in which this form relates to behavior of the system and the hypothesized mechanisms responsible for that particular behavior. These relationships involve parameters that can often be related to the expected frequency of occurrence of certain underlying events.

For example, if the hypothesized equation is

$$y = K;$$

where K is a parameter.

A single data point is enough to determine the value of K. Usually, K would be estimated as the mean of at least two observations. Furthermore, if the difference between these two observations is more than what could reasonably be ascribed to expected errors, additional observations would have to be made in order to test the hypothesis that *y* is constant.

If the hypothesized relation is the straight line.

$$y = a + bx$$

the values of *a* and *b* can be determined from any two data points. If the true relation is not a straight line, it cannot be determined from just two points. Furthermore, even if the true relation is a straight line but one (or both) of the points is appreciably in error, we get the wrong straight line i.e. an erronous value for *a* and *b*.

1.2.3 System

A system may be any entity to which something comes in as exogenous input and something goes out as output of the system. This output may again influence the exogenous input, an aspect which is completely ignored. System includes a component, state variables and system parameters.

The levels of the biological system are:

(1)   Molecule
(2)   Organell
(3)   Cell
(4)   Tissue
(5)   Organ
(6)   Individual
(7)   Population
(8)   Community
(9)   Ecosystem
(10) Biosphere

Over the years, the concepts of system analysis have gradually emerged into an accepted body of theory. Initially, system analysis was conceived as an integrating framework wherein complex systems–possibly involving several disciplines–could be studied (Boulding, K.E. "General System Theory—The Skeleton Science', Man. Sci. Vol. 2, pp. 197-208, 1956, as quoted in Dent and Anderson, 1971). This inter disciplinary function is even now of prime importance; major industrial, commercial or military projects cannot be successfully handled within the confines of a single discipline. The system view is a holistic one, which implies that an isolated study of parts of the system will not be adequate to understand the complete system. This is because the separate parts are linked in an interacting manner. A system implies a complexity of factors that interact; it implies an interaction between these factors and also that a conceptual boundary may be erected around the complex as a limit to its organizational autonomy (Dent, J.B. and J.R. Anderson, 1971). Fridgen et al., (2004) defined the concept of identification and management of a region within the geographic area confined by field boundaries.

1.2.4 Dynamic Process/Model/System

Dynamic systems are systems or processes whose state (state variables of the system defined on the basis of inputs, parameters, and each other) is constantly changing with time. Even though most of the discussion is in terms of the rate of change with respect to time, the same considerations apply in discussing the rates of change with respect to any other variable taken as independent. If the state of the system is specified by the values of n state variables, it is convenient to represent it as a point in n-dimensional space, which is termed the state space for the system. As the system changes in time, so does the position of the point that represents the system. The change of the system in time is, therefore, represented by the motion of a point in the n-dimensional state space. As the point moves, it traces out a path in the n-dimensional state space, which is referred to as the trajectory of the system. Thus, the system and its progress in time can be represented either by a path (trajectory) in state space or by an equation that describes the trajectory.

If the state is one- or two-dimensional, then one can draw the trajectory and visually exhibit the path in state space along which the system 'moves'. However, such a drawing does not display how fast the system moves along this path. It is like the map of a road on which an automobile is travelling. The map shows how the automobile may travel, but not how fast it moves. If we wish to show the rate of motion of a system in its state space (or of the automobile on the road), we must plot the position of the system (or of the automobile) versus time. Time then becomes $(n+1)^{st}$ dimension.

Unfortunately, the ability to portray a curve in a n-dimensional space is lost when n is greater than three. The portrayal of a curve in three dimensions is inconvenient, but may be done either by means of a projective drawing or by the physical construction of three-dimensional curves. In five or more dimensions, we have only the mathematical description, our power of imagination, and the possibility of constructing two-dimensional projections as aids to the imagination.

1.2.5 Continuous versus Discrete State Spaces

The term 'continuous state space' implies that the state variables are continuous, at least within the regions of interest. A number of state variables of interest in biology, including agriculture, are not continuous. For example, the number of individuals in a population (whether of plant, animals, microorganisms, or molecules) can only change by integer amounts. In such a case, as the point that represents the system moves through the state space, it moves in jumps rather than smoothly and continuously. A differential equation to describe the direction and rate of motion is not available unless the variables are continuous. However, because the formalism is so convenient, discrete spaces are often treated as continuous. The validity of such an approximation requires that the sizes of the discrete jumps be small in relation to the ability with which we can or wish to measure changes in the system. To put it another way, if the scale for which we observe the system is relative to the scale on which the individual jumps occur, then the path of the system through states space may appear to be continuous—just as a curve drawn on a television screen appears to be continuous, even though it is composed of many individual dots of light.

1.2.6 Stochastic versus Deterministic Descriptions

In a deterministic description, the behavior of the system is completely determined by its state and by the specified conditions. As a result, a deterministic description of a dynamic system and its evolution through time usually gives a description of a particular trajectory in a state space. On the other hand, in a stochastic description, we have—for each state that the system can be in—a distribution of probabilities on a set of possible behaviors, i.e. on the set of possible directions and rates of travel in the state space. The connection between deterministic and stochastic descriptions is made on the basis of the expected or average behavior. Although any real system must be considered to be subject to a variety of uncertainties when the relative uncertainty is small compared with the need for accuracy, one may take advantage of the comparative mathematical simplicity of the deterministic models.

In developing an explanatory model for a dynamic system, the purpose is to understand how the general laws that govern the behavior of the system arise from the laws that govern the constituent elementary event. The view point adopted is that these laws must always have a certain degree of random or uncertain characters. This randomness has to be taken into account unless the coefficient of variation is small, which usually means that the density of events is very large (Gold, 1977).

1.2.6.1 *Stochastic Models of Exponential Growth*

The deterministic equation of exponential population growth

$$N_t = N_0 e^{rt} \qquad (1.19)$$

predicts that a population with a stable age distribution of an unlimited environment will increase in the shape of a smooth, exponential curve.

As pointed out by Poole (1974), although the superiority of continuous time models to their discrete approximations is clear, unfortunately, in most cases they are extremely difficult to derive. Discrete time approximation is slightly less difficult. Sometimes it is possible to derive a discrete time model and, by using the methods discussed below, simulate the process for several time intervals. It is to be noted, however, that in the discrete time model, the time variable t is assumed to advance jumps of 1, although t is in reality a continuous variable. If, as later happens, the birth rate or death rate is postulated to be a function of time, it will also be a continuous variable. The discrete time approximation in these cases must assume the birth rate or death rate to be constant during a single interval of time, changing by a single jump from one time interval to the next. The situation is analogous to the discrete approximation of the percentage of the cohort alive during the mid period of time interval and the fecundity of the female.

The advantage of the stochastic model over the equivalent deterministic model is its greater reality. The greater complexity of the stochastic model, particularly in its derivation, tends in some cases to outweigh their advantages. This is particularly true if the variance in the number of individuals is small. There are cases, however, in which chance deviations can push the result of a process either one way or the other. Stochastic models have an advantage of being more realistic than deterministic models, predicting only what can happen and the probability of its happening, not what will happen.

The methods employed to arrive at an answer using a discrete time stochastic approximation to a continuous time situation are roughly the same as playing a game involving chance. The expected number is calculated and modified by a factor representing a random sample from the probability distribution to which the answer belongs.

The equations

$$E\ (N_{t+1}) = e^r\ N_t$$

$$\text{var}\ (N_{t+1}) = \frac{b+d}{b-d}\ (e^{b-d} - 1)\ (e^{b-d})N_t$$

denote the respected value of the population size at $t + 1$, plus the variance of the expected value. If the effect of chance variation is not taken into consideration, the increase in density of a population begin with 20 individuals with the parameters $b = 0.52$ and $d = 0.48$ as shown in column 1 of Table 1.2.

Table 1.2   A deterministic and two stochastic simulations of population growth in an unlimited environment
($b = 0.52$, $d = 0.48$, $N_0 = 20$)

T	Deterministic	Stochastic No. 1	Stochastic No. 2
0	20.000	20.000	20.000
1	20.816	14.641	27.129
2	21.665	18.352	32.422
3	22.549	20.381	22.832
4	23.469	21.678	18.889
5	24.426	18.159	17.287
6	25.423	25.091	10.795
7	26.460	28.838	7.478
8	27.540	23.823	10.319
9	28.663	15.339	10.012
10	29.833	17.861	15.277

The calculations in column 1 are strictly deterministic. To make the result stochastic, the following steps are taken:

1. Calculate the expected number at time t +1.
2. Calculate the variance of the estimate.
3. Calculate the standard deviation, i.e. the square root of the variance.
4. Pick a number at random from a table of random normal deviates and multiply it by the standard deviations.
5. Add the answer from step 4 to the answer from step 1. This is a possible size of the population at time t + 1.

This procedure assumes that the distribution of a possible answer is normally distributed. In this case, the distribution of the possible answer is approximately normal except at low population densities. It should be emphasized, however, that the distribution is not necessarily always normal, e.g. it might have an exponential distribution. An example with the parameters above is:

1. $E (N_{t+1}) = 20.816$
2. var $(N_{t+1}) = 21.232$
3. Standard deviations = 4.608
4. A random normal deviates, −1.38, times the standard deviation, 4.608 = − 6.359
5. 20.816 − 6.359 = 14.457

A table of random normal deviates in Beyer (1968, as quoted in Poole, 1974) was used. A random normal deviate is a number drawn at random from the standardized normal distribution. Unlike random numbers, where each number is equally likely to occur, random normal deviates are normally distributed. Even though the birth rate is greater than the death rate—although they are nearly equal—a chance negative deviation has caused the population to decrease rather than increase as expected.

The course of stochastic population growth can be simulated by repeatedly carrying out the above calculations. For example, this stochastic estimate of the population after the second interval of time can be arrived at by taking 14.641 as the population density at time t, recomputing the standard deviation, selecting another random normal deviate, and so on. A stochastic representation of population growth for 10 intervals of time is plotted against with the deterministic result in Figure 1.4.

Because deviations are random, if this procedure were repeated then the exact path of the growth of the population would not be the same as it was the first time. The calculation of a second set of 10 intervals of time is shown in column 3 of Table 1.2.

If long periods of time are involved in the simulation, as they often are, or if large numbers of replications of the simulation are needed to estimate the variance of the population at sometime t, the calculations become tedious. The calculations are easily programd for a computer and a subroutine can be used to generate random normal deviates.

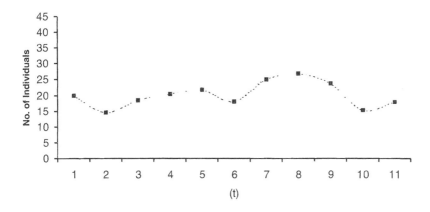

Fig. 1.4 Population growth in an artificial population with parameters b = .52, d = .48 and N_0 = 20. Solid line, the deterministic model; dashed line, a Monte Carlo simulation of the discrete time stochastic model of exponential population growth.

1.2.7 Modeling

Modeling is a creation of some representation of the important aspects of some system of interest. Modeling is an organized activity.

1.2.8 Model

Gold (1977) defined the word 'model'. 'Model' is used with more or less the same meaning as its every day meaning. A certain object (call it object M) is a model of another object (object S), provided the following conditions hold:

1. There is some collection of components of M, each of which corresponds to a component of S;
2. For certain relationships, the relation between the component of M is analogous to that between the corresponding components of S.

So, for example, the features of a marionette are intended to correspond to that of human being that it models, and certain relations between the features are the same in both. Similarly, the architect's blueprint is a model for a finished building.

A mathematical equation simply consists of a collection of symbols, some of which stand for variables (quantities with number and dimension) and some of which stand for operations on these variables (such as addition, multiplication, and differentiation). If the variables can be associated with physical entities for a given real world system and if the relation between the variables in the mathematical expression is analogous to the relation between the corresponding physical entities, then we may say that the mathematical expression is a model for the real world system.

Conditions (1) and (2) define the word 'model' in the sense that anything that has these two properties qualifies as a model and anything that lacks one or the other of them does not. Notice, however, that this definition does not require the model M to be an exact duplicate of object S. That is, condition (1) does not require every component of M to correspond to a component of S or vice versa. Nor does condition (2) require that every relationship between components of S be mirrored by an analogous relationship in the model. For example, the ratio of arm length to leg length may be the same for the marionette as for a human, but the difference between the arm length and leg length is not normally the same for the two. Furthermore, the mechanism by which the human being moves has no analogy in the mechanism by which the marionette operates. It should be clear that the two objects cannot correspond to each other in every detail unless they are identical objects; in which case, the concept of the model loses its usefulness.

In constructing a model, one of the first jobs is, therefore, to decide which characteristics of the object or system of interest are going to be represented in the model. In order to make such decisions, the purpose of making the model must be defined as clearly as possible.

We may divide mathematical models into two broad types that Gold called correlative and explanatory models. Beckner (1959, as quoted in Gold 1977) referred to them non-explanatory and explanatory models, while Poole (1974) called them inductive and deductive models. Regression models are inductive. An experiment is run or observations are made, the data are plotted, and a regression equation is fitted to the data, depending on the functional relationship suggested by the data. Models may also be derived deductively. In a deductive model, we make a logical hypothesis about the process being studied, formulate the hypothesis in the form of a mathematical model and then try to fit the model to the data. Sometimes a model may be derived by a combination of deductive and inductive methods. A deductive model is fit to the data and, based on the observed residuals of the model to the data, the model is modified.

1.2.9 Steps in Modeling

The system in operation in a computer is called simulation and a system as a blueprint on paper is called a model.

1.2.9.1 First Step: Define the Problem

To begin with, one must define the characteristic of the object or system of interest. If one wants to simulate the ratio of the arm length to the leg length of a marionette with that of a human being, then he should keep the same ratio of arm and leg in marionette as is in human being. If one wants to simulate the mechanism in movement of the marionette as of the human being, then he should keep in mind the mechanism of movement of the human being while simulating the movement mechanism of the marionette. So, a modeler should be clearly focused about the object of the modeling. Keeping in mind the objective of modeling, the system must be defined accordingly.

1.2.9.2 Second Step: Component Identification

Next, one should identify the major subsystems, processes, components, elements, etc., important to the operation of the system and identify the links between them. This step may be examplified as:

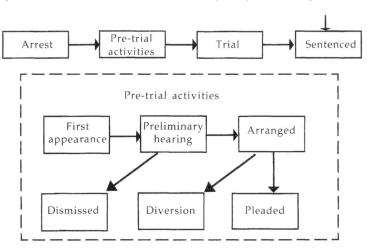

Fig. 1.5 A top down structured approach.

1.2.9.3 Third Step: Specify Component Behavior

Developing a quantitative or mathematical relationship between the inputs of each component and their outputs.
   (1)   What is the general form of the relationships?
   (2)   How can that form be represented numerically?
   (3)   How can mathematics be fitted to the data?

1.2.9.4 Fourth Step: Computer Implementation

This step includes all the stages necessary to convert a model description to a computer program which simulates the system and produces useful outputs. Program is a tool which reveals the manner in which these models perform.

1.2.9.5 Fifth Step: Validation

The validation is a comparison of model behavior with that of the real system. In a management model, one should answer the question 'What is the risk of making poor recommendation?' In the research model, one should answer the question 'Is there significant deviation between predicted and observed behavior?'

1.2.9.6 Sixth Step: Analysis

1.2.9.6.1 Sensitivity Analyses

Sensitivity analysis isolates those variables and parameters to which model outputs are most responsive. If there are two parameters A and B, by changing these two variables in the system with the same extent, and on the supposition that A parameter gives 20% change in outputs and B parameter gives only 2% change in outputs, than can we say that parameter A is more sensitive in changing the system output than the parameter B.

1.2.9.6.2 Stability Analyses

Stability analysis helps in delineating the conditions of the system, which may maintain the system stability rather than extinction of the system. System stability depends upon the interaction between biotic and abiotic stresses.

1.3 THREE PROBLEMS

Management, Research, Design.

1.3.1 System Management Problem

Here, a system is stated along with certain performance criteria and ways of controlling the system so that it functions in an optimal manner. Problem: Control of some insects.

   (i) Needs an analysis.

  (ii) Formulate alternative management approaches.

     (a) Current practice (Calender sprays).

     (b) Integrated bio-chemical control.

     (c) Host plant resistance.

 (iii) Feasibility analysis.

 (iv) Model to predict the performance of the system when managed by each of the alternative strategies.

  (v) Choose the management strategy which works the 'best' and implement it.

1.3.2 Pure Research Problem

Here, a partially understood system is stated with certain 'goodness to fit' criteria that determine the efficiency of the existing knowledge.

  (1) Formulate a hypothesis.

  (2) Make a prediction.

  (3) Conduct controlled experiments.

  (4) Compare results to prediction.

  (5) Accept or reject hypothesis.

Restated Procedure

   (i) Choose the topic to be investigated based on the current model.

  (ii) Formulate a hypothesis in the form of a suggested modification to this model.

 (iii) Make prediction based on the modified model.

 (iv) Conduct a controlled experiment monitoring any controlled inputs.

  (v) Compare the predicted versus observed behavior.

 (vi) Accept or reject hypothesis (acceptance is equivalent to retaining the modified and discarding the original).

1.3.3 System Design Problem

Here, one is given some sets of needs and performance criteria and has to create a system which meets all these needs.

(i) Needs analyses

    (a) Define the need precisely.

    (b) Evaluate the needs.

    (c) Evolve a criteria for judging the potential systems.

(ii) Formulate possible design alternatives.

    (a) Train the growers.

    (b) Work on a state-operated county program—high labor intensive.

    (c) Mobile mite control lab—high capital intensive.

    (d) Redirect state survey staff.

    (e) Current practice.

(iii) Feasibility analysis—rough screening to eliminate obviously unsuitable alternatives.

(iv) Modeling—build model to predict the behavior of each of the alternative systems.

(v) Choose the system to build based on performance criteria and predicted system behavior.

REFERENCES

Asare, D.K., Sammis, T.W., Assadian, H. and Fowler, J.F. (1992). Evaluating three cotton simulation models under different irrigation regimes. Agric. Water Manage. **22**: 391-407.

Beckner, M. (1959). The Biological Way of Thought. Berkeley, University of California Press.

Bell, M.A. and Fischer, R.A. (1994). Using yield prediction models to assess yield grain. A case study for wheat. Field Crops Res. **36**: 161-166.

Beyer, W.H. (ed.) (1968). Handbook of Tables for Probability and Statistics. 2nd ed. Cleveland: Chemical Rubber Co.

Boulding, K.E. (1956). General system theory—The skeleton science, Man. Sci. **2**: 197-208.

Bowen, H.D., Colwick, R.F. and Batchelder, D.G. (1973). Computer simulation of crop production: Potential and Hazards. Agric. Eng.**54** (10): 42-45.

Budagovsky, A.I., Nichiporovich, A.A. and Ross, J. (1964). Quantitative theory of photosynthesis and its use for solving scientific and applied problems of physical geography. (In Russian.) Izv. Acad. Sci., USSR Geogr. Ser. **6**: 13-27.

Budyko, M.I. (1964). An addition to the theory of the influence of climate factors on photosynthesis. (In Russian). Proc. Acad. Sci. USSR 158 (2): 331-334.

Budyko, M.I. and Gandin, L.S. (1964). About taking into account characteristics of physics of the atmosphere in agrometeorological research. (In Russian). Meteorol. Hydrol. **11**: 3-11.

Carberry, P.S. and Abrecht, D.G.(1991). Tailoring crop models to the semi-arid tropics. pp. 157-182. In: R.C. Muchow and J.A. Bellamy (ed.). Climatic Risk in Crop Production: Models and Management for the Semiarid Tropics and Subtropics. Wallingford, UK.: CAB Int.

Cowan, I.R. (1977). Stomatal behavior and environment. Adv. Bot. Res. **4:** 117-228.

Dent, J.B. and Anderson, J.R. (1971). Systems Analysis in Agricultural Management. 395p. Adelaide: A Wiley International Edition, printed at the Griffin Press, South Australia.

Duncan, W.G., Loomis, R.S., Williams, W.A. and Hanau, R. (1967). A model for simulating photosynthesis in plant communities. Hilgardia **38:** 181-205.

Fridgen, J.J., Kitchen, N.R., Sudduth, K.A., Drummond, S.T., Wiebold, W.J. and Fraisse, C.W. (2004). Management Zone Analyst (MZA): Software for subfield management zone delineation. Agron. J. **96** (1): 100-108.

Fritschen, L.J. and Gay, L.W. (1979). Environmental Instrumentation. 216p. New York: Springer-Verlag.

Gold, H.J. (1977). Mathematical Modeling of Biological System: An Introductory Guidebook, 357p. New York: John Wiley & Sons.

Hart, R.H. and Hanson, J.D. (1990). PASTORAL grazing simulator. J. Agron. Educ. **19:** 53-58.

Hatfield, J.L. (1993). Agronomic models. Agron. J. **85:** 713.

International Benchmark Site Network For Agrotechnology Transfer (1993). The IBSNAT decade. 178 p., IBSNAT, Dept. of Agronomy and Soil Science, College of Tropical Agriculture and Human Resources, University of Hawaii, Honolulu, Hawaii.

Ives, P.M. and Hearn, A.B. (1987). The SURATAC system for cotton pest management in Australia. pp. 251-268. In: P.S. Teng (ed.), Crop Loss Assessment and Pest Management. St. Paul. MN. APS Press.

Lemon, E.R., Stewart, D.W. and Shawcroft, R.W. (1971). The sun's work in a corn field. Science (Washington, DC) 174: 371-378.

Loague, K.M. and Freeze, R.A. (1985). A comparison of rainfall-runoff modeling techniques on small upland catchments. Water Resource Res. **21:** 229-248.

Loomis, R.S. and Williams, W.A. (1963). Maximum crop productivity: An estimate. Crop Sci. **3:** 67-72.

Meisner, C.A., Karnok, K.J. and Mc Crimmon, J.N. (1991). Using crop models in a beginning crop science laboratory. J. Agron. Educ. **20:** 157-158.

Oreskes, N., Shrader-Frechette, K. and Belitz, K. (1994). Verification, validation, and confirmation of numerical models in the earth sciences. Science (Washington DC) **263:** 641-646.

Pease, C.M. and Bull, J.J. (1992). Is Science logical? BioScience **42:** 293-298.

Penning De Vries, F.W.T. (1975). Use of assimilates in higher plants. pp. 459-480. In: J.P. Cooper (ed.) Photosynthesis and Productivity in Different Environments. Cambridge: Cambridge Univ. Press.

Poole, R.W. (1974). An Introduction to Quantitative Ecology, 532p., New York: McGraw-Hill.

Porter, J.R., Jamieson, P.D. and Wilson, D.R. (1993). Comparison of the wheat simulation models AFRCWHEAT 2, CERES-Wheat and SWHEAT for nonlimiting conditions of crop growth. Field Crops Res. **33:** 131-157.

Rabbinge, R. and Rijsdisk, F.H. (1983). Epipre: A Disease and Pest Management System for Winter Wheat, Taking Account of Micrometeorological factors. EPPO Bull. **13:** 297-305.

Ritchie, J.T., Godwin, D.C. and Otter-Nacke, S. (1985). CERES-Wheat: A Simulation Model of Wheat Growth and Development. Texas AM Univ. Press, College Station.

Sinclair, T.R. and Seligman, N.G. (1996). Crop modelling: From Infancy to Maturity. Agron. J. **88:** 698-704.

Sirtenko, O.D. (2001). Crop Modeling: Advances and Problems. Agron. J. **93:** 650-653.

Spitter, C.J.T. (1990). Crop growth models: Their usefulness and limitations. Acta Hortic. **267:** 349-362.

Waldren, R.P. (1984). CROPROD: A crop management computer model for under graduate agronomy courses. J. Agron. Educ. **13:** 53-56.

Weiss, A. (1994). From crop modeling to information systems for decision making. pp. 285-290. In: J.P.Griffiths (ed.) Handbook of Agricultural Meteorology. New York; Oxford Univ. Press.

Whisler, F.D., Acock, B., Baker, D.N., Fye, R.F., Hodges, H.F., Lambert, J.R. (1986). Crop simulation models in agronomic systems. Adv. Agron. **40:** 141-208.

Wilkerson, G.G., Jones, J.W., Boote, K.J., And Mishoe, J.W. (1985). SOYGRO V.5.0. Soybean Crop Growth and Yield Model. Technical documentation, Agric. Eng. Dep., Univ. of Florida, Gainesville.

Wit De, C.T. (1965). Photosynthesis of Leaf Canopies. Inst. Biol. Chem. Res. Field Crops Herb. Agric. Res. Rep. 663, Wageningen, Netherlands.

Wullschleger, S.D., Hanson, P.J. and Sage, R.F. (1992). PHOTOBIO: Modeling the stomatal and biochemical control of plant gas exchange. J. Nat. Resour. Life Sci. Educ. **21:** 141-145.

2

Development of Model Structure

2.1 VARIABLES AND THEIR CLASSIFICATION

2.1.1 Individual Observations

Such observations comprise measures and measurements taken on the smallest sampling unit. If we measure weight in 100 rats, then the weight of each rat is an individual observation; the weight of a hundred rats together represents the sample of observations.

2.1.2 Sample of Observations

A sample of observations is defined as a collection of individual observations recorded by a specified procedure. One individual observation is based on a single unit in a biological sense i.e., one rat. However, if we had studied the weight of a single rat over a period of time, the sample of individual observations would be the weights recorded of one rat at successive times. If we wish to measure the temperature in a study of ant colonies, when each colony is a basic sampling unit, each temperature reading for one colony is an individual observation, and the sample of observations is the temperature for all the colonies considered. A synonym for individual observation is item.

2.1.3 Variables

'Individual observation' and 'sample of observations' define only the structure but not the nature of the data in a study. The actual property measured by the individual observations is the character or variable. The more common term employed in general statistics is variable. However, in biology, the word character is frequently used synonymously. More than one character can be measured on each smallest unit. Thus, in a group of 25 mice, we might measure the blood pH and the erythrocyte count. Each of the 25 mice (a biological individual) is the smallest sampling unit; blood pH and red cell count would be the two characters studied; the pH readings and cell counts are the individual observations, and the result would be two samples of 25 observations on pH and erythrocyte count. Or we may speak of bivariate sample of 25 observations, each referring to a pH reading paired with an erythrocyte count.

2.1.4 Population

The biological definition of population refers to all the individuals of a given species found in a circumscribed area at a given time. In statistics, population always means the totality of individual observations about which inferences are to be made, existing anywhere in the world or atleast within a definitely specified sampling area which is limited in space and time.

2.1.5 Variables and Their Classification

The biological variables have already been referred to, but the variables are yet to be defined. A variable is defined as a property with respect to which individuals in a sample differ in some ascertainable way. If the property does not differ within a sample at hand or at least among the samples being studied, it cannot be of statistical interest. Being entirely uniform, such a property would also not be a variable from the etymological point of view. Lengths, height, weight, number of teeth, vitamin C content, and genotypes are examples of variables in ordinary, genetically and phenotypically diverse groups of organisms.

Variables can be divided as follows:

1. Measurement variables
    (a) Continuous variables
    (b) Discontinuous variables
2. Ranked variables (ordinals)
3. Attributes (nominals)

2.1.5.1 Measurement Variables

Measurement variables are such variables whose differing states can be expressed in a numerically ordered fashion. They are divisible into two kinds. The first of these are continuous variables, which, at least theoretically, can assume an infinite number of values between any two fixed points. For example, between the two length measurements 1.5 and 1.6 cm, there is an infinite number of lengths that could be measured if one were so inclined and have a precise enough method of calibration to obtain such measurements. Any given reading of a continuous variable, such as length of 1.57 cm is, therefore, an approximation to the exact reading, which in practice is unknowable. However, for purpose of computation, these approximations are usually sufficient and may even be made more approximate by rounding them off. Many of these variables studied in biology are continuous variables. Examples are lengths, areas, volumes, weights, angles, temperatures, periods of time, percentages, rates.

2.1.5.2 Discontinuous Variables

These variables are also known as meristic or discrete variables. These variables have only certain fixed numerical values, with no intermediate values possible in between. Thus, the number of segments in a certain insect appendage may be 4 or 5 or 6 but never 5½ or 4.3. Examples of discontinuous variables are numbers of a certain structure such as segments, bristles, teeth, or glands, the number of offspring, the number of colonies of microorganisms or animals, or the number of plants in a given quadrate.

Not all variables restricted to integral numerical values are meristic. An example will illustrate this point. If an animal behaviorist were to code the reactions of animals in a series of experiments as: (1) very aggressive; (2) aggressive; (3) neutral; (4) submissive; and (5) very submissive, one might be tempted to believe that these five different states of the variable were meristic because they assume integral values. However, they are clearly only arbitrary points (class marks) along a continuum of aggressiveness; the only reason why no values such as 1.5 occur is because the experimenter did not wish to subdivide the behavior classes too finely, either for reasons of convenience or because of inability to determine more than these five subdivisions of this spectrum of behavior with accuracy. Thus, the variable is clearly continuous rather than meristic, as it might have appeared at first sight.

2.1.6 Ranked Variables

Some variables cannot be measured but can certainly be ordered or ranked on the basis of their magnitude. Thus, in an experiment, one might record the rank order of emergence of ten pupae without specifying the exact time at which each pupa emerged. In such cases, we code the data as a ranked variable on the order of emergence. Special methods to deal with such variables have been developed. By expressing a variable as a series of ranks such as 1, 2, 3, 4, 5, we do not imply that the difference in magnitude between, say, ranks 1 and 2 is identical to or even proportional to the difference between, 2 and 3. Such an assumption is made for the measurement variables, discussed above.

2.1.7 Nominal Variables or Attributes

Variables that cannot be measured but must be expressed qualitatively are called attributes or nominal variables. Such properties, include qualities like black or white, pregnant or not pregnant, dead or alive, male or female. When such attributes are combined with frequencies, they can be treated statistically. In 80 mice, we may, for instance, state that four were black and the rest grey. When attributes are considered with frequencies into tables suitable for statistical analysis, they are reported as enumeration data. Thus, the enumeration data on the color in mice just mentioned would be arranged as follows:

Color	Frequency
Black	4
Grey	76
Total number of mice	80

In some cases, attributes can be changed into variables if desired. Thus, colors can be changed into wavelengths or color chart values, which are measurement variables. Certain other attributes that can be ranked or ordered may be coded to become ranked variables; for example, three attributes referring to a structure as 'poorly developed', 'well developed', and 'hypertrophied' could be conveniently coded as 1, 2, and 3. These variables imply the rank order of development, but not the relative magnitudes of these attribute states.

2.1.8 Variate

Variate means a single reading, score, or observation of a given variable. Thus, if we have measurements of the length of the tails of five mice, the tail length will be a continuous variable and each of the five readings of length will be a variate. Generally, the variables are identified by capital letters, the most common symbol being Y. Thus, Y may stand for the tail length of mice. A variate will refer to a given length measurement; Y_i is the measurement of tail length of the ith mouse and Y_4 is the measurement of tail length of the fourth mouse in the sample.

2.1.9 Derived Variable

The majority of variables in biometric work are observations recorded in the form of direct measurements or counts of biological material or even as readings that are the output of various types of instruments. However, there is an important class of variables in biological research, called the derived or computed variables, that are generally based on two or more independent measured variables whose relations are expressed in a certain way. These are referred as interval and ratio variables.

The difference in the temperature measurement in °C of two variables is called interval variable. The difference in temperature between 60°C and 20°C will equal 40°C, so the derived 40°C will be the derived interval variable.

A ratio expresses as a single value the relation that two variables have to one other. In its simplest form, it is expressed as 64 : 24, which may represent the number of wild type verses mutant individuals, the number of male versus female, a count of parasitized versus those not parasitized, and so on. The above examples imply ratios based on counts; a ratio based on a continuous variable might be similarly expressed as 1.2 : 1.8, which may represent the ratio of width to length in a sclerite of an insect or the ratio between the concentration of two minerals contained in water or soil. Ratios may also be expressed as fractions. Thus, the two ratios above could be expressed as 64/24 and 1.2/1.8. However, for computational purposes, it is most useful to express the ratio as a quotient. The few ratios cited above would, therefore, be 2.666 ... and 0.666, respectively. These are pure numbers, not expressed in measurement units of any kind. This form for ratios will be considered further below. Percentages are also a type of ratio; ratios and percentages are basic quantities in a lot of biological research. They are widely used and generally familiar.

Not all derived variables are in the form of ratio or percentages. The term index is used in a general sense for derived variables, although some may limit it to the ratio of an atomic variable divided by a larger, so-called standard one. For instance, in a study of the cranial dimensions of cats, Haltenorth (1937, as quoted in Sokal and Rohlf, 1981) divided all such measures on the basis of the length of the skull. Thus, each measurement was, in fact, a proportion of the basal length of the skull of the cat being measured. Conceived in the wide sense, an index could be the average of two measurements—either simply, such as 1/2 (length of A + lengths of B), or in weighted fashion, such as 1/3 [(2 × length of A) + length of B]. An index may refer to the summation of a series of numerically scored properties. Thus, if an animal is given six behavioral tests in which its score can range from 0 to 4, an index of its behavior might be the sum of the scores of the six tests. Similar indices have been described to determine the degree of hybridity in organisms.

Rates will be important in many experimental fields of biology. The amount of a substance liberated per unit weight or volume of biological material, weight gain per unit time, reproductive rates per unit population size and time (birth rates), and death rates would fall in this category. Many counts are really ratios or rates—the number of pulse beats observed in one minute or the number of birds of a given species found in some quadrate. In general, counts are ratios if the unit over which they are counted is not natural, such as an arbitrary time or space interval.

There are some serious drawbacks to the use of ratios and percentages in statistical work. In spite of these disadvantages, the use of these variables is deeply ingrained in scientific thought processes and is unlikely to be abandoned. Furthermore, is should be emphasized that ratios may be the only meaningful way to interpret and understand certain biological problems. If the biological process being investigated operates on the ratio of the variables studied, one must examine this ratio to understand the process. Thus, Sinnott and Hammond (1935, as quoted in Sokal and Rohlf, 1981) found that inheritance of the shapes of squashes, *Cucurbita pepo*, could be interpreted by a form index based on a length-width ratio, but not in terms of the independent dimensions of shape. Similarly, the evolution of shape in many burrowing animals is a function of the cross-sectional profile of the animal rather than of a single dimension. The selection affecting body proportions must be found to exist in the evolution of almost any organism when properly investigated (Sokal and Rohlf, 1981).

The disadvantages of using ratios are several: first as their relative inaccuracy. Let us return to the ratio 1.2/1.8 mentioned above. A measurement of 1.2 indicates a true range of measurement of the variable from 1.15 to 1.25; similarly, a measurement of 1.8 implies a range from 1.75 to 1.85. We realize, therefore, that the true ratio may vary anywhere from 1.15/1.85 to 1.25/1.75, or 0.622 and 0.714, respectively. We note a possible maximal error of 4.2% if 1.2 were an original measurement: $(1.25 - 1.2)/1.2$; the corresponding maximal error for the ratio is 7.0%: $(0.714 - 0.667)/0.667$. Furthermore, the best estimate of a ratio is usually not the mid point between the possible range. Thus, in our example, the midpoint between the implied limits is 0.668 and the true ratio is 0.666—only a slight difference, which may, however, be greater in other instances. In many cases, therefore, the ratio will not be as accurate as the measurements obtained directly. This liability can be overcome by empirical determinations of their variability through measures of their variance.

A second drawback to ratios and percentages is that they may not be approximately normally distributed, as required by many statistical tests. This difficulty can be frequently overcome by transformation of the variable. Another disadvantage of ratios is that they do not provide information on the form of the relationship between the two variables whose ratio is being taken. Often, more may be learned by studying the variable singly and their individual relationships to each other (bivariate and multivariate analysis).

Finally, it should be pointed out that when ratios involve enumeration data or meristic variables, they may occasionally give rise to curious distribution. By way of an example, we may cite the percentages obtained from an experiment performed some years ago by Sokal (as quoted in Sokal and Rohlf, 1981). In this experiment, ten drosophila eggs were put in a vial and the positions of the pupae noted after pupation. Some pupae pupated at the margin or the wall of the vials; these were called 'peripheral'. Others pupated away from the wall; these were called 'central'. Ideally, there should have been ten pupae if all the eggs had hatched and there had been no larvae mortality whatsoever. In fact, however, because of natural mortality and errors in preparing the vials, ten pupae were not always found. The logical minimum number of survivors on which a result could be reported was one pupae. The proportion of peripheral pupae was then calculated by dividing the number of such pupae in a vial by the total number of pupae found in that vial. Although the ratios obtained thus are a continuous variable in appearance, they are not so, in fact, because certain values can never be obtained. For instance, by limiting the maximum number of pupae to 10, we cannot obtain the percentages of peripheral population between 0 and 10% or between 90 and 100%. Some percentages are given by several ratios: 33% is obtained either by 3/9 or 2/6 or 1/3, but 57% can only be obtained by having 4 out of 7 pupae peripheral. In using ratios of meristic variates or of counts, such discontinuities and peculiarities of distributions must be taken into account. The problem in this instance was solved by transforming the data to probits.

2.1.10 Interval Variable

Numbers make their appearance and differences become meaningful in interval variable. For example, the difference between 50°C and 45°C is five. The interval variable has its zero scale arbitrary. One can take the difference in temperature in Celsius and Fahrenheit.

2.1.11 Ratio Variable

In ratio variable, multiplication and division become legal. The ratio of two variables which have mass as a dimension of 300 kg/150 kg is 20. The temperature in Kelvin is a ratio variable. One cannot take the ratio of the temperature in °C and °F scale.

2.1.12 Rate-Quantity Variable

Rate Variable: Rate variable describes how fast a quantity is changing at instant time. It is denoted as $r(t)$.

Quantity Variable

Quantity variable is defined as the amount of something present at one instant in time. It is denoted as q (t).

2.1.13 Example

System definition: A person is driving a car down an highway. It begins to rain.

2.1.13.1 Components

There may be four components in the system.
(1) Person; (2) Car; (3) Highway; and (4) Environment.

2.1.13.1.1 Person

This component may have the following variables:
(a) RAC (secs): Reaction time. This is a ratio variable.
(b) STR (degr.): Steering angle relation to straight ahead. This is a interval variable.
(c) DAC (lbs): Pressure on accelerator. This is a ratio variable.
(d) DBK (lbs): Pressure on brake. This is a ratio variable.

2.1.13.1.2 Car

A car component may have the following variables:
(a) SPD (mph) Speed (R)
(b) WHl (deg): Car angle relation to straight ahead (I)
(c) ACC (m/hr²): acceleration (R)
(d) TRD (Fresh, worn) Tread (Nominal)

2.1.13.1.3 Highway

This component may have the following variables:
(a) CRV (deg); Curve of highway: (I).
(b) Frc (?) Coefficient of friction (I)

2.1.13.1.4 Environment

(a) WSP (mph) wind speed (R)
(b) WDR ('Head', 'Tail', 'Cross') wind direction (N)
(c) RAN (in/sec) rainfall rate (R)
(d) Time (min) time (I)

Table 2.1 Classification of the variables of components of
car-driver-road and environment system

Car component

Variables

4 acceleration (mi/hr^2)	⎤ RQ	ACC
4 Speed (mi/hr)	⎦ ⎤ RQ	SPD
3 Position (mi)	⎦	POS
4 Fuel consumption rate (g/hr)	⎤ RQ	FCR
4 Gas in tank (g)	⎦	GAS
3 Angle of tires (deg)		TAG
2 Gear (1,2,3,4)		GER
1 Skid (Yes,No)		SKD
3 Lateral position (ft)		LTP
4 Lateral Speed (ft/Sec)	⎦ RQ	LTS

Parameter

1 Type (Heavy, Dodge, etc.)	TYP
4 Tank capacity (g)	TCP
4 Accelerator throw (in)	ACT
3 Max. time angle (deg)	MAT
3 Wheel lock to lock (deg)	LTL

Data type   1 = Nominal

2 = Ordinal

3 = Interval

4 = Ratio

R.Q. = Rate quantity data type

Abbreviation of units

mi	miles
hr	hour
g	gallon
deg	degree of angle
in	inches
sec	seconds
ft	feet
m	meters
min	minutes
ms	millisecond
°C	degree centigrade

Driver component

Variables

accelerator position (in)		ACP
break position (in)		BKP
3 wheel position (deg)		WHP
4 wheel turn rate (deg/sec)	⎤ RQ	WTR
4 distance can see (yds)	⎦	DSC

Parameters

3 reaction time (ms)	RET

(may be used in place of 'driver experience' which would require an
index)

Road component

Variables		
	Width (ft)	WTH
	4 Curve (deg/100m)	CRV
	4 Max. safe speed (mph)	MSS
	4 Max legal speed (mph)	MLV
	4 grade (dimensionless)	GDE
	rise/distance	GRD
	slickness (?)	SLK
	3 bank (deg)	BNK
	drainage rate (g/min/ft)	DRN
Parameters	1 RET Surface type (asphalt)	SUR

Environment

Variables		
	4 visibility (yds)	VIS
	3 Temperature (°C)	TMP
	2 rainfall (none, right, moderate, heavy)	RAN
	1 Light (day, night)	LIT
	3 Time (hr., min., sec.)	TIM

2.1.14 Exercise

Find out the data type from the following variable and parameter list.

Name	Meaning	Component	Units	Data Type
ACC	Acceleration	Car	mi/hr^2	
ACP	Acceleration position	Driver	in	
ACT	Accelerator throw	Car	in	
BKP	Brake position	Driver	in	
BNK	Banking	Road	deg	
CRV	Curvature	Road	deg/100m	
DRN	Drainage rate	Road	g/min/ft	
DSC	Distance visible	Driver	Yds	
FCR	Fuel consumption rate	Car	g/hr	
GAS	Gas in the tank	Car	g	
GER	Gear	Car	1,2,3,4	
GRD	Grade	Road	% (no units)	
LTT	Light	Environment	Day, Night	
LTL	Steering Wheel (Lock to Lock)	Car	deg	
LTP	Lateral position	Car	ft	
LTS	Lateral speed	Car	ft/sec	
MLV	Maximum legal speed	Road	mi/hr	
MSS	Maximum safe speed	Road	mi/hr	
MTA	Maximum tire angle	Car	deg	
POS	Position along the road	Car	mi	
RAN	Rainfall rate	Environment	N, L, M, H	

RET	Reaction times	Driver	ms
SKD	Skidding	Car	Yes, no
SLK	Slickness	Road	?
SPD	Speed	Car	mi/hr
SUR	Surface type	Road	Asphalt
TAC	Tire angle	Car	deg.
TCP	Tank capacity	Car	g
TIM	Time	Environment	hr, min, sec
TMP	Temperature	Environment	°C
TYP	Type	Car	Chevy, etc.
VIS	Visibility	Environment	Yds
WHP	Steering wheel position	Driver	Deg.
WTM	Width	Road	ft
WTR	Steering wheel		
TUR	Turning rate	Driver	Deg/sec

2.2 RELATIONSHIP BETWEEN VARIABLES

(1) **Define the system:** A person is driving a car on an highway in a rural area, when it starts to rain.

(2) **Identify the components**
   (a) Person
   (b) Car
   (c) Highway
   (d) Environment

(3) **Identify the variables and parameters in each component**

	Person	*Car*	*Highway*	*Environment*
STR	Steering Angle relative to Straight Head (degree)	WHL wheel Angle in relation to Highway (deg)	CRV Curve of the Highway (degree)	
RAC	Reaction time (sec)			

(4) **Relationship between variables**

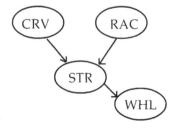

2.2.1 Causal Loop Diagrams

2.2.1.1 Direct Relations

A causative relation exists between two variables in a given model if a change in one variable leads directly to a change in the other.

2.2.1.2 Indirect Relations

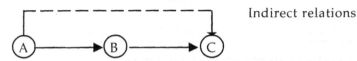

Indirect relations

Variable (A) affects variables (C) indirectly mediated by variables (B).

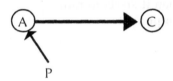

P parameter affects variable (A) and variable (A) affects variables (C)

Suppose there are five variables
(A), (B), (C), (D), (E).

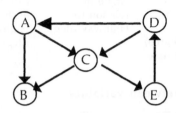

The above figure shows seven direct relations. And there will be (5×4) = 20 indirect relations.

2.2.1.3 Relationship Between Rate and Quantity Variable

Such a relationship is explained by way of a mathematical theory. Integration approach is much later.

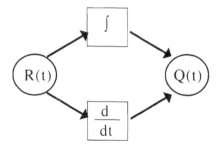

2.2.2 Types of Relationship between Variables

2.2.2.1 *Direct (Together) Relations*

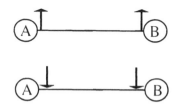

2.2.2.2 *Inverse Relations*

2.2.2.3 *Indeterminate Relations*

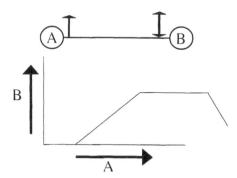

2.2.2.4 *Feedback Relationship*

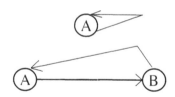

No variable can affect itself directly but it can certainly do so indirectly.

Variable (A) affects the variable (B) and, after some delay, the variable (B) may effect variable (A). This shows the feedback relations.

2.2.3 Example of Public Address System

Modeling of Public Address System:

This example, given by Gold (1977), is ideal in explaining the process of modeling for students.

2.2.3.1 *Step 1*

Formulation of the problem: The problem considered here concerns the build up of a high-pitched whistling sound in a public address system. One needs to know enough about the processing of signals in the system in order to get a clue as to how this build up can be eliminated.

2.2.3.2 *Step 2: Qualitative Description of the System*

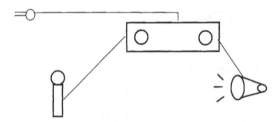

Fig. 2.1 Components of a public address system.

The system consists of a microphone, which converts sound energy into an electrical signal; an amplifier, which uses electrical energy to increase the strength of the signal from the microphone; and a loudspeaker, which converts this increased signal back to sound energy. Electrical signals and electrical energy are carried through wires, whereas the sound energy is carried through the air. The qualitative description of the system is shown in Fig. 2.1.

2.2.3.3 *Step 3: Definition of Relevant Components Subsystems, and Interactions*

The components have already been partially defined as part of the qualitative description of the system. Now one has to assume which of the components are important. In this example, a component is taken to be important only if it does something to the signal being processed. It is assumed that neither the electrical plug nor the wire that connects the physical components do anything to change the signal and that the fate of the signal can be adequately described by knowing what the three components (microphones, amplifier, and loudspeaker) do to the signal and how the signal passes from one component to the other. Note that the assumption that the connecting wires can be neglected would not be valid if transmission were over a very long distance. The assumptions concerning the choice of relevant components and their interconnections are shown in the component diagram Fig. 2.2. The arrows do not represent the wires; they simply indicate that some signal of interest issues from the component or the block to which the tail of an arrow is attached and is received by the block to which the arrowhead is attached. Arrows that do not originate in any component of the system represent the net input to the system. Arrows that do not terminate in any component of the system represent the net output of the system. When an arrow originates in another arrow, the signal has been split. Similarly, when one arrow terminates in another, the two signals have been combined. The component diagram is a schematic representation of the structure of the system to be considered. It is not altered unless there is a need to modify the structure.

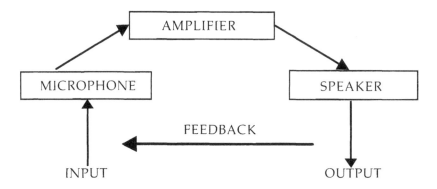

Fig. 2.2 Component and flow of signal for public address system.

2.2.3.4 *Step 4: Definition of Relevant Variables*

The next step is to define the variables (the quantities that may vary with time) whose values at any given time tell us what we want to know about the system. The problem, of course, is to understand what is relevant. A useful thumbrule here is to take the variables to be the inputs and outputs of the individual components. Just as we define the variables, it is convenient to assign them symbols so that we may refer to them later without any problem. A table might be constructed as follows:

Table 2.2 Variables and their symbols in a public address system

Variables	Symbol
Input to the system	$I(t)$
Input to the microphone	$I_M(t)$
Output from the microphone	$O_M(t)$
Input to the amplifier	$I_A(t)$
Output from the amplifier	$O_A(t)$
Input to the speaker	$I_S(t)$
Output from the speaker	$O_S(t)$
Output from the system	$O(t)$

In Table 2.2, (t) is meant to indicate that each of the variables may vary from one instant to the next.

When the variables are listed, it is usually desirable to specify their dimensions. The dimensions of a variable essentially reveal the kind of physical entity represented by the variable.

2.2.3.5 *Step 5: Representation of the Relations Between the Variables*

The manner in which each variable changes is influenced by one or more of the others. The pattern of relations between the variables may be represented by a type of diagram, sometimes called a signal flow graph, such as that of Fig. 2.3.

In the following diagram, the arrows represent directions of influence.

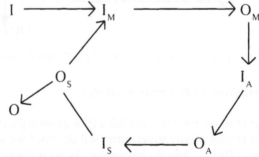

Fig. 2.3 Signal flow graph showing how one variable influences another in a public address system.

The signal flow graph may be summarized by a set of dependency statements of the type:

$$X = X \ (A, B, C, \ldots\ldots, Z) \tag{2.1}$$

X is one of the system variables

and the notation is meant to indicate that X is a function of the variables that are inside the parenthesis. Fig. 2.3 translates to the following set of dependency statements.

$$I_M = I_M \ (I, O_S) \tag{2.2}$$

$$O_M = O_M \ (I_M) \tag{2.3}$$

$$I_A = I_A \ (O_M) \tag{2.4}$$

$$O_A = O_A \ (I_A) \tag{2.5}$$

$$I_S = I_S \ (O_A) \tag{2.6}$$

$$O_S = O_S \ (I_S) \tag{2.7}$$

2.2.3.6 Step 6: Description of the Subsystems

The next step is to determine the specific form of the dependencies and use this information to convert the component diagram of Fig. 2.2 into the input output diagram in Fig. 2.4. To do so, the characteristics of the components of the system must be examined.

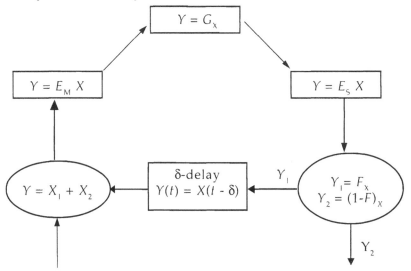

Fig. 2.4 Augmented input-output diagram for public address system.

A sound signal is converted to an electrical signal in the microphone. The microphone is not perfectly efficient; and a part of sound energy supplied to the microphone is lost. The remainder is converted to electrical signal. The total electrical signal output is proportional to the sound energy input.

$Y = E_M X$, where Y be output, X be input, E_M is the efficiency of the microphone and a number between 0 and one.

The use of linear equations (proportionality) is generally associated with increments that behave identically and independently. Each increment of input is identical and independent of the total size of the output, so the input-output equation for the amplifier becomes $Y = G_X$; where G is the amplification, or 'gain', and is generally greater than one. The input-output equation for the speaker would be $Y = E_S X$, where E_S is the efficiency of the speaker.

The next component, which is not shown in Fig. 2.4, is the atmosphere through which the sound is carried. The sound signal is of two types: the output for the speaker and the input of the microphone.

The speaker output must be split into the net output of the system and the part that feeds back to the microphone. A signal splitter has two outputs for any input.

$$Y_1 = F_x \qquad (2.8)$$

$$Y_2 = (1 - F)_x \qquad (2.9)$$

Notice that $F + (1 - F) = 1$, so that the signal splitter assigns a fraction F of the input to Y_1 and the remaining fraction $(1 - F)$ to Y_2.

The input to the microphone is the sum of the net input of the system and the feedback from the speaker. A summer, which receives two inputs but gives a single output.

$$Y = X_1 + X_2 \qquad (2.10)$$

An approximation is applied that the transfer of electric signal is instantaneous when compared with the much longer time that it takes for sound signal to get from the speaker to the microphone. Another conceptual component that shows a delay of δ time units between the splitter and the summer, is accounted for this time.

The input-output diagrams can be applied directly for computer simulation.

2.2.3.7 Step 7: The Model Equations

It is convenient to express the model in a set of equations by combining the information of the input-output diagram with the signal flow graph, or the dependency equations in order to get:

$$I_M(t) = I(t) + FO_S(t - \delta) \tag{2.11}$$

$$O_M(t) = E_M I_M(t) \tag{2.12}$$

$$I_A(t) = O_M(t) \tag{2.13}$$

$$O_A(t) = GI_A(t) \tag{2.14}$$

$$I_S(t) = O_A(t) \tag{2.15}$$

$$O_S(t) = E_S I_S(t) \tag{2.16}$$

$$O(t) = (1 - F) O_S(t) \tag{2.17}$$

2.2.3.8 Step 8: Studying the Behavior of the Model

What are the interesting aspects of the system behavior? Feedback has already been recognized as an important part of the problem, i.e. that the signal getting to the microphone is not limited to the system input I to see what the total input to the microphone is; it can be started with equation (2.11) and proceed by successive substitutions, to get

$$I_M(t) = I(t) + FE_S GE_M I_M(t - \delta) \tag{2.18}$$

The problem is the second term on the right hand side. An examination of the system parameters reveals that G is the easiest to adjust (a knob is usually provided) and that F may be adjusted by changing the physical arrangement of the components. Reducing F brings that $I_M \approx I$ ideal closer to reality. So does reducing G, but this also reduces O, which may not be tolerable.

Suppose we put a single short pulse into the microphone to establish $I_M(t - \delta)$ in Equation (2.18) and look at I_M immediately after the pulse, so that $I(t) = 0$. It should clear that if

$$\frac{I_M(t)}{I_M(t - \delta)} = FE_M GE_S > 1$$

then the signal continues to build up (even though $I = 0$) until the speaker cone rips, or someone shuts it off. If

$$FE_M GE_S < 1$$

then some echo effect may be produced, but the situation is at least tolerable. I_M eventually dies to zero.

This rather lengthy analysis tells us what most people already knew anyway. To eliminate feedback buildup in a public address system, one must either turn the volume (reduce G) or move the microphone away from the speaker (reduce F). Experimental test of the conclusion demonstrates its correctness and gives us some degree of confidence in the accuracy of our conceptual understanding of the system.

2.2.3.9 *Example of Feedback Relationship: Simple Public Address System*

The signal flow graph (Fig. 2.3) shows the feedback loop for the simple public address system which have been considered here. The feedback loop can be studied by looking at the relation between any two variables in the loop, say, I_M and O_S (the input to the microphone and the total output from the speaker). Equation 2.11 is

$$I_M(t) = I(t) + FO_S(t - \delta)$$

Now, suppose that the input I is fixed and constant and that, by some means or other, we fix the value of O_S at some constant value. From Equation (2.11) it is found that the resulting value for I_M would be

$$I_M = I + FO_S \qquad (2.19)$$

Next, Equations (2.11) through (2.17) for O_S in terms of I_M are solved,

$$O_S(t) = E_S I_S(t) = E_S O_A(t) = E_S GI_A(t) = E_S G O_M(t) = E_S GE_M I_M(t)$$

Therefore, if somehow we fix I_M (say, by adjusting F or I), the resulting value for O_S is

$$O_S = E_S GE_M I_M \qquad (2.20)$$

At the steady state (2.19) and (2.20) must be simultaneously satisfied. From (2.19) and (2.20), we get

$$\frac{dI_M}{dO_S} = F$$

$$\frac{dO_S}{dI_M} = E_S GE_M$$

These are of the same sign, so apparently the system has a positive feedback loop. The magnitude of the product of the derivatives might be taken as an index of the intensity of the feedback loop. The values of the product greater than one lead to the 'blow-up' of the system. For values less than one, the effect of the loop may be considered to be sufficiently damped so that the system does not spontaneously blow up.

2.2.3.10 *Example: Amplifier Circuit with Negative Feedback*

The next step in the development is to modify the circuit so that it has a negative feedback loop and to examine some of the consequences.

It can be imagined, therefore, that the system is equipped with a device that causes the returning signal to be subtracted from the incoming signal, rather than being added to it. A symbol arrow graph is shown in Fig. 2.5.

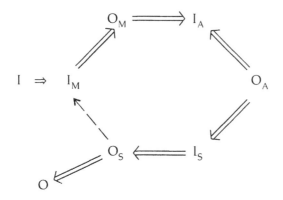

Fig. 2.5 Symbol arrow graph for public address system with negative feedback.
[⇒ increase; — — → decrease]

The only modification in the set of Equation (2.11) through (2.17) is change a sign in Equation 2.11.

It becomes

$$I_M(t) = I(t) - FO_S(t - \delta) \tag{2.21}$$

In place of Equation (2.18), it becomes

$$I_M(t) = I(t) - FE_S GE_M I_M(t - \delta)$$

This system would be at a steady state if $I_M(t) = I_M(t - \delta)$. So, assuming that this is the case, letting the ss superscript designate the steady state value and assuming I is kept constant, it becomes

$$I_M{}^{SS} = \frac{I}{I + FE_S GE_M} \tag{2.22}$$

which describes a physically realizable steady state for any positive value of the parameters.

Given the same procedure used to get Equations (2.19 and 2.20), we get

$$I_M = I - FO_S$$

$$O_S = E_S GE_M I_M$$

and

$$\left(\frac{dI_M}{dO_S}\right)\left(\frac{dO_S}{dI_M}\right) = -FE_S\, GE_M$$

This is always negative (assuming that the parameters are positive).

2.2.3.11 *Effect of Feedback on Response to Change in Input*

Till now, the point has been made that a negative feedback loop works to oppose disturbances, whereas a positive feedback loop tends to magnify them. In this subsection, it is looked more explicitly at the sensitivity of the steady-state position to change in the input variable and how the sensitivity is affected by the feedback loop. The amplifier system of the two previous sections are used to illustrate the differing affects of positive and negative feedback.

In these examples, solving for the steady-state value of O_S (the total speaker output) gives

$$O_S^{SS} = \frac{E_S\, GE_M}{1 - FE_S\, GE_M}\, I \quad \text{(Positive feedback)}$$

$$O_S^{SS} = \frac{E_S\, GE_M}{1 + FE_S\, GE_M}\, I \quad \text{(Negative feedback)}$$

The sensitivity of the steady-state speaker output to a change in input can be expressed by the derivative dO_S^{SS}/dI. To judge the effect of the feedback loop on this sensitivity, the value of this derivative can be compared under the assumption that the loop has been interrupted, i.e. open. In this example, the loop can be opened setting F to zero. The result is

$$\frac{\left(\dfrac{dO_S^{SS}}{dI}\right)\text{open loop}}{\left(\dfrac{dO_S^{SS}}{dI}\right)\text{closed loop}} = \begin{cases} 1 - FE_S GE_M, & \text{Positive feedback} \\ 1 + FE_S GE_M, & \text{Negative feedback} \end{cases} \tag{2.23}$$

Two things, in particular, should be noted above in Equation (2.23). Firstly, a positive feedback loop increases the sensitivity of the steady-state position to a change of input (ratio less than one), while a negative feedback loop decreases the sensitivity (ratio greater than one). In other words, the negative feedback loops serves to buffer or insulate the system against changes in the environment (as represented by the value of I). This is specially important for biological systems whose functioning often requires that certain internal state variables be reasonably independent of environment conditions.

The second thing to note is that in both cases, positive or negative feedback,

$$\frac{\left(\dfrac{dO_S^{SS}}{dI}\right)\text{open loop}}{\left(\dfrac{dO_S^{SS}}{dI}\right)\text{closed loop}} = 1 - \frac{dI_M}{dO_S}\frac{dO_S}{dI_M}$$

with the derivatives on the right hand side taken around the loop. This is indicative of a fairly general relation.

The following two systems are compared:

A. *Closed loop system*

$$I \longrightarrow X \rightleftarrows Y$$

$$I \longrightarrow X \longrightarrow Y$$

$$X = f(I, Y)$$
$$Y = g(X)$$

B. *Open loop system*

$$I \longrightarrow X \longrightarrow Y$$

$$X = f'(I)$$
$$Y = g(X)$$

with the following relation between the two systems

$$\frac{df'}{dI} = \frac{\partial f}{\partial I} \qquad\qquad (2.24)$$

Note that the same rule (the function g) gives the dependence of Y on X in the open- and closed loop systems. Different functions must be specified for X, because in one case X depends on Y, but in the other case it does not. Nevertheless, the assumption embodied in Equation (2.24) is that the direct dependence of X on the input I is the same with or without the feedback loop. Therefore, the relations that are going to be developed are not valid unless the assumption can be made.

An expression for dx/dI in the closed loop system is obtained using the chain rule.

$$\frac{dX}{dI} = \frac{\partial f}{\partial I} + \frac{\partial f}{\partial Y} \quad \frac{dY}{dI} \tag{2.25}$$

$$\frac{dY}{dI} = \frac{\partial g}{\partial x} \quad \frac{dx}{dI} \tag{2.26}$$

Substituting (2.26) into (2.25) and solving for dx/dI gives

$$\left(\frac{dX}{dI}\right)_{\text{closed loop}} = \frac{\partial f / \partial I}{1 - (\partial f / \partial Y)(\partial g / \partial X)} \tag{2.27}$$

For system (B),

$$\left(\frac{dX}{dI}\right)_{\text{open loop}} = \frac{df'}{dI} \tag{2.28}$$

The ratio of these gives, after algebraic rearrangement and use of assumption (2.24),

$$\frac{\left(\dfrac{dX}{dI}\right)_{\text{open loop}}}{\left(\dfrac{dX}{dI}\right)_{\text{closed loop}}} = 1 - \frac{\partial f}{\partial Y} \quad \frac{\partial g}{\partial X} \tag{2.29}$$

The derivatives on the right hand side of (2.29) are taken around the feedback loop. The value of this product of derivatives—or rather, its negative is a measure of the effectiveness of the feedback loop in the stabilization of the steady-state position against changes in the environment.

This is termed the homeostatic index of the loop (abbreviated H.I.); in the literature of system control theory, it is called the open loop gain. What we have just found is that under the assumption (2.24),

$$H.I. = \frac{(dX / dI)_{\text{open loop}}}{(dX / dI)_{\text{closed loop}}} - 1 \tag{2.30}$$

$$= - \text{ product of derivatives around the loop} \tag{2.31}$$

Once a model is constructed, the homeostatic index may be obtained from Equation (2.31). If the feedback loop involves more than two variables, direct application of the chain rule leads to exactly the same prescription: one should use the product of the derivatives taken around the loop. That is, if

$$X_K \longrightarrow X_{K-1} \longrightarrow - - \longrightarrow X_2 \longrightarrow X_1 \longrightarrow X_K$$

then

$$HI = - \frac{\partial X_1}{\partial X_2} \frac{\partial X_2}{\partial X_3} \cdots \frac{\partial X_{K-1}}{\partial X_K} \frac{\partial X_K}{\partial X_1}$$

the calculated value of H.I. can be compared with an experimentally determined value, provided it is possible to experimentally interrupt the feedback loops without altering the other functions.

For the two amplifier systems,

$$HI = \begin{cases} -FE_S\, GE_M & \text{Positive feed back} \\ +FE_S\, GE_M & \text{Negative feed back} \end{cases}$$

These two systems illustrate the following points:

(a) H.I. > 0 (negative feedback loop). In this case, the feedback loop stabilizes the position of the steady-state against changes in the environment.

(b) 0 > H.I. > − 1 (damped positive feedback loop). For H.I. between 0 and − 1, the feedback loop makes the steady-state position more sensitive to changes in the environment.

(c) H.I. ≤ − 1 (Positive feedback loop). For H.I. equals to or less than − 1, the sensitivity may be thought of as being so great as to eliminate the steady-state altogether.

2.3 STRUCTURAL (BLACK BOX) MODEL

The structural representation of a model focusses on the basic elements or processes which comprise a system.

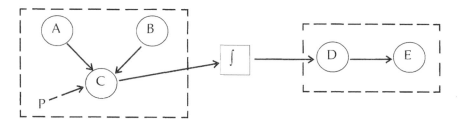

Fig. 2.6 Black box with variables leading into a variable.

1. For each set of arrows leading into a variable, make a box.
2. Each variable becomes an arrow.

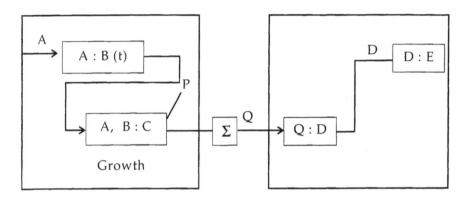

Fig. 2.7 Representing the equations with computer codes.

Fig. 2.7 represents a needed equation with computer codes. This can be given the name of some process.

Based on the structural representation of the model, variables may be classified as follows:

1. Those appearing only to left of: these are called the 'system' inputs or stimulus variables.
2. These which appear only on the right side of a: these are called output or response variables.
3. Those which appear on both sides, are called 'state variable'. In a dynamic system, state variables tell the condition of the system of a particular instant of time. But in a static system, the term 'intermediate' variable is more appropriate.

2.4 REFINEMENT IN STRUCTURAL MODELS

The system, which goes from more detail to less detail, is of less resolution. On the contrary, the system which goes from less detail to more details is of high resolution. The refinement in structural model is dependent on the development of the recent investigation on the system.

2.4.1 The Structure of Crop Simulation Models

Passioura (1996) has critically discussed the structure of crop simulation models (Passioura, J.B. 1996). We have adopted the word 'validation' for the process of comparing the output of a model with a new data set. The inevitable disagreement usually leads us to adjust the parameters in our models, rather than to examine their structure. Changes in structure typically come from explicit experimental work, aimed at testing well-defined hypotheses (Passioura, 1996).

Passioura (1996) further posed a question: What do I mean by structure? The following example illustrates the idea. Many of the mechanistically based simulation models use the photosynthetic rate as the control variable to determine the growth rate of a crop. Photosynthesis is determined by many other variables, one of the most important of which is stomatal conductance. If the crop starts to experience water stress, it is supposed that the leaf water potential falls in a way that can be calculated in terms of the transpiration rate and the hydraulic resistances within plant and soil and this, in turn, induces stomatal closure and reduced photosynthesis. This conceptual structure is essentially devised by Cowan (1965; as quoted in Passioura, 1996) many years ago, in which for the first time he combined the work of Gardner (1960; as quoted in Passioura, 1996) and others on the flow of water to roots with a simple algorithm to relate the stomatal conductance to leaf water potential.

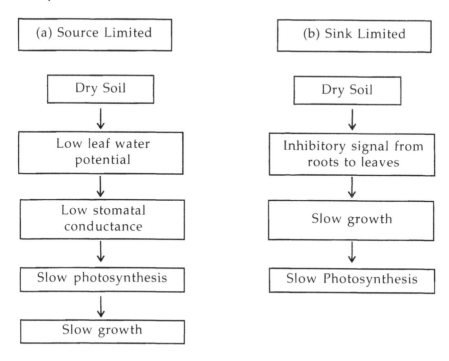

Fig. 2.8 Two scenarios of markedly different structures that depict how the growth of drought plants may be controlled (From Passioura, 1996).

This structure, depicted in Fig.2.8, may often be true in situations where leaf water potentials are induced in well-watered plants by very large evaporative demands, at least for the connection between leaf water potential and stomatal conductance, provided by the study of Saliendra et al. (1995) on a riparian species. However, a simulation model based on Fig. 2.8a can be made to fit the data even where the structure is not true. Applying the model does not challenge this strucutre.

An alternative view of the relation between photosynthesis and growth in a water-stressed plant is that the plant is able to sense that its environment is deteriorating, and determines its growth rate accordingly. In this scenario, the photosynthetic rate does not determine the growth rate of the plant. The reverse is true: the growth rate determines the rate of photosynthesis. In the language of carbon-partitioning- physiologists, the plant is sink limited rather than source limited. In fact, there is a spectrum of circumstances ranging from complete sink limitation to complete source limitation. Many pieces of evidence are available to suggest that plants growing in inhospitable soils are largely sink limited, and that the root system responds to the soil conditions by generating a signal, as depicted in Fig. 2.8b (Barlow, 1986; Passioura, 1988; Davies and Zhang, 1991; Masle, 1992; as quoted in Passioura, 1996). Now, if this structure is true, a model based on an entirely source limited model can never be relied to work, and the callibration of it against a given data set is, in de Wit's (1970; as quoted in Passioura, 1996) devastating words, 'the most cumbersome and subjective technique of curve-fitting that can be imagined'. It is notable that the well-established CERES family of crop models, which are predominantly functional rather than mechanistic, implicitly favor this second scenario in the sense that they relate transpiration and growth to soil water content rather than to leaf water potential.

An example from soil physics that matches the physiological example of Fig. 2.8 is that of infiltration and redistribution of soil water. Many models assume that the flow is one dimensional, and seek to improve the accuracy of their prediction by measuring the appropriate parameters—for example those refining hydraulic conductivity—with increasing spatial resolution. If the flow is occurring preferentially, for example in continuous macropores, or if there are perched water tables that result in lateral flow in the soil, then the one-dimensional model is inappropriate, and persisting with it while increasing the level of spatial detail is futile.

Reynolds and Acock (1985) have discussed the sources of error in relation to the complexity of models dissected the notational total error into two components, one arising from errors in estimating parameters, the other arising from a systematic bias resulting from over simplifying. They postulated that cumulative errors in the parameters grow with the number of the parameters as a model becomes more complex. And they postulated that this systematic bias (erroneous structure) decreases with increasing complexity (See Figure 3a of Passioura 1996). However, if the structure is fundamentally wrong, as it could be in the example of photosynthetically driven growth illustrated in Fig. 2.8, then no amount of complexity will improve the structural error. There will be an irreducible minimum error, as illustrated by the dotted asymptote in Fig. 3b of Passioura, 1996.

Occasionally though, the structure seems to be so incorrect that no amount of adjusting of the parameters enables the model to fit the data. When that happens, we have moved beyond the realm of validation and are in a position to discover something new (Passioura, 1996). A good example is the problem that the CERES models met with their routine in withdrawing water from the subsoil (J.T. Ritchie, personal communication, 1983; as quoted in Passioura, 1996). This routine greatly overestimated the rate of uptake by the roots, even using the measured root length density. The disagreement stimulated research into alternative structures for the routine: for example, that the roots were not uniformly distributed through the given layer of soil, but were clumped into pre-existing pores or cracks (Passioura, 1991; as quoted in Passioura, 1996).

Another example comes from Loomis et al., (1976; as quoted in Passioura, 1996) whose sugar beet (*Beta vulgaris* L.) model failed when they changed plant density, owing to its having the wrong structure for partitioning assimilate between root and shoot. This failure stimulated work on reciprocal grafts between beet (large root, small leaves) and chard (small roots, large leaves) that showed that the voracious appetite of a small fraction of the cells in the root of the beet largely determined the size of the axis (Rapoport and Loomis, 1986; as quoted in Passioura, 1996).

Even if the structure is right—as it might be in some of the leaching models when they are applied to soils in which the flow is essentially one dimensional—the models can rarely be applied with confidence to a field, because the parameters vary greatly in space. We have to assume average values of, say, hydraulic conductivity to apply the Richards equation, and because this equation is not linear, the averaging is an art rather than a well-defined procedure, and often works poorly (Passioura, 1996).

There is much that we do not know about the mechanistic structure of the workings of plant and their interactions with their environment. As crop physiologists and agronomists, we are faced with two main challenges: to illuminate those hidden structures—scientific challenges; and to make use of what we do know to improve the management of agriculture enterprises—an engineering challenge. It is important to distinguish between the two. While we remain ignorant of the essential structure, it is futile to develop mechanistic simulation models to help manage farms that are based on guesses about these structures.

Dobermann and Ping (2004) mentioned the primary and secondary variables for given yield mapping. Ordinary kriging, co-kriging, simple kriging with varying local means, and kriging with external drift were compared using grain yield as the primary variable and three different vegetation indices as secondary variables. They have analyzed and designed the structure of the MZA (Management Zone Analyst).

REFERENCES

Barlow, E.W.R. (1986). Water relations of expanding leaves. Aust. J. Plant Physiol. **13**: 45-58.

Cowan, I.R. (1965). Transport of water in the soil-plant-atmosphere continuum. J. Appl. Ecol.**2**: 221-239.

Dobermann, A. and Ping, J.L. (2004). Geostatistical integration of yield monitor data and remote sensing improves yield maps. Agron. J. **96** (1): 285-297.

Davies, W.J., and Zhang, J. (1991). Root signals and the regulation of growth and development of plants in drying soil. Annu. Rev. Plant Physiol. Plant. Mol. Biol. **42**: 55-76.

Gardner, W.R. (1960). Dynamic aspects of water availability to plants. Soil Sci. **89** : 63-67.

Gold, H.J. (1977). Mathematical Modeling of Biological System: An Introductory Guidebook, 357p. New York: John Wiley & Sons.

Haltenorth, T. (1937). Die verwandtschaftliche Stellung der Grosskatzen Zueinander. Z. Säugetierk, **12**: 97-240.

Loomis, R.S., NG., E., and Hunt, W.F. (1976). Productivity of root crops. pp. 269-286. In: R.H. Burris and C.C. Black (ed.) CO_2 Metabolism and the Productivity of Plants. Baltimore, MD:Univ. Park Press.

Masle, J. (1992). Will plant performance on soils prone to drought or with high mechanical impedance to root penetration be improved under elevated atmospheric CO_2 concentration? Aust. J. Bot. **40**: 491-500.

Passioura, J.B. (1988). Root signals control leaf expansion in wheat seedlings growing in drying soil. Aust. J. Plant Physiol. **15**: 687-693.

Passioura, J.B. (1991). Soil structure and plant growth. Aust. J. Soil Res.**29**: 717-728.

Passioura, J.B. (1996). Simulation models: Science, snake oil, education, or engineering? Agron. J. **88**: 690-694.

Rapoport, H.F., and Loomis, R.S. (1986). Structural aspects of root thickening in *Beta vulgaris* L.: Comparative thickening in sugarbeet and Chard. Bot. Gaz. (Chicago) **147**: 270-277.

Reynolds, J.F. and Acock, B. (1985). Predicting the response of plants to increasing Carbon dioxide: A critique of plant growth models. Ecol. Modell. **29**: 107-129.

Ritchie, J.T. (1983). Personal communication as quoted in Passioura, J.B. (1996).

Saliendra, N.Z., Sporry, J.S. and Comstock, J.P. (1995). Influence of leaf water status on stomatal response to humidity, hydraulic conductance, and soil drought in *Betula occidentalis*. **196**: 357-366.

Sinnott, E.W. and Hammond, D. (1935). Factorial balance in the determination of fruit shape in cucurbita. Amer. Nat., **64**: 509-524.

Sokal, R.R. and Rohlf, F.J. (1981). Biometry: The principles and practice of statistics in biological research. 2nd ed. 859p. New York: W.H. Freeman and Co.

Wit De, C.T. (1970). Dynamic concepts in biology. pp. 17-23. In: Prediction and management of photosynthetic productivity. Proc. IBP/PP.Tech. Meet., Trebon. Wageningen, Netherlands: PUDOC.

3

Specification of Component Behavior

Steps in Quantification

1. Determine the general form or 'Shape' of the relationship.
2. Find the equation which represents that particular shape.
3. Fit the equation to actual form.

3.1 ALGEBRAIC FORM

The category systems that are complicated but not steady state are open-ended in the sense that there is no limit to the degree of complexity that can be built into the system. Since all the necessary environmental factors can be included in the system, such as temperature, humidity, competitor species, parasites, diseases, dispersal and the results of strategies imposed by man (Watt, 1968).

$$\text{Number of adults} = f(X_1, X_2, ... X_n) \qquad (3.1)$$

There are two basic ways to break down the function represented by Eq. 3.1: (1) a series of components which are added; and (2) a series of components which are multiplied. A model may consist of both additive and multiplicative components. Each of these components is a function of several variables. Given an imaginary situation in which the variable to be predicted is a function of 10 other variables

$$Y = f(X_1, X_2, ..., X_{10})$$

The division of the process into components might make the model

$$Y = f(X_1, X_2, X_3) + f(X_4, X_5, X_6, X_7) + f(X_8) + f(X_9, X_{10})$$

The functions represented by each of these components are more responsible to modeling than the single 10 variable function.

Holling's component model of the functional response of a predator to prey density (Holling, 1966; as quoted in Poole (1974) is based on dividing the time period between the eating of one prey to the eating of the next into four components: (1) time taken in a digestive pause after a prey is eaten, TD; (2) time spent searching for a prey, T.S.; (3) time spent pursuing prey, TP, and (4) time spent eating prey, T.E. Each of the four components is modeled separately, and the four components are added to give the total time taken by a predator from the eating of one prey to the eating of the next, i.e., $T = TD + TS + TP + TE$.

A change in the population density can often be represented by a series of multiplicative components. In the hypothetical insect population, the development of eggs to adult may be broken down into three components: (1) the proportion of eggs surviving to become larvae; (2) the proportion of larvae surviving to become pupae and (3) the proportion of pupae surviving to become adults. These three components are equivalent to:

(1)   The probability E of an egg surviving to become a larva;

(2)   The probability L of a larva surviving to become a pupa; and

(3)   The probability P of a pupa surviving to become adult. Because the probability of two independent probabilistic events occurring simultaneously is equal to the product of their probabilities, the probability of an egg surviving to become an adult its $E \times L \times P$. Therefore, the number of adults A resulting from 100 eggs is $A = 100$ ELP. Each of the components E, L and P is a function of one or several environmental variables, so that the model might be

$$A = 100 \, [f(X_1, X_2, X_3)] \, [f(X_4, X_5)] \, [f(X_6, X_7, X_8, X_9)]$$

This type of model is open ended. If the model formulated is not accurate, it is possible to add more variables and formulate the components affected (refinement of structural model). Because of the complexity of the computation, and the several advantages of computer programming terminology, cyclic operations and either/or possibilities, to name two, this type of models are almost always intimately tied in with computer programming and simulation. Apart from their predictive value, the models may also be formulated to serve as a summary of the set of actions and interactions within a population.

The construction of this type of model of population change can be illustrated by a simplified, hypothetical example of the way one might go about making such a model and programming it for a computer. Imagine an animal population divided into three life history stages: egg, larvae and adult. The stages are not overlapping and reproduction occurs only in the adults. Each stage is also affected by a source of mortality.

The increase in the density of the population in one generation using the simple deterministic model is $N_{t+1} = e^r N_t$. Note that no distinction is made between the three stages of the animal, and also that r represents the intrinsic rate of increase but says nothing about why animals die or when and how they are born. In contrast, this type of model breaks this process down into several components such as: (1) mortality to egg, (2) mortality to larvae; (3) mortality of adults before laying eggs; and (4) number of eggs laid for adult. Let us suppose that at the beginning of the generation, there are 100 eggs, and that 45% of the eggs are destroyed every generation by an egg parasite. If we represent this variable egg as EGG, and parasitism by PARA. The model of egg survival is

1. Read, EGGS
2. PARA = 0.45
3. SECS = EGG – (PARA *EGG)

The first statement instructs the computer to read the value of EGG, the number of eggs from some source. The second statement sets the value of parasitism at a constant 45 percent, and the third statement says that the number of larvae, SECS, is equal to the number of eggs minus the number of eggs lost to parasitism. The asterisk indicates multiplication. If 30 percent of the larvae are lost to a predator every generation, PRD, the number of adults is

4. ADLT = SECS – (PRD * SECS)

If 10 adults are lost to a disease every generation

5. ADLT = ADLT – 10.0

Unlike a normal equation, statement 5 is legal in Fortran. The meaning is 'Take the value of ADLT substract 10.0, and call this new value ADLT'.

The possibility now arises that because of all this mortality, the population will become extinct and the computations must stop. If the population has not reached extinction, the process continues. The contingency is provided for by using an if statement.

6. IF (ADLT) 9.9.7

This statement reads, 'If this number of adult is less than 0 or equal to 0, go to statement 9'. The first two numbers represent less than 0 and equal to 0. Statement 9 is

9. **Stop**

and the computations stop. The population has become extinct. If the population does not become extinct and each adult lays 10 eggs, ADLT is greater than 0 and the program goes to statement 7 which is

7. EGG = 10.0 * ADLT
8. PRINT, EGG, ADLT
   GO TO 3

Statement 8 prints the number of adults and eggs, and the next statement starts the entire process over again with a new value for the number of eggs as computed by the program. The computations and cycles will continue till either the population becomes extinct or the computer is told to stop.

The calculations in real situations often become long and complicated. In order to follow this general flow of the computations, the program steps are usually illustrated in the form of a flow chart. This flow chart for the hypothetical example is shown in Fig. 3.1.

The shape of boxes depends on the type of program step or steps involved. In an example, statements are in diamonds, READ and PRINT statements in trapezoids, computations in rectangles, and start and stop instructions in ovals.

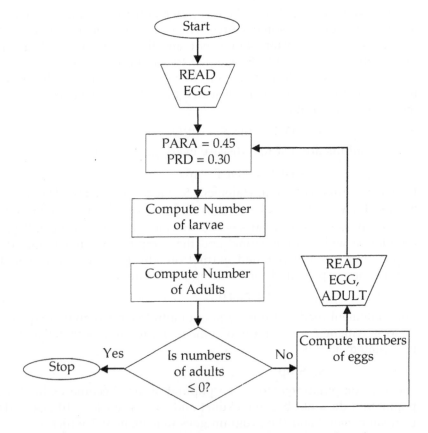

Fig. 3.1 A flowchart for the programmed model of the hypothetical insect populations in the text.

Formulating this type of model in summary consists of four steps:

1. Break the process down into its component steps.
2. Determine the variables significantly affecting different components.
3. Model the components and reassemble the model, including programming of the model for a computer if necessary.
4. Test the model with real or artificial data to determine if the model accurately mimics the process we originally set out to model. Application of the model to data will suggest improvement (refinement) in the component models.

3.1.1 Matrix Algebraic Form for Studying a Specific Behavior of Components

The use of matrices in population demography was largely developed in two articles by Leslie (1945, 1948, as quoted in Poole, 1974). Three basic statistics are required for the use of this technique:

$N_{x,t}$ = the number of females alive in the age group x to $x + 1$ at time t.

F_x = the probability that a female in the x to $x + 1$ age group at time t will be alive in the age group $x + 1$ to $x + 2$ at time $t + 1$.

F_x = the number of female offspring born in the interval t to $t + 1$ for female aged x to $x + 1$ at time t that will be alive in the age group 0 to 1 at time $t + 1$.

It is assumed that the population consists of discrete age grouping, in contrast to the reality of a continuously changing set of fertility and mortality statistics. For computational purposes, P_x is equal to L_{x+1}/L_x, and F_x is equal to m_x.

The basic information needed to study density changes and rates of increase or decrease is contained in a life table. A life table contains vital statistics as the probability of an individual of a certain age dying or, conversely, the average number of offspring produced by a female at a given age.

The most reliable method of determining these statistics is to begin with a group of individuals all born at the same time, a cohort, and follow the life of the cohort, noting the deaths of individuals and the birth of offspring until the demise of the last individual. It is convenient in the calculations to use a cohort of 100 or 1000 individuals. Rabinovich (1970, as quoted in Poole, 1974) formulated a life table for a muscide fly, *Synthesiomyia nudiseta* reared at 20°C. He began with 100 eggs and counted the number of individuals remaining after intervals of 5 days, with the exception of a first interval of only 2.5 days. The number of the original 100 individuals surviving at 5-day intervals until the last individual died approximately 95 days later, as shown in Fig. 3.2. The curve, called a survivorship curve, demonstrates a high initial mortality followed by a steady decline in numbers to the death of the last individual. The number of individuals of the initial cohort alive at age x is termed l_x. The value of l_{125}, the number of individuals alive at day 12.5, is 64. The statistic l_x will be given a slightly different definition when estimating r, the intrinsic rate of increase to define l_x as the percent of the cohort alive at the mid point of the time interval. The l_x values for the fly cohort are tabulated in Table 3.1.

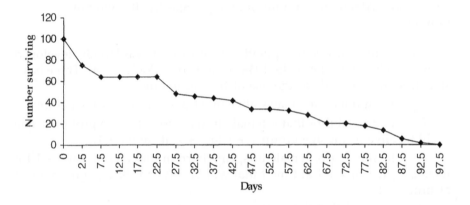

Fig. 3.2 Survivorship curve for a cohort of 100 individuals of the muscid fly
S. nudioeta reared at 20°C from egg to the death of the last individual
(After Rabinovich, 1970, as quoted in Poole, 1974).

The number of individuals dying during each interval of time is usually noted as d_x and is simply the number of individuals alive at age x minus the number alive at age $x + 1$; i.e. $d_x = l_x - l_{x+1}$. If $l_{25} = 75$ and $l_{75} = 64$, d_x for the 5-days interval between ages 2.5 and 7.5 days is 11. The d_x values for the cohort of flies are also listed in Table 3.1. The percentage of the cohort dying in the interval x to $x + 1$ of those alive at age x is q_x and is the value of d_x for the same interval of time divided by the number of flies alive at age x, l_x. The percentage of flies dying during the interval between 2.5 and 7.5 days, which were alive at 2.5 days, is 11/75, or 14.7%.

A slightly more complicated set of statistics to arrive at are the number of individuals alive during the interval between x and $x + 1$, L_x. Stating that individuals are dying continuously that is no exact answer except at an instant of time. Therefore L_x, in reality, represents the average number of individuals alive during the interval. Strictly speaking, L_1 is the total number of units of time during the interval x to x +1 lived by all the individuals of the cohort and is equal to

Table 3.1 The life table for a cohort of 100 individuals of the fly
S. *nudiseta* reared at 20°C from EGG to the death of the last individual

X days	q_1	l_1	d_1	L_1	T_1	e_1
0	0.250	100	25	87	768	19.20
2.5	0.147	75	11	69	681	45.40
7.5	0.000	64	0	64	612	47.81
12.5	0.000	64	0	64	548	42.81
17.5	0.000	64	0	64	484	37.81
22.5	0.250	64	16	56	420	32.81
27.5	0.041	48	2	47	364	37.92
32.5	0.043	46	2	45	317	34.46
37.5	0.045	44	2	43	272	30.91
42.5	0.190	42	8	38	229	27.26
47.5	0.000	34	0	34	191	28.09
52.5	0.058	34	2	33	157	23.08
57.5	0.125	32	4	30	124	19.38
62.5	0.285	28	8	24	94	16.79
67.5	0.000	20	0	20	70	17.50
72.5	0.100	20	2	19	50	12.50
17.5	0.222	18	4	16	31	8.61
82.5	0.000	14	8	10	15	5.35
87.5	0.571	6	4	4	5	4.15
92.5	1.000	2	2	1	1	2.00
97.5	...	0				

the area below the survivorship curve and between the two points x and x + 1. Again, strictly speaking, this area is found by integration of the l_1 curve between x and x + 1, or

$$L_1 = \int_x^{x+1} l_x \, dx$$

where dx is not the d_1 of the life table and only a part of the integral notation. These are two practical methods of determining the value of L_1 of some interval x to x + 1. If it can be assumed that individuals are dying at a constant rate during the interval of time, i.e., l_1 and l_{1+1} can be connected by a straight line, L_1 is the average of l_1 and l_{1+1}.

$$L_1 = \frac{l_x + l_{x+1}}{2}$$

If this cannot be assumed and the line connecting l_1 and l_{1+1} in the survivorship curve is known to be curvilinear, and not straight, an approximate value of L_1 may be arrived at by finding the value of l_1 on the l_1 curve corresponding to $x + 1/2$, the midpoint of the age interval shown in Fig. 3.3. Usually, nothing is known about the changes in rates of mortality during the age interval and Eq. 3.1 will be the appropriate way of determining the series of L_x statistics. The values of L_x are usually called the stationary age distribution.

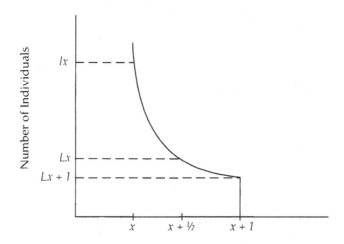

Fig. 3.3 Graphic method of finding L_1 when the decrease in number of survivors from time period x to $x + 1$ is not linear.

One further statistics is T_{x}, the number of units of times, 5 days for the fly cohort, lived by the cohort from age x, until all individuals are decreased. The values of T_1 are found by adding up the L_x column from x to the end. From the T_1 statistics, the life expectancy of an individual of age x from age x, e_1, can be determined.

$$e_x = \frac{T_x}{l_x}$$

The life expectancy of a 12.5-days-old muscid fly is $e_{12.5} = 548/64$, or 8.56 age intervals of 5 days, which is equivalent to 42.8 days.

One final statistics, m_{x}, is the average number of offspring produced per female of age x from x to $x + 1$. Like l_{x}, m_x is a continuously changing variable, and in its calculation, the standard procedure is to take the number of offspring produced per female in the interval x to $x + 1$ as m_x. The exact value of m_x cannot, in general, be determined and the above method of estimation assumes that all the births occur instantaneously at the midpoint of the time interval. Therefore, between $x = 4$ and $x = 5$,

$m_{4.5}$ is taken to be the average number of offspring produced by a single female between $x = 4$ and $x = 5$.

Table 3.2 A partial life table of the weevil *C. oryzae*

Age in weeks:			
x	l_x	m_x	$l_x m_x$
4.5	0.87	20.0	17.400
5.5	0.83	23.0	14.090
6.5	0.81	15.0	12.150
7.5	0.80	12.5	10.000
8.5	0.79	12.5	9.875
9.5	0.77	14.0	10.780
10.5	0.74	12.5	9.250
11.5	0.66	14.5	9.570
12.5	0.59	11.0	6.490
13.5	0.52	9.5	4.940
14.5	0.45	2.5	1.125
15.5	0.36	2.5	0.900
16.5	0.29	2.5	0.800
17.5	0.25	4.0	1.000
18.5	0.19	1.0	0.190

R0 = 113.560.
Source: From Birch, 1948.

In Table 3.1, l_x was defined as the number of individuals of a cohort alive at age x. However, l_x is always changing, although usually it is known only at discrete intervals of time. Therefore, in the computations that follow, it is necessary to ignore the continuous nature of l_x and m_x and treat them as numbers changing in jumps. In other words, mortality and fertility are assumed to occur instantaneously at the midpoint of the time interval. In contrast to the definition of l_x given in Table 3.1, it is standard notation in the literature when estimating r, intrinsic rate of increase to define l_x in Table 3.2, as the percent of the cohort alive at the midpoint of the time interval. In terms of the Table 3.1, l_x as redefined is L_x divided by the initial size of the cohort. Birch (1948, as quoted in Poole, 1974) has estimated the l_x and m_x statistics as defined above for a cohort of the weevil *Calandra oryzae* in unlimited environment of wheat at 20°C and 14% relative humidity (Table 3.2).

In any one generation, total number of individuals produced per female per generation is equal to the fecundity of the female m_x, as modified by the probability of the living to be that old, l_x. A female in the age group 8 to 9 weeks could produce 12.5 offspring (Table 3.2) but has a 79% chance of reaching that age. Therefore, the total number of offspring produced per female during the interval of time is

$$l_x m_x \ (8\text{-}9) = (0.79) \ (12.5) = 9.88 \text{ offspring}$$

The total number of female offspring produced per female during a single generation is equal to the sum of the $l_x m_x$ products, or

$$R_0 = \sum_{x=0}^{n} l_x m_x$$

In studies of population density using the matrix algebra, the following three statistics are very important:

- $n_{x,t}$ = the number of females alive in the age group x to $x + 1$ at time t.
- P_x = The probability that a female in the x to $x + 1$ age group at time t will be alive in the age group $x + 1$ to $x + 2$ at time $t + 1$.
- F_x = number of female offspring born in the interval t to $t + 1$ per female aged x to $x + 1$ at time t that will be alive in the age group 0 to 1 at time $t + 1$.

Using these three statistics, a series of equations can be created accounting for the change in number of individuals from one time, t, to the next $t + 1$. Given F_x (fertility rate) and the number of females in each age group ($n_{x,t}$) at time t, the number of females alive in the 0 to 1 age group after one interval of time $t + 1$ is equal to the sum of the numbers of females produced by each age group.

$$\sum_{x=0}^{n} F_x \, n_{x,0} = n_{0,1}$$

As a simple example with four age groups.

$$n_{0,1} = F_0 n_{0,0} + F_1 n_{1,0} + F_2 n_{2,0} + F_3 n_{3,0}$$

If the four F_x values are 2, 6, 8, and 2 and the number of individuals in each of the four age groups at time t are 10, 6, 8 and 4, the number of individuals in the 0 to 1 age group in the next generation will be

$$n_{0,1} = (2) \ (10) + (6) \ (6) + (8) \ (8) + (2) \ (4) = 128 \text{ individuals}$$

The number of females in the age interval 1 to 2 at time $t + 1$ is equal to the number of individuals in the age group 0 to 1 at time t times the probability that the female will survive to the next age group, P_x, or

$$p_0 n_{0,0} = n_{1,1}$$

The argument is the same for all higher age groups there are, therefore, a set of $n + 1$ linear equations in which n is the last age group listed in the life table.

$$\sum F_x \, n_{x,0} = n_{0,1}$$

$$P_0 n_{0.0} = n_{1.1}$$

$$P_1 n_{1.0} = n_{2.1}$$

$$P_2 n_{2.0} = n_{3.1}$$

$$P_{n-1} x_{n-1\,0} = n_{n.1}$$

These equations can be represented in matrix forms as

$$Mn_0 = n_1$$

where n_0 is a column vector giving the distribution of individuals in age groups at time 0, and n_1 is the column vector for the population after one interval of time.

Written as matrices, this set of equations becomes

$$
\begin{bmatrix}
F_0 & F_1 & F_2 & F_3 & \cdots & F_{n-1} & F_n \\
P_0 & 0 & 0 & 0 & & 0 & 0 \\
0 & P_1 & 0 & 0 & & 0 & 0 \\
0 & 0 & P_2 & 0 & & 0 & 0 \\
0 & 0 & 0 & P_3 & & 0 & 0 \\
\cdot & \cdot & \cdot & \cdot & & & \\
\cdot & \cdot & \cdot & \cdot & & & \\
\cdot & \cdot & \cdot & \cdot & & & \\
0 & 0 & 0 & 0 & & P_{n-1} & 0
\end{bmatrix}
\begin{bmatrix}
n_{0.0} \\
n_{1.0} \\
n_{2.0} \\
n_{3.0} \\
n_{4.0} \\
\cdot \\
\cdot \\
\cdot \\
n_{n.0}
\end{bmatrix}
=
\begin{bmatrix}
n_{0.1} \\
n_{1.1} \\
n_{2.1} \\
n_{3.1} \\
n_{4.1} \\
\cdot \\
\cdot \\
\cdot \\
n_{n.1}
\end{bmatrix}
$$

If the matrix M is post-multiplied by the column vector n_0, the equations listed above are restored. The matrix M is of order $n + 1$ and is square.

As a simple example with four age groups along with a mortality and fertility schedule as in the matrix M below and a distribution of individual into age groups as in the column vector n_0, the number of individuals in each of the four age groups after one time interval is given by the column vector n_1.

$$
\begin{bmatrix}
0 & 1 & 2 & 3 \\
0.6 & 0 & 0 & 0 \\
0 & 0.4 & 0 & 0 \\
0 & 0 & 0.2 & 0
\end{bmatrix}
\begin{bmatrix}
10 \\
5 \\
2 \\
1
\end{bmatrix}
=
\begin{bmatrix}
12 \\
6 \\
2 \\
0.4
\end{bmatrix}
$$

The matrix model does not assume a stable age distribution. Leslie (1945) gives an example of the use of the matrix model in simulating population growth in a laboratory population of the Norway rat, *Rattus norwegicus*.

Because of the relationship

$$M \, n_0 = n_1$$

Then

$$M \, n_1 = n_2 \qquad \text{or} \qquad M^2 n_0 = n_2$$

and

$$M \, n_2 = n_3 \qquad \text{or} \qquad M^3 n_0 = n_3$$

In general

$$M^r n_0 = n_r \tag{3.2}$$

The number and distribution of individuals after r intervals of time is equal to the matrix M raised to the rth power times the original n_0 column vector, as long as the mortality and fertility statistics defining the rate of population growth remains the same.

To illustrate the use of Eq. (3.2), suppose that the M matrix is

$$\begin{bmatrix} 0.00 & 6.43 & 14.00 & 18.00 \\ 0.78 & 0 & 0 & 0 \\ 0 & 0.71 & 0 & 0 \\ 0 & 0 & 0.60 & 0 \end{bmatrix}$$

and the initial population distributions is

$$\begin{bmatrix} 0 \\ 0 \\ 20 \\ 0 \end{bmatrix}$$

This example is actually possible because many experiments begin with an initial population of adults which are of the same age group. After one interval of time the number and distribution of individuals is

$$n_1 = \begin{bmatrix} 280 \\ 0 \\ 0 \\ 12 \end{bmatrix}$$

and after two, three, and four interval

$$n_2 = \begin{bmatrix} 216 \\ 218 \\ 0 \\ 0 \end{bmatrix}$$

$$n_3 = \begin{bmatrix} 1402 \\ 168 \\ 155 \\ 0 \end{bmatrix}$$

$$n_4 = \begin{bmatrix} 3250 \\ 1094 \\ 119 \\ 93 \end{bmatrix}$$

The age group distribution is changing, although with time the population will reach a stable age distribution. Of more interest is the change in numbers of individual with time. The total number of individuals after each interval of time is plotted in Figure 3.4.

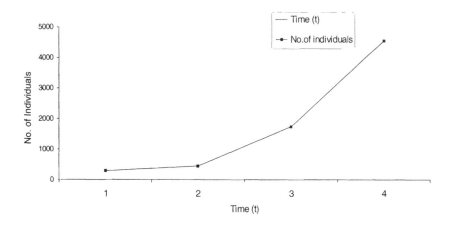

Fig. 3.4 The population increase in a hypothetical population with relatively high survival probabilities for each age group.

Instead of a smooth experimental curve, this curve oscillates slightly. These oscillations will damp themselves out as the age distribution becomes stable. This example demonstrates the utility of the matrix equations. The exponential equation treats all individuals of the population as identical without regard to their age or assumes that the age groups form a stable age distribution. In most real cases, the individuals of a population do not have a stable age distribution.

These oscillations are even more pronounced if the mortality of the early age group is high. If the P_2 values in the M matrix are changed to $P_0 = 0.20$, $P_1 = 0.10$ and $P_3 = 0.05$, then

$$n_1 = \begin{bmatrix} 280 \\ 0 \\ 0 \\ 1 \end{bmatrix}$$

$$n_2 = \begin{bmatrix} 18 \\ 56 \\ 0 \\ 0 \end{bmatrix}$$

$$n_3 = \begin{bmatrix} 360 \\ 4 \\ 6 \\ 0 \end{bmatrix}$$

$$n_4 = \begin{bmatrix} 109 \\ 72 \\ 0 \\ 0 \end{bmatrix}$$

$$n_5 = \begin{bmatrix} 463 \\ 22 \\ 7 \\ 0 \end{bmatrix}$$

The population continues to increase, even at these high mortalities, but the oscillations are pronounced (Fig. 3.5).

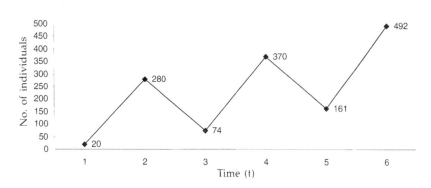

Fig. 3.5 The population oscillations in a hypothetical population with an unstable age distribution and low survival probabilities of the age groups.

Given enough time, the population will achieve a stable age distribution and stop oscillating.

The dominant latent root of the matrix M is equal to the finite rate of increase of a population with a stable age distribution. The rate of increase changes and the numbers oscillate in a population without a stable age distribution until the population achieves a stable age distribution.

The dominant latent root of the matrix M may be found by the process of diagonal expansion. However, Leslie (1948) has proposed a simpler method of calculating the latent root of M starting with a hypothetical matrix.

$$\begin{bmatrix} 0 & 45/7 & 18 & 18 \\ 7/9 & 0 & 0 & 0 \\ 0 & 5/7 & 0 & 0 \\ 0 & 0 & 3/5 & 0 \end{bmatrix}$$

Form a diagonal matrix H with diagonal elements $h_{1.1} = P_0,\ P_1,\ P_2,\ h_{2.2} = P_1 P_2,\ h_{3.3} = P_2$ and $h_{4.4} = 1$

$$\begin{bmatrix} 1/3 & 0 & 0 & 0 \\ 0 & 3/7 & 0 & 0 \\ 0 & 0 & 3/5 & 0 \\ 0 & 0 & 0 & 1 \end{bmatrix}$$

and then form a new matrix B by taking

$$B = HM\ H^{-1}$$

when H^{-1} is the inverse of the matrix H. The inverse of a diagonal matrix such as H is another diagonal matrix with the elements of the diagonal the reciprocal of the elements of the original matrix, in this case

$$H^{-1} = \begin{bmatrix} 3 & 0 & 0 & 0 \\ 0 & 7/3 & 0 & 0 \\ 0 & 0 & 5/3 & 0 \\ 0 & 0 & 0 & 1 \end{bmatrix}$$

In the example

$$B = \begin{bmatrix} 0 & 5 & 10 & 6 \\ 1 & 0 & 0 & 0 \\ 0 & 1 & 0 & 0 \\ 0 & 0 & 1 & 0 \end{bmatrix}$$

Instead of decimals in the sub-diagonal of the matrix as in M, in B there are all 1s. The top row of B gives the coefficients of the equation for the latent roots of M. The equation for the example is:

$$\lambda^4 - 5\lambda^2 - 10\lambda - 6 = 0$$

Because there are four rows and columns, there are four latent roots. Although in this case the roots are relatively easy to find, in most actual situations:

(1) Some of the roots will be complex, i.e. contain imaginary components, and (2) the roots will be too many to find with elementary algebra. Therefore, a numerical computer program is ordinarily needed to find the roots of the equation.

In the above case, the largest positive root, the dominant latent root is 3. Therefore, the finite rate of increase is 3. If the finite rate of increase exists, the population will eventually reach a stable age distribution, provided the largest positive root is larger than the absolute values of all the remaining roots.

Returning to the definition of latent roots and vectors, there is not just one latent vector per root, but an infinite number. However, the elements of every latent vector of a given root are always in the same proportion which, after all, is what a stable age distribution is all about. The stable age distribution of a population is calculated as

$$Mn_s = \lambda_1 n_s \tag{3.3}$$

where n_s is the stable age distribution of the population, and λ_1 is the dominant latent root of the matrix M. To help the computations the matrix B is used again. Equation (3.3), using the B matrix, becomes

$$Bv_s = \lambda_1 v_s$$

where v_s is a latent vector of B but not of M. The vector v_s is calculated as

$$v_{0,s} = \lambda_1 v_{1,s}$$
$$v_{1,s} = \lambda_1 v_{2,s}$$
$$v_{2,s} = \lambda_1 v_{3,s} \text{, etc.}$$

By arbitrarily taking any one of the age groups as 1, the rest can be calculated. In the example, the dominant latent root is 3, and if the last age group is taken as 1, $v_{2,s} = \lambda_1 v_{3,s}$, or $v_{2,s} = (3)(1) = 3$, and so forth, giving the vector v_s as

$$v_s = \begin{bmatrix} 27 \\ 9 \\ 3 \\ 1 \end{bmatrix}$$

This vector is converted to the stable age distribution of the matrix M by the equation

$$n_s = H^{-1} v_s$$

where n_s is the stable age distribution of the population. In the example

$$n_s = \begin{bmatrix} 3 & 0 & 0 & 0 \\ 0 & 7/3 & 0 & 0 \\ 0 & 0 & 5/3 & 0 \\ 0 & 0 & 0 & 1 \end{bmatrix} \begin{bmatrix} 27 \\ 9 \\ 3 \\ 1 \end{bmatrix} = \begin{bmatrix} 81 \\ 21 \\ 5 \\ 1 \end{bmatrix}$$

If the dominant latent root of the M matrix of fertility and mortality statistics is greater than 1, the population will increase, and if the root is less than 1, then population will become extinct. For a game or pest animal for which hunting will cause a known mortality in each age group, the matrix M can be created and the dominant latent root calculated. If the dominant latent root is less than 1, the regime of hunting will cause the animal to become extinct.

3.1.1.1. USE OF MATRIX ALGEBRA IN PRINCIPAL COMPONENT ANALYSIS

Principal component analysis is one of the simplest multivariate methods. The object of the analysis is to take P variables $X_1, X_2, ..., X_p$

and find combinations of these to produce indices Z_1, Z_2, ..., Z_p that are correlated. The lack of correlation is a useful property because it means that the indices are measuring different 'dimensions' in the data. However, the indices are also ordered so that Z_1 displays the largest amount of variation; Z_2 displays the second largest amount of variation, and so on. That is, var $(Z_1) \geq$ var $(Z_2) \geq ... \geq$ var (Z_p), where var (Z_1) denotes the variance of Z_1 in the data set being considered. The Z_1 are called the principal components. When doing a principal components analysis, there is always the hope that the variance of most of the indices will be so low as to be negligible. In that case, the variation in the data set can be adequately described by the few Z variables with variances that are not negligible. Some degree of economy is then achieved and the variation in the p original X variables is accounted for by a smaller number of Z variables.

It must be stressed that a principal components analysis does not always work in the sense that a large number of original variables are reduced to a small number of transformed variables. Indeed, if the original variables are uncorrelated, then the analysis does absolutely nothing. The best results are obtained when the original variables are very highly correlated, positively or negatively. If that is the case then it is quite conceivable that 20 or 30 original variables can be adequately represented by two or three principal components. If this desirable state of affairs does occur, then the important principal components will be of some interest in measuring the underlying 'dimension' in the data. However, it will also be of value to know that there is a good deal of redundancy in the original variables, with most of them measuring similar things (Manly, 1994).

Before launching into a description of the calculations involved in a principal components analysis, it may be of some value to briefly scan the outcome of the analysis when it is applied to the data in Table 3.3 on 20 measurements of six cultivars of chickpea. Many measurements are correlated, as shown in Table 3.4. This is, therefore, good material for the analysis in question. It turns out, as we shall see, that the first principal component has a variance of 7.816 and second principal component has a variance of 7.393, whereas all the other components have variances very much less than these (2.943, 1.507, .441, and .005) (Table 3.5). This means that the first and second principal components are by far the most important of the twenty components in representing the variations in the measurements of the five cultivars.

First principal component for variety Katila:

$Z_1 = 0.286\ X_1 + 0.288\ X_2 - 0.127\ X_3 + 0.054\ X_4 + 0.0001\ X_5 - 0.121\ X_6 - 0.342\ X_7 - 0.343\ X_8 - 0.314\ X_9 - 0.316\ X_{10} - 0.065\ X_{11} + 0.282\ X_{12} +$

$$0.16\ X_{13} - 0.23\ X_{14} - 0.059\ X_{15} + 0.285\ X_{16} + 0.47\ X_{17} - 0.255\ X_{18} - 0.153\ X_{19} + 1.79\ X_{20}$$

where X_1, X_2, ... , X_{20} represent here the measurements in Table 3.3 after they have been standardized to have zero means and unit standard deviations. The standardized data are given in the Table 3.6.

Second component for variety Katila:

$$Z_2 = -1.71\ X_1 + -0.147\ X_2 + -0.308\ X_3 + -0.327\ X_4 + -0.359\ X_5 + -0.340\ X_6 + -0.206\ X_7 -0.320\ X_8 + -0.050\ X_9 + -0.157\ X_{10} + 0.179\ X_{11} + -0.129\ X_{12} + 0.065\ X_{13} -0.101\ X_{14} + 0.347\ X_{15} + -0.151\ X_{16} + 0.152\ X_{17} + -0.223\ X_{18} + 0.248\ X_{19} + -0.294\ X_{20}$$

Table 3.3 Phenological, partitioning, pod filling and photosynthetic traits, 2000-01, of different cultivars of chickpea (unpublished data).

Traits	Katila	Vijay	RSG 143-1	Phule G5	ICC 4958	K 850	Mean	Standard deviation
HT	24.49	26.08	25.10	25.95	26.31	25.66	25.59	0.689
HI	0.24	0.33	0.30	0.30	0.36	0.29	0.303	0.04
BM/Plant	3.25	3.56	3.78	3.75	3.84	5.31	3.915	0.716
Sy/Plant	0.17	1.16	1.12	1.11	1.39	1.56	1.185	0.270
A-PM (d)	14	14	18	18	19	20	17.166	2.562
A-PM (dd)	298	271	376	343	363	426	346.168	55.726
S-A(d)	83	77	86	72	72	87	79.5	6.774
S-A (dd)	1112	1013	1104	936	923	1179	1044.5	103.647
S-PM (d)	97	91	104	90	91	107	96.666	7.339
S-PM (dd)	1410	1284	1480	1279	1286	1605	1390.666	133.488
Pod/Plant	6.9	7.5	4.8	4.9	5.4	6.5	6.0	1.124
SFR mg/d	10	10	10	11	12	10	10.5	0.836
SFR mg/dd	0.6	0.5	0.4	0.6	0.6	0.5	0.533	0.081
RWR	0.32	0.25	0.29	0.32	0.28	0.38	0.306	0.044
Straw WR	0.36	0.33	0.33	0.31	0.28	0.26	0.312	0.036
AI	0.31	0.48	0.42	0.42	0.56	0.41	0.433	0.082
HWR	0.08	0.10	0.08	0.07	0.08	0.08	0.081	0.0098
Wu/Pot	5.5	3.6	6.6	5.3	5.1	7.0	5.516	1.205
No.seed/Pod	2	1	1	1	1	1	1.166	0.408
SW/Pod	127	147	163	235	248	215	189.166	50.113

In the same way, these first two components are calculated for all the six varieties. In Table 3.7, the value of Z_1 and Z_2 in different 6 varieties are shown.

Figures 3.6 a,b and 3.7 have demonstrated the classification of varieties and agrophysiological traits. These figures are drawn from the data given in Table 3.7 and Table 3.5, respectively. The correlation coefficient are given in Table 3.4.

Table 3.4 Correlation coefficients between twenty variables calculated from the data of Table 3.3

Variables	X_1	X_2	X_3	X_4	X_5	X_6	X_7	X_8	X_9	X_{10}	X_{11}	X_{12}	X_{13}	X_{14}	X_{15}	X_{16}	X_{17}	X_{18}	X_{19}	X_{20}
X_1	1	0.898	0.248	0.674	0.380	0.094	-0.687	-0.665	-0.497	-0.473	-0.119	0.620	0.163	-0.289	-0.650	0.878	0.182	-0.41	-0.73	0.7
X_2		1	0.75	0.605	0.344	0.075	-0.601	-0.657	-0.431	-0.475	-0.222	0.657	-0.041	-0.548	-0.518	1	0.289	-0.4	-0.77	0.57
X_3			1	0.841	0.740	0.828	0.410	0.465	0.636	0.705	-0.011	-0.105	-0.265	0.733	-0.857	0.086	-0.155	0.62	-0.45	0.46
X_4				1	0.784	0.711	0.001	0.018	0.276	0.310	-0.122	0.296	-0.201	0.302	-0.975	0.612	0.019	0.28	-0.75	0.69
X_5					1	0.938	0.037	0.002	0.386	0.388	-0.638	0.419	-0.128	0.520	-0.870	0.358	-0.57	0.66	-0.60	0.82
X_6						1	0.375	0.343	0.675	0.679	-0.507	0.131	-0.284	0.692	-0.778	0.096	-0.54	0.86	-0.42	0.59
X_7							1	0.948	0.941	0.919	0.216	-0.801	-0.660	0.490	0.088	-0.597	0.07	0.67	0.25	-0.52
X_8								1	0.903	0.917	0.345	-0.811	-0.528	0.584	0.045	-0.645	0.08	0.63	0.31	-0.5
X_9									1	0.980	-0.027	-0.586	-0.650	0.633	-0.224	-0.422	-0.12	0.85	0.02	-0.18
X_{10}										1	0.062	-0.577	-0.530	0.737	-0.287	-0.458	-0.14	0.84	0.06	-0.14
X_{11}											1	-0.488	0.087	-0.061	0.217	-0.206	0.74	-0.43	0.39	-0.56
X_{12}												1	0.590	-0.217	-0.431	0.695	-0.36	-0.2	-0.29	0.78
X_{13}													1	0.93	-0.022	0.015	-0.33	-0.35	0.4	0.33
X_{14}														1	-0.461	-0.514	-0.58	0.78	0.14	0.26
X_{15}															1	-0.54	0.21	-0.36	0.65	-0.83
X_{16}																1	-0.54	-0.37	-0.72	0.58
X_{17}																	1	0.26	-0.61	-0.52
X_{18}																		1	-0.006	0.18
X_{19}																			1	-0.60
X_{20}																				1

Table 3.5 The eigen values and eigen vectors of the correlation matrix for twenty measurements on 5 chickpea genotypes. The eigen values are the variances of the principal components. The eigen vectors give the coefficient of the standardized variables. Eigen vector, coefficient of

Variance	X_1	X_2	X_3	X_4	X_5	X_6	X_7	X_8	X_9	X_{10}	X_{11}	X_{12}	X_{13}	X_{14}	X_{15}	X_{16}	X_{17}	X_{18}	X_{19}	X_{20}
7.8160	0.286	0.288	-0.127	0.054	0.0001	-0.121	-0.342	-0.343	-0.314	-0.316	-0.065	0.282	0.160	-0.230	-0.059	-0.285	0.047	-0.255	-0.153	0.179
7.3930	-0.171	-0.147	-0.308	-0.327	-0.359	-0.340	-0.206	-0.020	-0.150	-0.157	0.179	-0.129	0.065	-0.101	0.347	-0.151	0.152	-0.223	-0.248	-0.294
2.9430	-0.173	-0.235	-0.130	-0.218	0.099	0.070	-0.156	-0.142	-0.108	-0.085	-0.298	0.235	0.377	0.221	0.066	-0.212	-0.517	0.163	0.251	0.181
1.5070	-0.133	0.117	0.269	-0.172	0.093	0.075	0.080	-0.083	0.107	-0.034	-0.554	-0.012	-0.486	-0.335	0.253	0.076	-0.129	0.161	-0.224	-0.086
0.4410	0.263	-0.235	0.17	-0.063	-0.052	-0.133	-0.173	-0.105	-0.177	-0.134	-0.014	-0.47	-0.151	0.248	0.056	-0.337	-0.147	-0.149	-0.486	0.087
0.0048	0.48	0.008	0.006	0.094	-0.124	-0.084	0.145	0.074	0.019	-0.028	0.291	-0.029	-0.124	-0.398	-0.322	-0.006	-0.733	-0.146	-0.0002	-0.133
0	0.028	0.075	-0.047	0.059	0.108	-0.137	0.189	-0.071	0.111	-0.217	0.009	0.163	0.278	-0.326	-0.226	-0.599	0.228	0.344	-0.241	-0.093
0	0.024	-0.399	-0.012	-0.143	-0.129	-0.092	0.150	0.009	-0.032	-0.373	-0.182	-0.282	-0.031	-0.077	-0.470	0.351	0.168	0.249	0.184	0.203
0	0.324	-0.271	0.097	-0.016	-0.245	0.132	-0.547	0.141	0.263	0.350	-0.182	0.037	0.059	-0.387	-0.043	-0.104	0.085	0.032	0.121	-0.012
0	-0.254																			
0	-0.243																			
0	-0.025																			

Table 3.6 Standardized values of Table 3.3

Traits	Katila	Vijay	RSG 143-1	Phule G5	ICC 4958	K 850
X_1	−1.619	0.704	−0.728	0.514	1.040	0.090
X_2	−1.575	0.675	−0.075	−0.075	1.425	−0.325
X_3	−0.928	−0.495	−0.188	−0.230	−0.104	1.948
X_4	−1.537	−0.092	−0.240	−2.77	0.759	1.388
X_5	−1.235	−1.235	0.325	0.325	0.715	1.106
X_6	−0.8643	−1.3488	0.535	−0.0568	0.3021	1.4326
X_7	0.516	−0.369	0.959	−1.121	−1.121	1.107
X_8	0.651	−0.303	0.574	−1.046	−1.172	1.297
X_9	0.051	−0.772	0.999	−0.908	−0.772	1.408
X_{10}	0.144	−0.779	0.669	−0.836	−0.784	1.605
X_{11}	0.800	1.334	−1.067	−0.978	−0.533	0.444
X_{12}	−0.598	−0.598	−0.598	0.598	1.794	−0.598
X_{13}	0.827	−0.407	−1.641	0.827	0.827	−0.407
X_{14}	0.318	−1.272	−0.363	0.318	−0.590	1.681
X_{15}	1.361	0.527	0.527	−0.027	−0.861	−1.416
X_{16}	−1.463	0.573	−0.158	−0.158	1.548	−0.280
X_{17}	−0.102	1.938	−0.102	−1.122	−0.102	−0.102
X_{18}	−0.013	−1.59	0.899	−0.179	−0.346	1.231
X_{19}	2.044	−0.406	−0.406	−0.406	−0.406	−0.406
X_{20}	−1.240	^−0.841	−0.522	0.914	1.174	0.515

Table 3.7 The values of Z_1 and Z_2 for agrophysiological variables in chickpea genotypes

Varieties	Z_1	Z_2
Katila	−2.43	3.75
Vijay	1.85	2.55
RSG 143-1	−1.95	−0.43
Phule G-5	1.9	−0.44
ICC 4958	3.65	−1.75
K 850	−2.995	−3.61

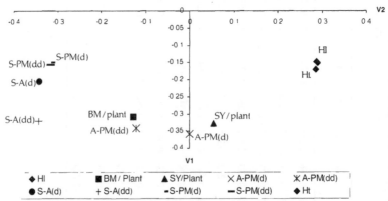

Fig. 3.6a Classification of agro-pheno-morpho-physiological traits in chickpea (*Cicer arietinum*) through principal component analysis.

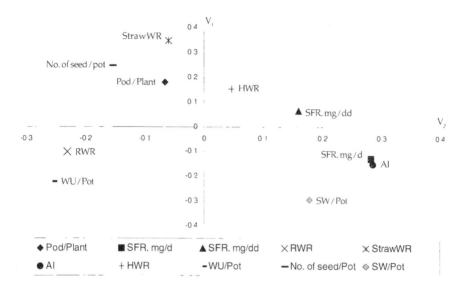

Fig. 3.6b Classification of agro-pheno-morpho-physiological traits in chickpea (*Cicer arietinum*) through principal component analysis.

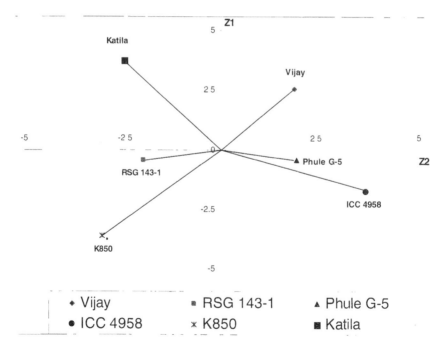

Fig. 3.7 Varietal Classification of Chickpea (*Cicer arietinum*) through principal component analysis.

3.1.1.2 USE OF MATRIX ALGEBRA IN LINEAR PROGRAMMING FOR OPTIMIZATION OF THE SYSTEM

Farming system is a complex involving allocating the resources as well as development management, activities and decision, which—in an unit of farm or combination of farm units, — result in agricultural production, marketing and processing of produce.

Table 3.8 shows the constraints (land and water) and activities (crops).

Table 3.8 The constraints land and water and activities (crops) (Singh *et al.*, 1978)

Constraints	Activities	
	Rs 7398/ha. as net profit from wheat	Rs 3238/ha. as net profit from barley
Land 70 ha.	1	1
15 Nov. to 30 Nov. 269.82 ha.cm.	1.7	1.7
December 257.88 ha.cm.	2.92	2.92
January 327.68 ha.cm.	7.10	6.62
February 345.81 ha.cm.	13.24	7.40
March + upto 8 April 303.42 ha.cm.	13.12	2.14

Let us maximize the net profit from wheat and barley.

Let X_1 = Wheat

X_2 = Barley

Restriction:

(i) $X_1 + X_2 \leq 70$

(ii) $1.7\,X_1 + 1.7X_2 \leq 269.82$

(iii) $2.92\,X_1 + 2.92\,X_2 \leq 257.88$

(iv) $7.10\,X_1 + 6.62\,X_2 \leq 327.66$

(v) $13.24\,X_1 + 7.40\,X_2 \leq 345.81$

(vi) $13.12\,X_1 + 2.14\,X_2 \leq 303.42$

After adding slack variables $X_3, X_4, X_5, X_6, X_7, X_8,$ we have

	X_3	X_4	X_5	X_6	X_7	X_8		
$X_1 + X_2 +$	X_3	0	0	0	0	0	=	70
$1.7\,X_1 + 1.7\,X_2 +$	0	X_4	0	0	0	0	=	268.82
$2.92\,X_1 + 2.92\,X_2 +$	0	0	X_5	0	0	0	=	257.88
$7.10\,X_1 + 6.62\,X_2 +$	0	0	0	X_6	0	0	=	327.66
$13.24\,X_1 + 7.40\,X_2 + 0$	0	0	0	0	X_7	0	=	345.81
$13.12\,X_1 + 2.14\,X_2 + 0$	0	0	0	0	0	X_8	=	303.42

Our objective is to maximize the net profit of this system

$7398\ X_1 + 3238\ X_2 + OX_3 + OX_4 + OX_5 + OX_6 + OX_7 + OX_8$ subject to the restriction land and available water.

We shall be using the simplex method to solve this problem (Heady & Candler, 1973):

First iteration:

			↑							
			7398	3238	0	0	0	0	0	0
			X_1	X_2	X_3	X_4	X_5	X_6	X_7	X_8
0	X_3	70	1	1	1	0	0	0	0	0
0	X_4	269.82	1.7	1.7	0	1	0	0	0	0
0	X_5	257.88	2.92	2.92	0	0	1	0	0	0
0	X_6	327.66	7.10	6.62	0	0	0	1	0	0
0	X_7	345.81	13.24	7.40	0	0	0	0	1	0
← 0	X_8	303.42	(13.12)	2.14	0	0	0	0	0	1
	$Zi =$		0	0	0	0	0	0	0	0
	$Zi - Ci$		−7398	−3238	0	0	0	0	0	0

Replace X_8 by X_1

Using Ring Round the Rosy method: $\begin{bmatrix} 70 & 1 \\ 303.42 & 13.12 \end{bmatrix}$

Second iteration:

Available land of first row of first iteration $70-1\left[\dfrac{303.42}{13.12}\right] = 46.87$

X_1 column of first row $= \begin{bmatrix} 1 & 1 \\ 13.12 & 13.12 \end{bmatrix}$

$$1-1\left(\frac{13.12}{13.12}\right)=1-1=0$$

X_2 column of first row $= \begin{bmatrix} 1 & 1 \\ 2.14 & 13.12 \end{bmatrix}$

$$1-1\left(\frac{2.14}{13.12}\right)=1-0.16=.84$$

X_3 column of first row $= \begin{bmatrix} 1 & 1 \\ 0 & 13.12 \end{bmatrix}$

$$1-1\left(\frac{0}{13.12}\right)=1-0=1$$

X_4 column of first row $= \begin{bmatrix} 0 & 1 \\ 0 & 13.12 \end{bmatrix}$

$$0-1\left(\frac{0}{13.12}\right)=0-0=0$$

X_5 column of first row $= \begin{bmatrix} 0 & 1 \\ 0 & 13.12 \end{bmatrix}$

$$0-1\left(\frac{0}{13.12}\right)=0$$

X_6 column of first row $= \begin{bmatrix} 0 & 1 \\ 0 & 13.12 \end{bmatrix}$

$$1-1\left(\frac{0}{13.12}\right)=0$$

X_7 column of first row $= \begin{bmatrix} 0 & 1 \\ 0 & 13.12 \end{bmatrix}$

$$0-1\left(\frac{0}{13.2}\right)=0$$

X_8 column of first row $= \begin{bmatrix} 0 & 1 \\ 1 & 13.2 \end{bmatrix}$

$$0-1\left(\frac{0}{13.12}\right)=-0.08$$

Available water of second row of first iteration

$$= \begin{bmatrix} 269.82 & 1.7 \\ 303.42 & 13.12 \end{bmatrix}$$

$$269.82 - 1.7 \left(\frac{303.42}{13.12} \right)$$

$$269.92 - 39.32 = 230.50$$

X_1 column of second row $= \begin{bmatrix} 1.7 & 1.7 \\ 13.12 & 13.12 \end{bmatrix}$

$$1.7 - 1.7 \left(\frac{13.12}{13.12} \right) = 0 - 0 = 0$$

$$1.7 - 1.7 = 0$$

X_2 column of second row $= \begin{bmatrix} 1.7 & 1.7 \\ 2.14 & 13.12 \end{bmatrix}$

$$1.7 - 1.7 \left(\frac{2.14}{13.12} \right)$$

$$= 1.7 - 2.7 \ (0.16) = 1.43$$

X_3 column of second row $= \begin{bmatrix} 0 & 1.7 \\ 0 & 13.12 \end{bmatrix}$

$$0 - 1.7 \left(\frac{0}{13.12} \right) = 0$$

X_4 column of second row $= \begin{bmatrix} 1 & 1.7 \\ 0 & 13.12 \end{bmatrix}$

$$1 - 1.7 \left(\frac{0}{13.12} \right) = 1$$

X_5 column of second row $= \begin{bmatrix} 0 & 1.7 \\ 0 & 13.12 \end{bmatrix}$

$$0 - 1.7 \left(\frac{0}{13.12} \right) = 0$$

X_6 column of second row $= \begin{bmatrix} 0 & 1.7 \\ 0 & 13.12 \end{bmatrix}$

$$= 0$$

X_7 column of second row $= \begin{bmatrix} 0 & 1.7 \\ 0 & 13.12 \end{bmatrix}$

$$0 - 1.7 \left(\frac{0}{13.12} \right) = 0$$

X_8 column of second row $= \begin{bmatrix} 0 & 1.7 \\ 1 & 13.12 \end{bmatrix}$

$$0 - 1.7\,(.08) = -0.136$$

Water available of third row of first iteration

$$\begin{bmatrix} 257.88 & 2.92 \\ 303.42 & 13.12 \end{bmatrix}$$

$$257.88 - 2.92 \left(\frac{303.42}{13.12} \right) = 190.34$$

X_1 column of third row $= \begin{bmatrix} 2.92 & 2.92 \\ 13.12 & 13.12 \end{bmatrix}$

$$2.92 - 2.92 \left(\frac{13.12}{13.12} \right) = 0$$

X_2 column of third row $= \begin{bmatrix} 2.92 & 2.92 \\ 2.14 & 13.12 \end{bmatrix}$

$$2.92 - 2.92 \left(\frac{2.14}{13.12} \right) = 2.43$$

X_3 column of third row $= \begin{bmatrix} 0 & 2.92 \\ 0 & 13.12 \end{bmatrix} = 0$

X_4 column of third row $= \begin{bmatrix} 0 & 2.92 \\ 0 & 13.12 \end{bmatrix} = 0$

X_5 column of third row $= \begin{bmatrix} 1 & 2.92 \\ 0 & 13.12 \end{bmatrix} = 1 - 0 = 1$

X_6 column of third row $= \begin{bmatrix} 0 & 2.92 \\ 0 & 13.12 \end{bmatrix} = 0$

X_7 column of third row $= \begin{bmatrix} 0 & 2.92 \\ 0 & 13.12 \end{bmatrix} = 0$

X_8 column of third row $= \begin{bmatrix} 0 & 2.92 \\ 0 & 13.12 \end{bmatrix} = 0 - 2.92\,(0.08) = -0.234$

Water available of fourth row of first iteration =

$$\begin{bmatrix} 327.66 & 7.10 \\ 303.42 & 13.12 \end{bmatrix}$$

$$= 327.66 - 7.1\left(\frac{303.42}{13.12}\right) = 163.46$$

X_1 column of fourth row $= \begin{bmatrix} 7.1 & 7.1 \\ 13.12 & 13.12 \end{bmatrix}$

$$7.1 - 7.1\left(\frac{13.12}{13.12}\right) = 0$$

X_2 column of fourth row $= \begin{bmatrix} 6.62 & 7.1 \\ 2.14 & 13.12 \end{bmatrix}$

$$6.62 - 7.1\left(\frac{2.14}{13.12}\right) = 5.48$$

X_3 column of fourth row $= \begin{bmatrix} 0 & 7.10 \\ 0 & 13.12 \end{bmatrix} = 0$

X_4 column of fourth row $= \begin{bmatrix} 0 & 7.10 \\ 0 & 13.12 \end{bmatrix} = 0$

$$0 - 7.1\,(0/13.12) = 0$$

X_5 column of fourth row $= \begin{bmatrix} 0 & 7.10 \\ 0 & 13.12 \end{bmatrix}$

$$0 - 7.1\,(0/13.12) = 0$$

X_6 column of fourth row $= \begin{bmatrix} 1 & 7.1 \\ 0 & 13.12 \end{bmatrix} = 0$

$$1 - 7.1\,(0/13.12) = 1$$

X_7 column of fourth row $= \begin{bmatrix} 0 & 7.1 \\ 0 & 13.12 \end{bmatrix}$

$$0 - 7.1\ (0/13.12) = 0$$

X_8 column of fourth row $= \begin{bmatrix} 0 & 7.1 \\ 1 & 13.12 \end{bmatrix}$

$$0 - 7.1\ (0.08) = -0.57$$

Available water of fifth row of first iteration =

$$\begin{bmatrix} 345.81 & 13.24 \\ 303.42 & 13.12 \end{bmatrix}$$

$$= 345.81 - 13.24 \left(\frac{303.42}{13.12} \right) = 39.61$$

X_1 column of fifth row $= \begin{bmatrix} 13.24 & 13.24 \\ 13.12 & 13.12 \end{bmatrix}$

$$13.24 - 13.24 \left[\frac{13.12}{13.12} \right]$$

$$13.24 - 13.24 = 0$$

X_2 column of fifth row $= \begin{bmatrix} 7.4 & 13.24 \\ 2.14 & 13.12 \end{bmatrix}$

$$7.4 - 13.24 \left[\frac{2.14}{13.12} \right] = 5.24$$

X_7 column of fifth row $= \begin{bmatrix} 1 & 13.24 \\ 0 & 13.12 \end{bmatrix}$

$$1 - 13.24 \left(\frac{0}{13.12} \right) = 1 - 0 = 1$$

X_8 column of fifth row $= \begin{bmatrix} 0 & 13.24 \\ 1 & 13.12 \end{bmatrix}$

$$= 0 - 13.24\ (0.08) = -1.06$$

X_8 is replaced by X_1
Available land at sixth row

$$303.42/13/12 = 23.13$$

X_1 of sixth row $= 13.12/13.12 = 1$
X_2 of sixth row $= 2.14/13.12 = 0.16$
X_3 of sixth row $= 0/13.12 = 0$
X_4 of sixth row $= 0/13.2 = 0$
X_8 of sixth row $= 1/13.2 = 0.08$

Second iteration :

$\uparrow$

			7398	3238	0	0	0	0	0	0
			X_1	X_2	X_3	X_4	X_5	X_6	X_7	X_8
0	X_3	46.87	0	0.84	1	0	0	0	0	-0.08
0	X_4	230.51	0	1.42	0	1	0	0	0	-0.14
0	X_5	190.35	0	2.43	0	0	1	0	0	-0.23
0	X_6	163.46	0	5.48	0	0	0	1	0	-0.57
← 0	X_7	39.61	0	(5.24)	0	0	0	0	1	-1.06
→ 7398	X_1	23.13	1	0.16	0	0	0	0	0	0.08
$Zi =$		171115.74	7398	1183.68	0	0	0	0	0	591.84
$Zi - Ci$			0	-2054.32	0	0	0	0	0	+591.84

Delete X_7 by X_2

Third iteration :

Available land of first row of second iteration

$$= \begin{bmatrix} 46.87 & 0.84 \\ 39.61 & 5.24 \end{bmatrix}$$

$$46.87 - 0.84 \left(\frac{39.61}{5.24} \right) = 40.52$$

Water available of second row of second iteration

$$= \begin{bmatrix} 230.51 & 1.42 \\ 39.61 & 5.24 \end{bmatrix}$$

$$230.51 - 1.42 \left(\frac{39.61}{5.24} \right) = 219.78$$

Water available for third row of second iteration

$$= \begin{bmatrix} 190.35 & 2.43 \\ 39.61 & 5.24 \end{bmatrix}$$

$$190.35 - 2.43 \left(\frac{39.61}{5.24} \right) = 171.98$$

Water available for fourth row of second iteration

$$= \begin{bmatrix} 163.46 & 5.48 \\ 39.61 & 5.24 \end{bmatrix}$$

$$163.46 - 5.48 \left(\frac{39.61}{5.24} \right) = 122.03$$

Water available for sixth row of second iteration

$$= \begin{bmatrix} 23.13 & 0.16 \\ 39.61 & 5.24 \end{bmatrix}$$

$$23.13 - 0.16 \left(\frac{39.61}{5.24} \right) = 21.9$$

Third iteration :

X_1 column for first row $= \begin{bmatrix} 0 & 0.84 \\ 0 & 5.24 \end{bmatrix}$

$$0 - 0.80 \left(\frac{0}{5.24} \right) = 0$$

X_2 column for first row $= \begin{bmatrix} 0.84 & 0.84 \\ 5.24 & 5.24 \end{bmatrix}$

$$0.84 - 0.84 \left(\frac{5.24}{5.24} \right) = 0$$

X_2 column for sixth row $= \begin{bmatrix} 0.16 & 0.16 \\ 5.24 & 5.24 \end{bmatrix}$

$$0.16 - 0.16\left(\frac{5.24}{5.24}\right) = 0$$

X_3 column for first row $= \begin{bmatrix} 1 & 0.84 \\ 0 & 5.24 \end{bmatrix}$

$$1 - 0.84\left(\frac{0}{5.24}\right) = 1$$

X_3 for second row $\qquad = \begin{bmatrix} 0 & 1.42 \\ 0 & 5.24 \end{bmatrix}$

$$0 - 1.42\,(0) = 0$$

X_4 for first row $\qquad = \begin{bmatrix} 0 & 0.84 \\ 0 & 5.24 \end{bmatrix} = 0$

X_5 for third row $\qquad = \begin{bmatrix} 1 & 2.44 \\ 0 & 5.24 \end{bmatrix} = 1$

X_8 for first row $\qquad = \begin{bmatrix} -0.08 & 0.84 \\ -1.06 & 7.24 \end{bmatrix}$

$$-0.8 - 0.84\left(\frac{-1.06}{5.24}\right)$$

$$= -0.08 + 0.1699$$

$$= +0.0899$$

X_8 for second row $\qquad = \begin{bmatrix} -0.14 & 1.42 \\ -1.06 & 5.24 \end{bmatrix}$

$$-0.14 - 1.42\left(\frac{-1.06}{5.24}\right)$$

$$= -0.14 + 0.287$$
$$= 0.147 = 0.15$$

X_8 for fourth row $\qquad = \begin{bmatrix} -0.57 & 5.48 \\ -1.06 & 5.24 \end{bmatrix}$

$$-0.57 - 5.48\left(\frac{-1.06}{5.24}\right)$$

$$-0.57 + 1.109$$
$$= 0.54$$

X_8 for sixth row = $\begin{bmatrix} 0.08 & 0.16 \\ -1.06 & 5.24 \end{bmatrix}$

$$0.08 - 0.16\left(\frac{-1.06}{5.24}\right)$$

$$0.08 + 0.0323$$
$$= 0.112$$

Third iteration :

			7398	3238	0	0	0	0	0	0
			X_1	X_2	X_3	X_4	X_5	X_6	X_7	X_8
0	X_3	40.52	0	0	1	0	0	0	0	0.09
0	X_4	219.78	0	0	0	1	0	0	0	0.15
0	X_5	171.98	0	0	0	0	1	0	0	0.26
0	X_6	122.03	0	0	0	0	0	1	0	0.54
→ 3238	X_2	7.56	0	1	0	0	0	0	0.19	−0.2
7398	X_1	21.9	1	0	0	0	0	0	0	0.112
Zi = 186495.48		7398	3238						+615.22	+116
$Zi - Ci$		0	0	0	0	0	0	0	+	+

So the optional solution is
  Grow wheat in 21.9 ha.
  Barley 7.56 ha.
  Surplus land 40.52 ha.
  Surplus water in Pd I = 219.78 ha.cm.
  Surplus water in Pd II = 171.91 ha.cm.
  Surplus water in Pd III = 122.03 ha.cm.
  Water in Pd IV and V is completely used

3.1.1.2.1 Remark

 (1) The pivotal value is the value in the outgoing row and outgoing column intersection.

 (2) To calculate the coefficients of incoming row for the next iteration, the outgoing row coefficients are divided by the pivotal value.

(3) To calculate the coefficient of rows other than the incoming row, subtract from the quantity on the row in the previous iteration section the product of the quantity on the incoming row in the new section (new iteration) and the input-output coefficients of outgoing column in the previous section (iteration).

3.1.1.3 USE OF MATRIX ALGEBRA FOR DISTANCE MEASUREMENTS

The simplest measure of the square of the distance between two Texa (Varieties) i and j is:

$$D_{ij}^2 = [(X_{1i} - X_{1j})^2 + (X_{2i} - X_{2j})^2 + ... + (X_{ni} - X_{nj})^2$$

$$= \sum_{K=1} (X_{ki} - X_{kj})^2$$

where X_{kj} is character k of Taxon j. Usually, the characters are standardized so that various scales of measurement such as millimeter, meter, counts, etc., will not be confounded. The standardization most commonly used is the division of each character by its among taxa standard error. Other standardizations can also be employed.

Such a simple measure of distance works quite well for characters that are reasonably non-correlated. However, when the value of one character can be reasonably predicted on the basis of the knowledge of the values of some of the other characters, the equation results in distances which differ more amongst themselves than is justified on the basis of the data. Such a situation can be avoided if the characters are adjusted or transformed so that they are not correlated before the distances are calculated. The resulting distances are called generalized distances. Such procedures involved require a knowledge of matrix algebra (Emigh and Goodman (1985).

For six cultivars of chickpea and twenty characters $(X_1, X_2, ...,X_{20})$, suppose we have the measurements given in Table 3.3. To standardize these scores, we can subtract the mean from each value of X_i and then divide by each value of its standard deviation. Thus, our standardized scores appear as in Table 3.6. We will then calculate the distance from both characters:

$$D_{I, II}^2 = (-1.62 - 0.7)^2$$
$$+ (-1.57 - 0.66)^2$$
$$+ (-0.93 - (-0.5)^2$$
$$+ (-1.54 - (-0.09)^2$$
$$+ (-1.24 - (-1.24)^2$$
$$+ (-.86 - (-1.35)^2$$

$$+ (0.52 \ - (-0.37)^2$$
$$+ (0.65 \ - (-0.3)^2$$
$$+ (.04 \quad - (-0.77)^2$$
$$+ (0.14 \ - (-0.8)^2$$
$$+ (0.8 \quad - 1.33)^2$$
$$+ (-0.6 \ - (-0.6)^2$$
$$+ (0.82 \ - (-0.41)^2$$
$$+ (0.3 \quad - (-1.27)^2$$
$$+ (1.32 \ - 0.5)^2$$
$$+ (-1.49 - 0.56)^2$$
$$+ (-0.17 - 1.86)^2$$
$$+ (-0.01 - (-1.59)^2$$
$$+ (2.04 \ - (-0.41)^2$$
$$+ (1.24 \ - (-0.84)^2$$
$$= 38.07$$

$$D_{I,II} = \sqrt{38.07}$$
$$D_{I, II} = 6.17$$

These distance are shown in Table 3.9 with the help of a computer program written in Quick Basic—a general purpose language run in IBM compatible PC. The dendogram is shown in Fig. 3.8.

Table 3.9 Showing the distances (unpublished data)

	Katila	Vijay	RSG 143-1	Phule G5	ICC 4958	K 850
	(1)	(2)	(3)	(4)	(5)	(6)
1. Katila	0	6.17	5.54	6.26	8.29	7.73
2. Vijay	6.17	0	5.96	5.64	5.75	8.10
3. RSG 143-1	5.54	5.96	0	5.21	6.34	4.86
4. Phule G5	6.26	5.64	5.21	0	3.31	6.64
5. ICC4958	8.29	5.75	6.34	3.31	0	7.15
6. K 850	7.73	8.10	4.87	6.64	7.15	0

Figures 3.6 a, b and 3.7 have demonstrated the classification of varieties and agrophysiological traits.

3.1.1.3.1 Calculation of Group Distances to Make a Dendogram

(a) Distance between 4 and 5 = 3.31 (Table 3.9)
(b) Distance between 3 and 6 = 4.86
(c) Distance between 1 and 2 = 6.17

(d) Distance between (4,5) and (3, 6) = 5.21 + 6.64 + 6.34 + 7.14 = 25.33/4 = 6.33

(e) Distance between (4, 5, 3, 6) and (1, 2) = + 6.26 + 5.64 + 8.28 + 5.75 + 5.54 + 5.96 + 7.73 + 8.10 = 53.26/8 = 6.66.

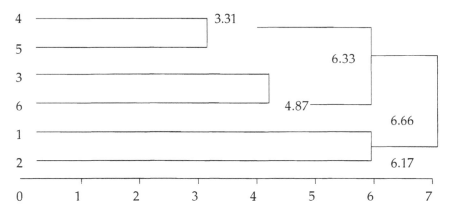

Fig. 3.8 Dendogram–Variety.

3.2. INTEGRAL-DIFFERENTIAL FORM

Pine (1975) decreed a great deal of importance to the differential and integration and its resulting brain child. Human beings have reached moon and are now trying to land on farthest planets. In biology, aging is a destructive mechanism built into us. If we can discover this mechanism, we can dismantle it. All these achievements can be made as a result of varying analyses.

Here an example is provided of using the differential and integration which helps in understanding the physical event with 100% perfection. Although this example is not particularly important, it does involve the principles which will help us in predicting and controlling events. This example needs only a thermometer and a wristwatch.

3.2.1. Example for Formulating a Differential Equations

Mathematical formulation of Newton's Law of cooling:

$$\frac{dT}{dt} = K\,(T{-}F)$$

where T is the temperature of the object, in this case, the thermometer. The initial temperature coldest is $0^{\circ}C$, t is time in minutes. dT/dt is the rate of change of temperature of the object with respect to time, i.e. the general slope; F is the temperature of the medium, in this case, the room temperature, a constant $21.66\,^{\circ}C$.

K is a constant of proportionality, to be computed.

Mathematical manipulation

$$\frac{dT}{T-F} = K dt$$

$$\int \frac{dT}{T-F} = \int K dt$$

$$\ln (T - F) = Kt + c$$

where $\ln (T - F)$ is the $\int y.\ d\ (T - F)$ form

$$\frac{dT}{T-F} \text{ is } y \text{ form. } \int \frac{dT}{T-F} = \ln (T - F)$$

If y form is a/x, then $\int y.\ dx$ form $= a \ln x$.

Kt is the result of the integration of the constant K and dt

c is used when the limits are not specified.

This c applies to both sides.

If we now bring each side to the power of e, then ln disappears.

$$e^{\ln (T-F)} = T - F$$

$$T - F = e^{Kt-c} = e^{Kt}.\ e^c$$

Then $T - F = c\ e^{Kt}$

$T = F + c\ e^{Kt}$ ← This is the general formula.

If $t = 0$

then $e^{K(o)} = e^o = 1$ $\begin{bmatrix} \text{anything (except zero), to the} \\ \text{zero power is equal to one.} \end{bmatrix}$

Inserting $T = 0°C$, and $F = 21.66°C$ so that

$0 = 21.66 + c$ and $c = -21.66°C$

giving us

$T = 21.66 - 21.66\ e^{Kt}$

By conducting an accurate experiment, transferring a thermometer having a reading of 0°C to a room having a constant temperature of 21.66°C, after two minutes of time, we observed the reading in thermometer to be 9.72°C.

$$9.72 = 21.66 - 21.66\ e^{K(2)}$$

or

$$9.72 - 21.66 = -21.66\ e^{K(2)}$$

$$-11.94 = -21.66\ e^{2K}$$

$$\frac{-11.94}{-21.66} = e^{2K}$$

$$0.551 = e^{2K}$$

$$\ln 0.551 = e^{\ln 2K}$$

$$\ln 0.551 = 2K$$

$$-0.59602 = 2K$$

$$K = \frac{-0.5960204}{2}$$

$$= -0.298$$

$T = 21.66 - 21.66\ e^{-.298t}$ ← This is the completed specified formula

We can predict the reading in thermometer after 3 minutes as:

$$T = 21.66 - 21.66\ e^{-.298(3)}$$

$$= 21.66 - 21.66\ e^{-0.894}$$

$$= 21.66 - 21.66\ (0.409)$$

$$= 21.66 - 8.8592955$$

$$T = 12.80°C$$

We can validate this predicted value of temperature after 3 minutes after conducting the experiment accurately. If we conduct the experiment accurately, then we observe the temperature of 12.80°C after 3 minutes without any error.

The resulting equation

$$T = 21.66 - 21.66\ e^{-.298t}$$

represents the Newton's law of cooling.

This Newton's law of cooling was discovered only due to thanks of differentiation and anti-differentiation. The calculus is used to locate all the different formulae for almost all and any event in nature.

We will now derive the equation governing another law in nature, i.e. the absorption law of Lambert.

3.2.2. The Absorption Law of Lambert

The rate of change of the amount of light (L) in a thin transparent layer with respect to the thickness (h) is proportional to the amount of light on the layer.

$$\frac{dL}{dh} = KL$$

$$\frac{dL}{L} = Kdh$$

$$\int \frac{dL}{L} = \int Kdh$$

$$\ln L = Kh + c$$

$$e^{\ln L} = e^{Kh + c}$$

$$L = e^{Kh + c}$$

$$L = e^{Kh} \cdot e^c$$

$$L = c\, e^{Kh}$$

To evaluate c and K, when $h = 0$, then $L = c$. So, c is the initial condition of L and can be designated as L_0. So,

$$L = L_0 e^{Kh}$$

Crop production modelers are using the absorption law of Lambert to estimate the extinction coefficient of the crop canopies.

There are countless books on calculus that will define the various methods and procedures for all sort of physical and biological situations.

The algebraic and calculus forms are complementary to each other. These two forms do not work in isolation but together.

3.3 PARAMETER ESTIMATION

3.3.1 Statistical Procedure

3.3.1.1 *Finding the Best Parameter Values for Linear Equations*

Gold (1977) gave a very good derivation for this aspect.

$$y_i = f(x_i ; a, b)$$

$$= a + bx_i \qquad (3.4)$$

3.3.1.1.1 *Useful Characteristic of Extrema*

Suppose the best value for a and b are a^* and b^*. If these values are made greater or smaller than these best values, then the d(y, $\hat{y}$), distance between y and $\hat{y}$ will increase. This is demonstrated in Fig. 3.9, which illustrates that the slope of the curve of $d(y - \hat{y})$ versus a (or of the curve of $d(y, \hat{y})$ versus b) is zero at the best values. The value $d(y, \hat{y})$ should be positive or, at least non-negative.

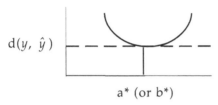

$$d(y, \hat{y})$$

a* (or b*)

Fig. 3.9 $d(y, \hat{y})$ has a minimum value at $a = a^*$ (or $b = b^*$).

The expression for derivatives of $d^2(y, \hat{y})$ with respect to a and b should equal zero and then solve for a and b.

$$\left(\frac{\partial d^2(y, \hat{y})}{\partial a} \right)_{a = a^*} = 0 \tag{3.5}$$

$$\left(\frac{\partial d^2(y, \hat{y})}{\partial b} \right)_{b = b^*} = 0 \tag{3.6}$$

If we have a dependent variable which is a function of many variables as $f(x_1, ..., x_n)$, then all of the $\partial f/\partial x_i$ have to be zero at the extremum. One should know that the point with zero derivative is an inflection point or an extremum and if it is extremum and whether it is a maximum or a minimum and whether it is unique (see figure 3.10).

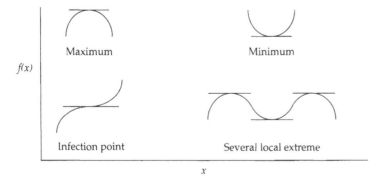

Fig. 3.10 Many cases for which df/dx can be zero.

3.3.1.1.2 *Expressions for Parameters a and b*

The derivatives of equations (3.5) and (3.6) can be evaluated using the chain rule.

3.3.1.1.2.1 *Derivative of a Function of a Function: The Chain Rule*

If $y = y(x)$, where x is a function of another variable, $x = x(z)$.

$$\frac{dy[x(z)]}{dz} = \lim_{\Delta z \to 0} \frac{y[x(z + \Delta z)] - y[x(z)]}{\Delta z}$$

$$\frac{dy}{dz} = \lim_{\Delta z \to 0} \frac{y(x + \Delta x) - y(x)}{\Delta x} \cdot \frac{\Delta x}{\Delta z}$$

$$\frac{dy}{dz} = \frac{dy}{dx}\frac{dx}{dz} \tag{3.7}$$

If a function f is a function of y, which is a further function x, which is a function of z, then

$$\frac{df(y(x(z)))}{dz} = \frac{df}{dy} \cdot \frac{dy}{dx} \cdot \frac{dx}{dz} \tag{3.8}$$

When one deals with functions, partial derivatives must be used. If we have a function $f(x, y)$, where $x = x(z)$ and $y = y(z)$. We have

$$\frac{df(x, y)}{dz} = \frac{\partial f}{\partial x} \cdot \frac{\partial x}{\partial z} \frac{\partial f}{\partial y} \cdot \frac{\partial y}{\partial z} \tag{3.9}$$

This can be extended to any finite number of arguments.

$$\frac{df(x_1, x_2, \ldots\ldots, x_n)}{dz} = \sum_{i=1}^{n} \frac{\partial f}{\partial x_i} \cdot \frac{\partial x_i}{\partial z} \tag{3.10}$$

Then we go back to (3.8)

$$\frac{\partial d^2(y, \hat{y})}{\partial b} = \frac{\partial \left[\sum_{i=1}^{n} (y_i - \hat{y}_i)^2 \right]}{\partial b}$$

$$= \frac{\partial \left[\sum\limits_{i=1}^{n} (y_i - \hat{y}_i)^2 \right]}{\partial (y_i - \hat{y}_i)} \cdot \frac{\partial (y_i - \hat{y}_i)}{\partial b} \qquad (3.11)$$

Since

$$\frac{\partial \sum\limits_{i=1}^{n} (y_i - \hat{y}_i)^2}{\partial (y_i - \hat{y}_i)} = \sum\limits_{i=1}^{n} \frac{\partial (y_i - \hat{y}_i)^2}{\partial (y_i - \hat{y}_i)}$$

$$= 2 \sum (y_i - \hat{y}_i)$$

So,

$$\frac{\partial d^2 (y, \hat{y})}{\partial b} = 2 \sum\limits_{i=1}^{n} (y_i - \hat{y}_i) \frac{\partial (y_i - \hat{y}_i)}{\partial b} \qquad (3.12)$$

Now

$$\hat{y}_i = a + bx$$

So,

$$\frac{\partial (y_i - \hat{y}_i)}{\partial b} = \frac{\partial y_i}{\partial b} - \frac{\partial (bx_i)}{\partial b} - \frac{\partial a}{\partial b}$$

Neither y_i nor x_i nor a contain any explicit dependance on b. This is simply

$$\frac{\partial (y_i - \hat{y}_i)}{\partial b} = 0 - x_i - 0 = -x_i$$

Putting this into (3.12)

$$\frac{\partial d^2 (y, \hat{y})}{\partial b} = 2 \sum\limits_{i=1}^{n} (y_i - a - bx_i)(-x_i)$$

$$= 2 \sum\limits_{i} (ax_i + bx_i^2 - x_i y_i) \qquad (3.13)$$

Setting (3.13) to zero gives

$$a^* \sum\limits_{i} x_i + b^* \sum\limits_{i} x_i^2 - \sum\limits_{i} x_i y_i = 0 \qquad (3.14)$$

Equation (3.14) has two unknowns $a*$ and $b*$. So, we need another equation before proceeding. It is provided by 3.5. Now, using the chain rule:

$$\frac{\partial d^2\,(y,\hat{y})}{\partial a} = 2\sum_i (y_i - \hat{y}_i)\cdot\frac{\partial(y_i - \hat{y}_i)}{\partial a} \tag{3.15}$$

Since

$$\hat{y}_i = a + bx_i$$

$$\frac{\partial(y_i - \hat{y}_i)}{\partial a} = \frac{\partial y_i}{\partial a} - \frac{\partial bx_i}{\partial a} - \frac{\partial a}{\partial a}$$

Since y_i, x_i and b do not depend on a

$$\frac{\partial(y - \hat{y})}{\partial a} = 0 - 0 - 1 = -1$$

Putting this into (3.15)

$$\frac{\partial d^2\,(y,\hat{y})}{\partial a} = 2\sum_i (y_i - a - bx_i)\,(-1) = 2\sum_i (a + bx_i - y_i)$$

Setting this equal to zero, gives

$$na* + b* \; 2\sum_{i=1}^{n} x_i - 2\sum_{i=1}^{n} y_i = 0 \tag{3.16}$$

Equations (3.14) and (3.16) are what we need to get $a*$ and $b*$. If Equation (3.16) is divided through by n,

$$a* + b* \; \bar{x} - \bar{y} = 0 \tag{3.17}$$

where the bar indicates arithmatic average

$$a* = y - b* \; \bar{x}$$

Substituting into equation (3.14)

$$(\bar{y} - b* \; \bar{x})\,\Sigma x_i + b*\,\Sigma x_i^2 - \Sigma x_i\,y_i = 0$$

$$\bar{y}\,\Sigma x_i - b*\bar{x}\,\Sigma x_i + b*\,\Sigma x_i^2 = \Sigma x_i\,y_i$$

$$b*\,(\Sigma x_i^2 - \bar{x}\,\Sigma x_i) = \Sigma x_i y_i - \bar{y}\,\Sigma x_i$$

$$b^* = \frac{\sum x_i y_i - \bar{y} \sum x_i}{\sum x_i^2 - \bar{x} \sum_i x_i} \tag{3.18}$$

Substituting the expression back into Equation (3.17) and rearranging to solve for

$$a^* + b^* \ \bar{x} - \bar{y} = 0$$

$$a^* + \left(\frac{\sum\limits_i x_i y_i - \bar{y} \sum\limits_i x_i}{\sum\limits_i x_i^2 - \bar{x} \sum\limits_i x_i} \right) (\bar{x}) - \bar{y} = 0$$

$$a^* + \frac{\bar{x} \sum\limits_i x_i y_i - \bar{x} \ \bar{y} \sum\limits_i x_i}{\sum\limits_i x_i^2 - \bar{x} \sum\limits_i x_i} - \frac{\bar{y}}{1} = 0$$

$$a^* = \frac{\bar{y}}{1} - \frac{\bar{x} \left(\sum\limits_i x_i y_i - \left(\bar{y} \sum\limits_i x_i \right) \right)}{\sum\limits_i x_i^2 - \bar{x} \sum\limits_i x_i}$$

$$= \frac{\bar{y} \left(\sum\limits_i x_i^2 - \bar{x} \sum\limits_i x_i \right) - \left(\bar{x} \left(\sum\limits_i x_i y_i - \bar{y} \sum\limits_i x_i \right) \right)}{\sum\limits_i x_i^2 - \bar{x} \sum\limits_i x_i}$$

$$= \frac{\bar{y} \sum\limits_i x_i^2 - \bar{x} \ \bar{y} \sum\limits_i x_i - \bar{x} \sum\limits_i x_i y_i + \bar{x} \ \bar{y} \sum\limits_i x_i}{\sum\limits_i x_i^2 - \bar{x} \sum\limits_i x_i}$$

$$= \frac{\bar{y} \sum\limits_i x_i^2 - \bar{x} \sum\limits_i x_i y_1}{\sum\limits_i x_i^2 - \bar{x} \sum\limits_i x_i} \tag{3.19}$$

In practice, one would simply substitute the numerical value for b^* from Equation (3.18) and solve for a^* directly from Equation (3.17).

$$b^* = \frac{\sum\limits_i x_i y_1 - \bar{y} \sum\limits_i x_i}{\sum\limits_i x_i^2 - \bar{x} \sum\limits_i x_i}$$

Given the values of $x = 1, 2, 3$ and $y = 2.1, 2.9, 4.05$
suppose

$$\sum_i x_i y_i = 20.05$$

$$\bar{y} = 3.016$$

$$\sum_i x_i = 6$$

$$\sum_i x_i^2 = 14$$

$$\bar{x} = 2$$

Substituting these values

$$b^* = \frac{20.05 - 3.016\,(6)}{14 - 2(6)}$$

$$= \frac{20.05 - 18.096}{14 - 12} = \frac{1.954}{2} = 0.977$$

$$a^* + b^* \; \bar{x} - \bar{y} = 0$$

$$a^* + 0.977\,(2) - 3.016 = 0$$

$$a^* = 3.016 - 0.977\,(2)$$

$$= 3.016 - 1.954 = 1.062$$

$$a^* = 1.062$$

a^* may also be calculated by using Equation (3.19)

$$a^* = \frac{\bar{y} \sum\limits_i x_i^2 - \bar{x} \sum\limits_i x_i y_i}{\sum\limits_i x_i^2 - \bar{x} \sum\limits_i x_i}$$

$$= \frac{3.016\,(14) - (2)\,20.05}{14 - 2\,(6)}$$

$$= \frac{-40.10 - 42.224}{2} = \frac{2.124}{2} = 1.062$$

These values of a^* and b^* are not the true values of the parameters. These are the best estimates in a least square sense. It is assumed that all the errors are embodied in y.

3.3.1.1.2.2 Graphical Representation

$$\hat{y} = a^* + b^* x_0 \qquad\qquad (3.20)$$

If the data set is being described by a best-fitting straight line, the line can be drawn directly on the data plot by calculating the value of two points that lie directly on the line. We can do this by plugging in any convenient value x_0 into the Equation (3.20) to get a point $(x_0, \hat{y} x_0)$. A second point may be obtained in the same way.

The two points, $(\bar{x}, \bar{y})$ and $(-a^*/b^*, 0)$, are convenient because they require no additional computation. To verify that these points do lie exactly on the line, one needs to only plug $\bar{x}$ and $-a^*/b^*$ into Equation (3.20).

$$\hat{y}_{\bar{x}} = a^* + b^* \bar{x}$$

The equation (3.17) already tells us that the right hand side expression is $\hat{y}_{\bar{x}}$. In the same way, for $x = -a^*/b^*$, it may be found that $\hat{y} = 0$. So, the line between the points $(-a^*/b^*, 0)$ and $(\bar{x}, \bar{y})$ is the 'least squares' line of best fit.

3.3.1.2 How Good is the Best Fitting Curve ?

After finding the parameters, it is to see whether it gives the smallest value of $d(y - \hat{y})$ for a given type of equation. Now, the question arises: is it small enough to be satisfactory? If it is small, it is good, but, if it is large, then we should try for another equation, possibly with more parameters. The term 'small' or 'large' should be interpreted relative to the size of the expected uncertainties.

If there is no difference between observed and predictive values, then the $d(x, \hat{y})$ would be zero, but it never happens. It shows that there is some variability unaccounted for by the equation. This unaccounted variability is of two types:

(i) The first type of variability is related to the randomness. This means that if we measure the many values of Y at the same value of X, we get the variability amongst the values of Y. This variability is associated with the uncertainty in the attempt to determine a value of Y for a given value of X.

(ii) If the equation is unable to accurately describe the underlying relation between x and y, this is the second source of error.

If the value of d $(y - \hat{y})$ is about equal of source (i), then the data are in agreement with the type of equation chosen. Otherwise, we should search for a better equation.

Now the question arises: how can we compare the value of d $(y, \hat{y})$ with values expected from the source of error (i)? It is most convenient to reduce the total distance to average distance per data point.

The usual method to find out the average distance is to divide the total contributed quantities by the number of contributors. In this case, the total is not the d $(y, \hat{y})$, but d^2 $(y, \hat{y})$. This was because of the additivity problem. So, this is the d$^2(y - \hat{y})$, which is divided by the number of contributors.

If we are fitting the equation with u parameters to u data points, we get the value d $(y - \hat{y})$ equals to zero. So, the data points should be at least u + 1 to get any contribution to the estimate of error. If we divide the error by u + 1, then it underestimates the error expected in a single observation. If there are n data points and u parameters to be estimated, then there will be n-u contributors to the error.

The quantity of mean square deviation is the measure of the goodness of fit.

$$\text{MSdev.} = \frac{d^2\ (y, \hat{y})}{n - u}$$

$$= \frac{\sum_{i=1}^{n} (y_i, \hat{y}_i)^2}{n - u} \tag{3.21}$$

and the root mean square deviation (standard error of estimate)

$$\text{RMSdev.} = \left[\frac{d^2\ (y, \hat{y})^2}{n - 1} \right]^{\frac{1}{2}}$$

$$= \left[\frac{\displaystyle\sum_{i=1}^{n} (y, \hat{y})^2}{n - u} \right]^{\frac{1}{2}} \tag{3.22}$$

3.3.1.3 Random Versus Systematic Deviations

The deviation may be either completely random as in Fig. 3.11a or systematic as in Fig. 3.11b.

If the deviations are as in Fig. 3.11a, then the source (i) is the dominant source of error. On the contrary, if the deviations are as in fig. 3.11b, then the contribution from source (ii) is appreciable, even though the RMSdev. may be small.

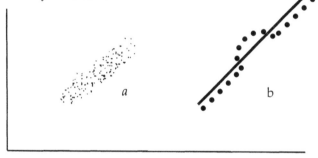

Fig 3.11   Random (a) and systematic (b) deviations from a curve.

The other method of judging the deviation properties is to plot the deviation directly. The difference $(\hat{y}_i - y_i)$ is called a residual value. If all the deviation is from source (i), then the residuals will be scattered in some random fashion about zero. If the residual values are in systematic fashion in the plot, then the error is from source (ii). Fig. 3.12 shows two such plots (adapted from Gold, 1977). Clearly, model A should be preferred.

3.3.1.4 Linear Approximations for Quick Estimating a Good Fitting Curve

The approximate method approach for estimating the parameters of a relation is to break the data into groups according to the value of the independent variables of at least as many groups as we have parameters. If we want to fit the straight line, we should divide the data into two groups as in Fig. 3.13a. Any straight line must go through ($\bar{x}$, $\bar{y}$), where $\bar{x}$ and $\bar{y}$ are determined by the data used. So if we use the data in

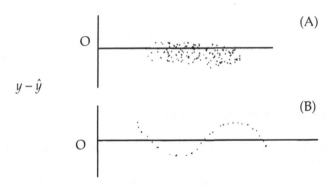

Fig. 3.12 Plots of residuals for two explanatory models for *x* and *y* variables.

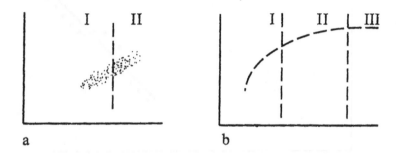

Fig. 3.13 Grouping of data point (adopted from Gold, 1977).

region I, the estimated line should go through the point ($\bar{x}_1$, $\bar{y}_1$). Similarly, if we use the data in region II, the line would have to go through ($\bar{x}_2$, $\bar{y}_2$). Now, if we assume that the same straight line goes through both regions, then the line that connects ($\bar{x}_1$, $\bar{y}_1$) with ($\bar{x}_2$, $\bar{y}_2$) is an estimate of the line that describes the data.

If the curve used is not a straight line, then the data should be divided into many small regions. The curvature within each region should be small. Within such a region, the curve could be approximated by a straight line (Fig. 3.14). Using the equation of a straight line, one can get an approximate value of $f(x^*)$ for any x^* near X_A. If we know the value of $f(X_A)$ and the value of $(df/dx)_A$

$$f(x^*) = f(x_A) + \left(\frac{df}{dx}\right)_A \cdot (x^* - x_A) + 0(x^* - x_A) \qquad (3.23)$$

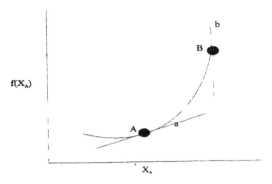

Fig. 3.14 Approximation of a curve by straight line (Adapted from Gold, 1977).

Equation (3.23) indicates that approximation becomes more precise as (X^*-X_A) becomes smaller, finally becoming exact in the limit as $|X^*-X_A|$ approaches zero. The straight line in region I (Fig. 3.13a) would go through $(\bar{x}_1, \bar{y}_1)$. To get an approximation to the values of the parameters K_1, K_2, ...K_u, one needs a minimum of u regions. The structural equation would be $g\,(y_1, x_1, k_1)$ and the reduced equation would be

$$\bar{y} = f\,(\bar{x}_1, K_1, ...K_u).$$

This relation gives u simultaneous equations, which may be solved for the parameters.

If we choose more regions, then the parameters may give some degree of freedom for testing the adequacy of the equation.

3.3.1.5 Weighting of Data

The procedure for finding the weights is as follows:

(a)   For each value of the independent variable x_1, make an estimate of error associated with the measurement of $\bar{y}_1$. Call the squares of this estimate ε_1^2 (equivalent to variance).

(b)   The weight to be used are proportional to $1/\varepsilon_1^2$;

$$W_1 = \frac{C}{\varepsilon_i^2} \tag{3.24}$$

(c)   It is often convenient to adjust the weights so that they add up to n, the total number of observations.

This leads to the following series of mathematical statements:

$$n = \sum_{i=1}^{n} w_i$$

$$= \sum_{i=1}^{n} \frac{C}{\varepsilon_1^2}$$

$$= C \sum_{i=1}^{n} \frac{1}{\varepsilon_1^2}$$

So, if we wish the w_1 to add to n, each estimate value of $1/\varepsilon_1^2$ is multiplied by

$$C = n \bigg/ \sum_{i=1}^{n} \frac{1}{\varepsilon_i^2}$$

3.3.1.5.1 Examples

x_i	y_i	$\hat{y}_i$	$y-\hat{y}$	$(\hat{y}-y)^2 = \varepsilon_i^2$	$1/\varepsilon_i^2$
1	2.1	2.099	0.001	0.000001	1000000
2	2.9	3.016	–0.116	0.13456	74.316
3	4.05	3.993	0.057	0.003249	307.787
					1000382.096

$$C = \frac{n}{\sum\limits_{i=1}^{n} 1/\varepsilon_i^2} = \frac{3}{1000382} = 0.000003$$

$$W_1 = \frac{C}{\varepsilon_1^2} = \frac{0.000003}{0.000001} = 3$$

$$W_2 = \frac{C}{\varepsilon_2^2} = \frac{0.000003}{0.13456} = 0.00002229$$

$$W_3 = \frac{C}{\varepsilon_3^2} = \frac{0.000003}{0.003249} = 0.0009233610$$

$$n = C \sum_{i=1}^{n} 1/\varepsilon_1^2$$

$$n = .000003 \,(1000382) = 3.001146$$

$$n = 3$$

$$n = \sum_{i=1}^{n} w_i$$

$$= 3 + .0000229 + 0.0003077 \approx 3$$

For a given vector of weight, w, $(w_1, w_2, \ldots, w_n)$ the value to be minimized is

$$d^2w(y, \hat{y}) = \sum_{i=1}^{n} w_i (\hat{y}_i - y_i)^2 \tag{3.25}$$

for $w = (1, 1, \ldots, 1)$, this is exactly the same as before. For the case of a straight line relation, the result is

$$b_w^* = \frac{\sum_i w_i x_i \hat{y}_i - \hat{y}_w \sum_i w_i x_i}{\sum_i w_i x_i^2 - \bar{x}_w \sum_i w_i x_i} \tag{3.26}$$

Example

x_i	y_i	w_i	$x_i - y_i$	$w_i x_i y_i$	$y_i - w_i$
1	2.1	3	2.1	6.3	6.3
2	2.9	0.00002	5.8	0.000116	0.000058
3	4.05	0.0009	12.15	0.010935	0.003645
				6.311051	6.303703

$$\bar{y}_w = 2.1012343$$

Example

$w_i x_i$	x^2	$w_i x_i^2$
3	1	3
0.00004	4	0.00008
0.00270	9	0.00810

$$\sum w_i x_i = 3.00274 \qquad \sum_i w_i x_i^2 = 3.00818$$

$$\bar{x}_w = 1.00091$$

$$b^* = \frac{6.311051 - 2.101234\,(3.00274)}{3.00818 - 1.00091\,(3.00274)}$$

$$= \frac{6.311051 - 6.3094594}{3.00818 - 3.0054725}$$

$$= \frac{0.0015916}{0.0027075} = 0.5878485$$

$$b'_w = 0.587$$

$$a_w^* = \bar{y}_w - b^* \bar{x}_w$$

Where $\bar{x}_w$ and $\bar{y}_w$, are weighted means

= 2.101 − 0.587 (1.00)

= 2.101 − 0.587 = 1.514

= 1.514

b*	a*	
0.977	1.062	without weighing (See Page 103)
0.587	1.514	with weighing

$$\bar{x}_w = \frac{1}{n} \sum_i w_i x_i$$

$$\bar{y}_w = \frac{1}{n} \sum_i w_i y_i$$

and

$$\sum w_i = n$$

b*	a*	
0.977	1.062	without weighing
0.587	1.514	with weighing

The weighting of data may be important when the data are not used in their original form even if all the measurements are equally accurate. The error structure of the transformed data may be different from that of the original data.

3.3.1.6 Error Due to Data Transformation

A straight line equation is the easiest to handle. For example, we might wish to use the equation

$$y = e^{Kx} \tag{3.27}$$

in the form

$$\ln y = Kx \tag{3.28}$$

and to plot $\ln y$ versus x. The main concern here is how the errors in y relates to the error in $\ln y$.

Another example is equation

$$v = \frac{V_{max} C_s}{K + C_s} \tag{3.29}$$

One might wish to use this equation in the form

$$\frac{1}{V} = \frac{K}{V_{max}} \times \frac{1}{C_s} + \frac{1}{V_{max}} \tag{3.30}$$

A plot of $1/V$ versus $1/C_s$ should then be a straight line with slope K/V_{max} and intercept $1/V_{max}$. One's concern is how the errors in V and C, relate to errors in $1/V$ and $1/C_s$.

The Equations 3.27 and 3.28 are identical, so it should make no difference which we work with. Similarly, Equations 3.29 and 3.30 would make no difference if the observations were error-free.

3.3.1.6.1 Example: Error Due to Data Transformation

The following data were obtained for an enzyme that catalyzes the reaction $S\text{-----}\rightarrow P$ (Segel, 1975). Determine K_m and V_{max} and also do the plotting.

$[S]$	V	$1/[S]$	$\dfrac{1}{V}$
$(M) \times 10^5$	$(M \times Min^{-1})10^9$ or $(\mu\ moles \times Litre^{-1} \times min^{-1})$	$(M^{-1}) \times 10^{-4}$	$(M \times Min^{-1})^{-1} \times 10^{-7}$
0.833	13.8	12.00	7.24
1.00	16.0	10.00	6.25
1.25	19.0	8.00	5.26
1.67	23.6	6.00	4.24
2.00	26.7	5.00	3.74
2.50	30.8	4.00	3.25
3.00	34.3	3.33	2.91
3.33	36.3	3.00	2.75
4.00	40.0	2.50	2.50
5.00	44.4	2.00	2.25
6.00	48.0	1.67	2.08
8.00	53.3	1.25	1.88
10.00	57.1	1.00	1.75
20.00	66.7	0.50	1.50

$$\Sigma \frac{1}{[S]} = 60.25$$

$$\left(\frac{\bar{1}}{[S]} \right) = 4.3035$$

$$\Sigma \frac{1}{V} = 47.6$$

$$\left(\frac{\bar{1}}{V} \right) = 3.4$$

(1/S or x_i) . (1/V or y_i)	(1/S)²
86.88	144
62.50	100
42.08	64
25.44	36
18.70	25
13.00	16
9.6903	10.89
8.25	9.00
6.25	6.25
4.50	4.00
3.4736	2.7889
2.35	1.5625
1.75	1.00
0.75	0.25
Σ1/S. 1/V = 285.6139	Σ(1/S)²= 420.7414

$$b^* = \left[\sum_i x_i y_i - \bar{y} \sum_i x_i \right] \bigg/ \left(\sum_i x_i^2 - \bar{x} \sum_i x_i \right)$$

$$\frac{1}{b^*} = \frac{285.6139 - 3.4\,(60.25)}{420.7414 - 4.3035\,(60.25)}$$

$$= \frac{285.6139 - 204.85}{420.7414 - 259.28588}$$

$$= \frac{80.7639}{161.45553} = 0.5002238 \approx 0.5$$

$$\frac{1}{b^*} = 0.5 = \text{slope} = \frac{K_M}{V_{max}}$$

$$\frac{1}{V} = \frac{K}{V_{max}} \times \frac{1}{C_s} + \frac{1}{V_{max}}$$

$$\frac{1}{V_{max}} = \frac{1}{a^*} = \frac{1}{V_{max}}$$

$$\left(\frac{\overline{1}}{V}\right) = \frac{1}{a^*} + \frac{1}{b^*}\left(\frac{\overline{1}}{C_s}\right) = 3.4 = \frac{1}{a^*} + 0.5(4.3035)$$

$$\frac{1}{a^*} = 3.4 - 2.15175 = 1.24825$$

$$a^* \text{ or } V_{max} = 0.8011215$$

$$\left(\frac{\overline{1}}{V}\right)_{arith} = \frac{1}{a} + \frac{1}{b}\left(\frac{\overline{1}}{[S]}\right)_{arith} \tag{3.31}$$

$$3.4 = 1.25 + 1/b \,(4.3035)$$
$$= 1.25 + 0.5 \,(4.3035)$$
$$3.4 = 3.40175$$
$$3.4 = 1.25 + K_M/V_{max} \,(4.3035)$$
$$3.4 = 1/V_{max} + K_M/V_{max} \,(4.3035)$$
$$3.4 = 1.25 + 1.25 \, K_M \,(4.3035)$$
$$3.4 - 1.25 = 5.379375 \, K_M$$
$$2.15 = 5.379375 \, K_M$$
$$K_M = 2.15/5.379375 = 0.3996 \approx 0.4$$
$$K_M = 0.4 \times 10^4 \, M$$
$$K_M = 4 \times 10^5 \, M$$
$$K_M = 4 \times 10^5 \, M$$

$$\left(\frac{\overline{1}}{V}\right)_{arith} = \frac{1}{a} + \frac{1}{b}\left(\frac{\overline{1}}{[S]}\right)_{arith}$$

The harmonic mean of a sequence of members $S_1, S_2, \ldots\ldots S_n$ is

$$\bar{S}_{harm} = \frac{n}{1/S_1 + 1/S_2 + \ldots 1/S_n}$$

If we now take the reciprocal of Eq. (3.31), we see that Eq. (3.29) expresses a relation between the harmonic means,

$$\bar{V}_{harm} = \frac{V_{max}\,\bar{S}_{harm}}{K + \bar{S}_{harm}}$$

$$\frac{14}{47.6} = \frac{(0.801)\,(14/60.25)}{K + 14/60.25}$$

$$K + \frac{14}{60.25} = \frac{(0.801)\,(14/60.25)}{14/47.6}$$

$$K = \frac{0.801*\ 14/60.25}{14/47.6} - \frac{14}{60.25}$$

$$K = \frac{0.801*\ 14/60.25}{0.2941} - \frac{14}{60.25}$$

$$K = 2.7235*\frac{14}{60.25} - \frac{14}{60.25}$$

$$K = \frac{38.129}{60.25} - \frac{14}{60.25}$$

$$K = \frac{24.129}{60.25} = 0.4\times10^4\ M$$

$$K_M = 4\times10^5\ M$$

$$\frac{\hat{1}}{V_i} = \frac{K}{V_{max}}\times\frac{1}{C_s} + \frac{1}{V_{max}}$$

$$= 0.5\ (12) + 1.25$$

$$= 6 + 1.25$$

$$= 7.25$$

$$\frac{\hat{1}}{V_2} = 0.5\,(10) + 1.25$$

$$= 5 + 1.25 = 6.25$$

$$\frac{\hat{1}}{V_3} = 0.5\,(8) + 1.25$$

$$= 4 + 1.25$$
$$= 5.25$$

$$\frac{\hat{1}}{V_4} = 0.5\,(6) + 1.25$$

$$= 3 + 1.25$$
$$= 4.25$$

$$\frac{\hat{1}}{V_5} = 0.5\,(5) + 1.25$$

$$= 2.5 + 1.25$$
$$= 3.75$$

$$\frac{\hat{1}}{V_6} = 0.5\,(4) + 1.25$$

$$= 2 + 1.25$$
$$= 3.25$$

$$\frac{\hat{1}}{V_7} = 0.5\,(3.33) + 1.25$$

$$= 1.665 + 1.25$$
$$= 2.915$$

$$\frac{\hat{1}}{V_8} = 0.5\,(3) + 1.25$$

$$= 1.5 + 1.25$$
$$= 2.75$$

$$\frac{\hat{1}}{V_9} = 0.5\,(2.5) + 1.25$$

$$= 1.25 + 1.25$$
$$= 2.50$$

$$\frac{\hat{1}}{V_{10}} = 0.5\,(2) + 1.25$$

$$= 1 + 1.25$$
$$= 2.25$$

$$\frac{\hat{1}}{V_{11}} = 0.5\,(1.67) + 1.25$$

$$= 0.835 + 1.25$$
$$= 2.085$$

$$\frac{\hat{1}}{V_{12}} = 0.5\,(1.25) + 1.25$$

$$= 0.625 + 1.25$$
$$= 1.875$$

$$\frac{\hat{1}}{V_{13}} = 0.5\,(1) + 1.25$$

$$= 1.75$$

$$\frac{\hat{1}}{V_{14}} = 0.5\,(0.5) + 1.25$$

$$= 0.25 + 1.25$$
$$= 1.50$$

Table 3.10 Validation of the model

Observed $1/V_i$	Predicted $1/V_i$
7.24	7.25
6.25	6.25
5.26	5.25
4.24	4.25
3.74	3.75
3.25	3.25
2.91	2.915
2.75	2.75
2.50	2.50
2.25	2.25
2.08	2.085
1.88	1.875
1.75	1.75
1.50	1.50

The above derivation shows that predicted and observed reaction velocity are exactly the same, indicating that the measurement of variables are accurate and that the difference between measured and predicted is zero and, thus, error term is zero. Under such conditions, there is no need to weigh the data.

3.3.1.6.1.1 Graphical Representation

Figure 3.15 shows the straight line relationship between $1/[S]$ versus $1/V$. Data are given in Table 3.11.

Table 3.11 Showing the values of $1/[S]$ and $1/V$

$1/[S]$ M^{-1} * 10^{-4}	$1/V$ $(M * min^{-1})^{-1}$ * 10^{-7}
12.00	7.24
10.00	6.25
8.00	5.26
6.00	4.24
5.00	3.74
4.00	3.25
3.33	2.91
3.00	2.75
2.50	2.50
2.00	2.25
1.67	2.08
1.25	1.88
1.00	1.75
0.50	1.50

3.3.1.7 Correlation between Variables

In a correlation between the variables, we do assume that both variables are measured and subject to error.

Here we look at two extreme types of situation. The first situation is that the true value (expected value) of y does not at all depend on x. In such case, the measurements of y would normally cluster about the mean $\bar{y}$. Deviation from $\bar{y}$ would be random 'noise'. If we choose to plot the individual values of y_i against the value of x_i that happens to be measured at the same time, we might get a scatter such as in Fig. 3.12.

Table 3.12 Showing the values of x and y variable

x	y	$y-\bar{y}$	$(y-\bar{y})^2$	$\hat{y}$	$(\hat{y}-\bar{y})^2$
1	2	-0.0125	0.0001562	2.13	0.0144
2	2.1	0.0875	0.0076562	2.10	0.0081
3	2.2	0.1875	0.0351562	2.10	0.0081
4	1.8	-0.2125	0.0451562	2.03	0.0004
5	2.2	0.1875	0.0351562	2.00	0.0001
6	2.3	0.2875	0.0833765	2.00	0.0001
7	1.7	0.3125	0.0976562	2.00	0.0001
8	1.8	-0.2125	0.0451562	2.00	0.0001

$\bar{y}$ =2.0125   Randomness   0.3494699 $\sum(\hat{y}-\bar{y})^2$ 0.0314

Σy = 16.1

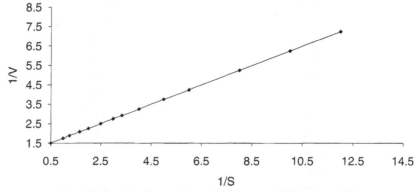

Fig. 3.15 Straight line relationship between 1/[S] versus 1/V.
Data are given in Table 3.11.

The variations in y are unrelated to any change in x; knowing x_i it gives no information about the value of y_i. In this case, the randomness or 'noise' in y can be measured by the distance of the vector y from a vector, each of whose components is equal to $\bar{y}$.

$$\bar{y} = (\bar{y}, \bar{y}, ., \bar{y})$$

The randomness would be measured by

$$d^2(y,\bar{y}) = \sum_{i=1}^{n} (y_i, \bar{y})^2$$

The variance in the data of Table 3.12 is 0.3494699 (randomness or total variation).

* The explained variation is 0.0314.

We might have a series of solutions of different concentrations of a given chemical compound and study the relation between the absorbance of light versus concentration. The absorbance $= \ln I_0/I$ where $I_0 =$ Total incident light; $I =$ light after transmission.

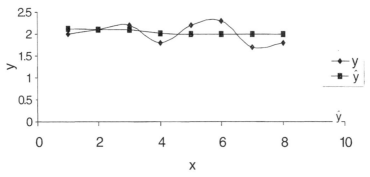

Fig. 3.16 Showing no relation between x and y

Table 3.13 Dependence of absorbance on concentration

A $\ln\left(\dfrac{I}{I_0}\right)$	$Antilog\ \ln\left(\dfrac{I}{I_0}\right)$	C $gm/liter$
0.1438	1.155	1
0.2876	1.333	2
0.5762	1.779	4
0.7765	2.176	5.4
0.8627	2.373	6

$$\bar{y} = 1.7632$$

Both of these variables can be measured to extremely high accuracy. A plot one against another looks as in Fig. 3.17 (Table 3.13).

Table 3.14 Randomness or variations in y in relation to change in x

x	y	$y - \bar{y}$	$(y - \bar{y})^2$
1	0.1438	-0.38556	0.14865
2	0.2876	-0.24176	0.05844
4	0.5762	0.04684	0.002193
5.4	0.7765	0.24714	0.061078
6	0.8627	0.33334	0.111115

$\bar{x} = 3.68$	$\bar{y} = 0.52936$		$\Sigma = 0.381476$
$\Sigma = 18.4$	$\Sigma = 2.6468$		

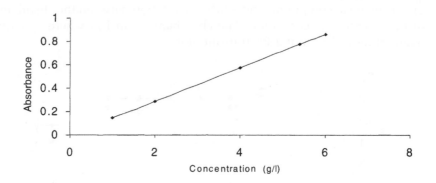

Fig. 3.17 A plot of absorbance on concentration

Table 3.15 showing y**,** $x_i y_i$ **and** $(x_i)^2$

x (gm/L)	$y \left(In\dfrac{I_0}{I} \right)$	$x_i\, y_i$	$(x_i)^2$
1	0.1438	0.1438	1
2	0.2876	0.5752	4
4	0.5762	2.3048	16
5.4	0.7765	4.1931	29.16
6	0.8627	5.1762	36
		$\Sigma = 12.3931$	$\Sigma = 86.16$

The value of $d^2\,(y,\ \bar{y}\,)$ predicted by the hypothesis might be written as

$$d^2\,(\hat{y}, \bar{y}) = \sum_{i=1}^{n} (\hat{y}_i - \bar{y})^2 \qquad (3.32)$$

$$\hat{y} = a^* + b^*\, x_0$$

$$\hat{y} = a^*_{\,geom.} + \frac{In\,(I_0 / I)}{C} \cdot x_0$$

$$b^* = \frac{\left(\displaystyle\sum xy_i - \bar{y} \sum_i x_i \right)}{\left(\displaystyle\sum_i x_i^2 - \bar{x} \sum_i x_i \right)}$$

$$b* = \frac{12.3931 - 0.52936(18.4)}{86.16 - 3.68(18.4)}$$

$$b* = \frac{2.652876}{18.848} = 0.1438029$$

$$b* = K$$

$$a* = \bar{y} - b* \bar{x}$$

$$a* = 0.52936 - 0.144 (3.68)$$

$$a* = 0.52936 - 0.52992$$

$$a* = 0$$

$$\ln \hat{y}_1 = \ln k * x = 0.144 * 1 = 0.144$$

$$\ln \hat{y}_2 = 0.144 * 2 = .288$$

$$\ln \hat{y}_3 = 0.144 \times 4 = .576$$

$$\ln \hat{y}_4 = 0.144 \times 5.4 = .778$$

$$\ln \hat{y}_5 = 0.144 \times 6 = .864$$

$$\hat{y}_{geom} = e^{Kx}$$

$$\hat{y}_1 = e^{KC} = e^{144(1)} = 1.155$$

$$\hat{y}_4 = e^{.144 \times 4} = e^{.576} = 1.779$$

$$\hat{y}_{5.4} = e^{.144 \times 5.4} = e^{0.7776} = 2.176$$

$$\hat{y}_6 = e^{.144 \times 6} = e^{.864} = 2.373$$

Table 3.16a Showing the explained and total variation

$\bar{y}$	$\hat{y}$	y	$\hat{y} - \bar{y}$	$(\hat{y} - \bar{y})^2$
	1.155	1.155	−0.6082	4.05695
1.7632	1.333	1.333	−0.4302	0.1851
	1.779	1.779	0.0158	0.0002496
	2.176	2.176	0.4128	0.1704
	2.373	2.373	0.6098	0.3719
				$\Sigma = 4.7845996$

$y-\bar{y}$	$(y-\bar{y})^2$
−0.6082	4.05695
−0.4302	0.1851
0.0158	0.0002496
0.4128	0.1704
0.6098	0.3719
	4.7845996

$$\eta^2 = \frac{\text{Explained variation}}{\text{Total variation (randomness)}} = \frac{(\hat{y}-\bar{y})^2}{(y-\bar{y})^2}$$

$$\eta^2 = \frac{4.7845}{4.7845} = 1$$

η^2 = coefficient of determination

Table 3.16b Showing the explained and total variation

$\ln \hat{y}_i$	$\ln \bar{y}$	$\ln \hat{y}_i - \ln \bar{y}$	$(\ln \hat{y}_i - \ln \bar{y})^2$
0.144	0.52936	−0.38536	0.1485
0.288		−0.24136	0.05825
0.576		0.04664	0.0002175
0.778		0.24864	0.06182
0.864		0.33464	0.11198
			0.3807675

$\ln y_i$	$\ln \bar{y}$	$\ln y_i - \ln \bar{y}$	$(\ln y_i - \ln \bar{y})^2$
0.144	0.52936	−0.38536	0.1485
0.288		−0.24136	0.05825
0.576		0.04664	0.0002175
0.778		0.24864	0.06182
0.864		0.33464	0.11198
			0.3807675

$$\eta^2 = \frac{(\ln \hat{y} - \bar{y})^2}{(\ln y - \bar{y})^2} = \frac{0.3807}{0.3807} = 1$$

The randomness would be measured by

$$d^2 (y, \bar{y}) = \sum_{i=1}^{n} (y_i - \bar{y})^2$$

Table 3.12 shows that this value equals 0.3495. This is a total variation (signal + noise).

The value of d^2 ($\hat{y}$, $\bar{y}$) predicted by the hypothesis might be written as

$$d^2(\hat{y},\bar{y})=\sum_{i=1}^{n}(\hat{y}_1-\bar{y})^2$$

In the case represented in Table 3.12, this is $0.0314{\approx}0$. In the case represented in Tables 3.16a and 3.16b, it is equal to d^2 ($\hat{y}$, $\bar{y}$). It is useful to speak of the correlation ratio (coefficient of determination) of y on x,

$$\eta^2 = \frac{d^2(\hat{y},\bar{y})}{d^2(y,\bar{y})}$$

$$= \frac{\sum\limits_{i=1}^{n}(\hat{y}_i,\bar{y})^2}{\sum\limits_{i=1}^{n}(y_i,\bar{y})^2} \tag{3.33a}$$

as a measure of the dependence of y on x. It varies from zero (as in Table 3.12) to one (as in Tables 3.16a and 3.16b). It is often referred to as the ratio of explained variation to total variation.

For the special case of a linear relation between y and x (that is $y = a + bx$), equation (3.33a) becomes the square of the ordinary correlation coefficient,

$$r = \frac{\sum\limits_{i}(x_i - \bar{x})(y_i - \bar{y})}{\left[\sum\limits_{i}(x_i - \bar{x})^2 \sum\limits_{i}(y_i - \bar{y})^2\right]^{1/2}} \tag{3.33b}$$

In the form given by equation (3.33b), the correlates coefficient has a negative or positive sign, depending upon the slope of the line relating y and x.

3.3.1.7.1 Example

Table 3.17 Showing the statistics of r

x	y	$x-\bar{x}$	$y-\bar{y}$	$(x-\bar{x})(y-\bar{y})$
1	0.144	-2	-0.3854	0.7708
2	0.288	-1	-0.2414	0.2414
3	0.576	0	0.0466	0
4	0.776	1	0.2466	0.2466
5	0.863	2	0.3396	0.6672

$\bar{x} = 3$, $\bar{y} = 0.5294$, $\Sigma x = 15$, $\Sigma y = 2.647$, $\Sigma x^2 = 55$, $\Sigma y^2 = 1.782401$, $\Sigma = (x-\bar{x})(y-\bar{y}) = 1.926$

$(x-\bar{x})^2$	$(y-\bar{y})^2$
4	0.1485331
1	0.0582739
0	0.00217156
1	0.0608115
4	0.1112889
$(\Sigma(x-\bar{x})^2 =10)$	$(\Sigma(y-\bar{y})^2 = 0.3811485)$

$$\sqrt{3.811484} = 1.9523025$$

$$r = \frac{1.926}{1.926} \approx 1$$

$$r^2 = 1$$

3.3.1.8 Forced Correlation

3.3.1.8.1 Example

Table 3.18 showing the values of x, y, z and $\hat{z}$

x_i	y_i	z_i	$\hat{z}_i$
1	2	2	5.64
2	2.1	4.2	6.67
3	2.2	6.6	7.60
4	1.8	7.2	7.53
5	2.2	11.0	9.36
6	2.3	13.8	10.29
7	1.7	11.9	11.22
8	1.8	14.4	12.15

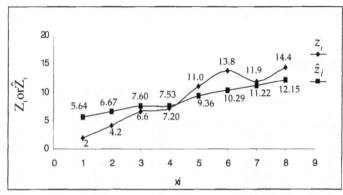

Fig. 3.18 The same data as used in Fig. 3.16 plotted after creating a variable called z_i. It illustrates the creation of a valid but meaningless correlation.

We take the case in which y and x really have no relation. Now, whatever value is attributed to y, just multiply x to it, and call the result z.

$$z_1 = x_1^* y_1$$

The variable Z is going to be a function of X, whether or not Y is. Moreover, the variable Z makes a larger journey away from its mean than the variable Y does, and *all* the increased variability is directly accounted for in terms of X. The signal to noise ratio and the correlation ratio are thereby increased.

Graphically, the situation may be represented as in Fig. 3.16, where we see a plot of y verses x, where

$$y_1 = \text{const.} + \varepsilon_1$$

This means that y is unrelated to x, and the correlation ratio is zero. In Fig. 3.18, we have a plot of z versus x, where $z_1 = y_1 + x_1$. The result is that

$$z_1 = x_1 + \text{const.} + \varepsilon_1$$

The errors are the same for y_1 and z_1. Only the meaningful variability is changed. We can say that z and x are correlated, whereas y and x are not. We can further increase the correlation still further increasing the steepness of the line by letting,

$$z = y + Kx$$

where K is any number we choose.

The degree of correlation suggested by the plot can be altered by changing the scale of the plot. As a general rule, it is a good idea to choose the scales of the coordinates so that the range of y observations and the range of x observations are represented by about the same length of paper.

3.3.1.9. Statistical Procedure for Parameters Estimation of Normal Distribution Curve

The gaussion or normal curve is described by the equation :

$$y = \frac{1}{\sigma\sqrt{2\pi}}\ e^{-(x-\mu)^2/2\sigma^2} \tag{3.34}$$

In this equation, y represents the relative frequency of some variable quantity, x. The values for π and e are constant; π is the familiar ratio of the circumference to the diameter of any circle, 3.1416; e is the base to the Napier or natural logarithms.

The important features are the two parameters μ and σ. The first, μ, is the arithmetic mean; the second, σ, is called the standard deviation.

The standard deviation is a measure of the spread of the data about the mean. Since the values for π and e are fixed forever and remain constant, the entire curve is completely defined by the two parameters the mean and standard deviation.

Table 3.18 shows the data on the seed filling rate (unpublished data).

Table 3.18 Showing the seed filling rate (mg/pod seed.day) in chickpea cultivar ICC4958 during the season 1999-2000 at HAU, Hisar (Pot experiment) The total observations are 41.

S. No.	SFR	S. No.	SFR
1.	5.105	22.	12.9
2.	7.695	23.	13.07
3.	8.37	24.	13.475
4.	9.016	25.	13.56
5.	9.142	26.	13.65
6.	9.927	27.	13.865
7.	10.031	28.	13.945
8.	10.158	29.	14.038
9.	10.485	30.	14.117
10.	10.846	31.	14.235
11.	10.996	32.	14.447
12.	11.259	33.	14.47
13.	11.42	34.	14.506
14.	11.7	35.	14.819
15.	11.7	36.	14.979
16.	11.925	37.	15.38
17.	12.409	38.	16.5
18.	12.435	39.	16.775
19.	12.523	40.	17.406
20.	12.656	41.	20.895
21.	12.813		

$$I = \frac{(M - m) - (n - 1)p + (Y + y)}{n} \quad \text{(Lewis 1971)}$$

Where I is the size of the class interval,

M is the highest value in the data.

m is the lowest value in the data.

p is the link value for precision.

n is the number of groups to be formed.

$(M - m)$ is the range of the data.

$(m - y)$ is the lowest value.

$(M + Y)$ is the highest value.

$$I = \frac{15.79-(10-1)(0.001)+Y+y}{10}$$

$$I = \frac{15.781+Y+y}{10} = 1.578 + Y + y = 1.578 + 0.004 + 0.003 = 0.585$$

Now, the mathematical formulation of a normal distribution curve is:

$$y = \frac{N}{\sigma\sqrt{2\pi}} e^{-\frac{1}{2}\left(\frac{d}{\sigma}\right)^2}$$

Where y = frequency density

    σ = standard deviation of population.

    d = deviation from the mean

    μ = mean of population

$$Z = \frac{e^{-\frac{1}{2}\left(\frac{d}{\sigma}\right)^2}}{\sqrt{2\pi}}$$

So, $y = z\,(cN/\sigma)$

where c = value of class interval

The value of z can also be taken from Table A-2 quoted in Goulden, 1962, against the values of d/σ.

From the Tables 3.19 and 3.20 and Figure 3.19, it is clear that, there is no significant difference between observed and theoretical normal frequencies. Hence, the data given in Table 3.18 follows the normal distribution.

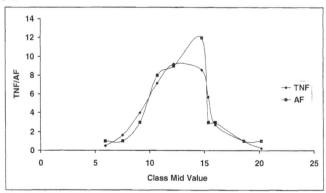

Fig. 3.19 The relationship between midvalues and TNF/AF. Data are given in Table 3.19

Table 3.19   The class range, class midvalue (cmd) and class limit, d (deviation from mean), d/s (the value of $s^* = 2.725$), z (from Table A-2, Goulden, 1962, page 442), y (frequency density at the point d), N/2 $(1+\alpha)$ (from Table A-1, Goulden, 1962, page 441), theoretical normal frequency and actual frequency. ($y = z$ (cN/s), where c is equal to class interval)

C.R.	CMD	C.L.	d	d/s	z	y=z(cN/s)	N/2 (1 + a)	TNF	AF
			12.321	4.521	0.0001	0.002	41.00	0.0000	
			10.735	3.939	0.0002	0.004	41.00	0.0164	
			9.149	3.357	0.0014	0.03	40.9836	0.984	
5.102 - 6.687	5.894		7.563	2.775	0.0086	0.19	40.8852	0.4715	1
6.688 - 8.273	7.480	6.687	5.978	2.194	0.0363	0.85	40.4137	1.6154	1
8.274 - 9.859	9.066	8.273	4.391	1.611	0.1092	2.59	38.7983	4.0098	3
9.860 - 11.445	10.652	9.859	2.805	1.029	0.2347	5.59	34.7885	7.1709	8
11.446 - 13.031	12.238	11.445	1.219	0.447	0.3605	8.59	27.6176	9.2373**	9
		12.664	0.000	0.000	0.3989	9.51	20.5000		
13.032 - 14.617	14.824	13.031	-0.367	-0.135	0.3951	9.41	22.9837	8.5485	12
14.618 - 16.203	15.410	14.617	-1.953	-0.717	0.3079	7.33	31.3322	5.699	3
16.204 - 17.789	16.996	16.203	-3.539	-1.299	0.1714	4.07	37.0312	2.7347	3
17.790 - 19.375	18.582	17.789	-5.125	-1.881	0.0681	1.61	39.7659	0.9512	0
19.376 - 20.961	20.168	19.375	-6.711	-2.463	0.0194	0.45	40.7171	0.2337	1
		20.961	-8.297	-3.045	0.0039	0.07	40.9508		
		-9.882	-3.626	0.0005	0.01	40.9959	0.0451		
		-11.467	-4.208	0.0001	0.002	41.0000	0.0041		
								41	41

*The value of standard deviation is calculated as per Goulden, 1962, page 21

**In the central class, the two differences are added.

Table 3.20 Showing *d/s* (standard deviation from mean), *y* (observed frequency), $\hat{y}$ (theoretical normal frequency) and χ^2 test for seed filling rate data

d/s	*y*	$\hat{y}$	$(\hat{y}-y)$	$(\hat{y}-y)^2$	$(\hat{y}-y)^2/\hat{y}$	Table χ^2 with 9df
2.775	1	0.47	-0.53	0.2809	0.5976 =	16.92
2.194	1	1.61	0.61	0.3721	0.2361	
1.611	3	4.00	1.00	1.0000	0.2500	
1.029	8	7.17	-0.83	0.6889	0.0960	
0.447	-	-	-	-	-	
0.000	9	9.23	0.23	0.0529	0.0057	
-0.135	-	-	-	-	-	
-0.717	12	8.54	-3.46	11.9716	1.4018	
-1.299	3	5.69	2.69	7.2361	1.2717	
-1.881	3	2.73	0.27	0.0729	0.0267	
-2.463	0	0.95	0.95	0.9025	0.95	
-3.045	1	0.23	-0.77	0.5929	2.5778	
				Caculated X^2	7.4084	

3.3.1.9.1 *Practical Uses of Normal Distribution Curve and Table of Normal Distribution (Double Tail)*

The mean, $\bar{x}$, and standard deviation, s, have been calculated after having analysed a large sample of data. One may determine what proportion of the population is included between the values ($\bar{x} + x$) and ($\bar{x} - x$) shown as the unshaded area in Figure 3.20. The deviations are converted into units of standard deviation, U, as follows :

$$U = \pm \frac{\bar{x}-x}{s} \tag{3.35}$$

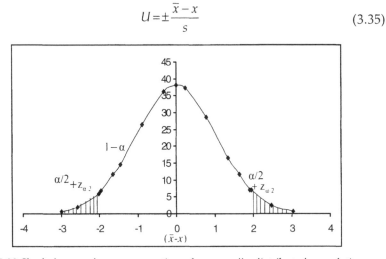

Fig. 3.20 Shaded areas show a proportion of a normally distributed population deviating from the mean by more than $\pm (\bar{x} - x)$.

The proportionate areas corresponding to U are given in a standard table, where the values give the proportion of the total area under the normal distribution curve lying in the shaded portions corresponding to various values of U. Since the total area is equal to unity, the unshaded area is calculated as $(1-p)$.

Here we quote an example cited by Lewis (1966). Let us imagine that somebody is conducting an experiment using spherical seeds. For the purpose of this study, he wants all of the seeds used in the experiment to deviate from the mean diameter by no more than 5% of the mean. The mean diameter is 10 mm with a standard deviation of 2 mm. The person conducting the experiment separates the seeds that are too large by passing the seeds through a sieve with holes 10.5 mm in diameter (i.e. 10 mm + 5% of 10 mm). The seeds to be used are separated by passing the remainder through a sieve with holes 9.5 mm in diameter (10 mm – 5% of 10 mm). From the available information, we are asked to determine how many useful specimens he will have if he starts with 1000 seeds.

First, he calculates the value of U :

$$U = \frac{(10 - 9.5)}{2} = 0.25$$

Turning to the standard table, in the column headed U, he finds 0.2. The next column to the right, with the heading 0.00, gives the proportion of the shaded area represented by a value of $U = 0.2 + 0.00$ or $U = 0.20$. The value here is 0.8415. However, he wants the area or proportion corresponding to $U = 0.25$. In the column headed 0.05 (at the level of the entry 0.2 in the column headed U), he finds the entry 0.8026 which may be approximately rounded off to 0.803. This is the entry for $U = 0.2 + 0.05$ or $U = 0.25$. The proportion of seeds retained is $(1-0.803)$ or 0.197 which, multiplied by the original 1000 seeds, gives the answer, 197 seeds.

Suppose the experimenter takes 1000 of these same seeds. How many will be retained by a sieve with holes 6 mm in diameter? In this example, only the seeds which are 4 mm less than the mean will be discarded, but all of the remainder, including those 4mm larger than the mean, will be retained. The value of U is obtained as :

$$U = \frac{(10 - 6)}{2} = 2$$

The standard table is 'double tailed', but for this particular example, we require a 'single tailed' value. In other words, the proportionate area p, represents both of the shaded areas represented at the ends of the distribution curve in Fig. 3.20. These areas are equal, and since we only require the area of the shaded tail on the left, he takes half of the tabulated value. In this example, that table shows that for $U = 2$, the proportion of the population included under both shaded areas is 0.045. However, only half of this value represents the seeds that have diameters less than 6mm and will fall through the sieve. Thus, 0.0225 (1000) represents the number of seeds he could expect to exclude. This would not exceed 23 and, therefore, he would expects the sieve to retain 977 seeds, i.e. 1000–23 = 977.

3.3.1.9.1.1 *Example (Quirin 1978)*

A random sample of 100 brand A cigarettes yields an average of $\bar{x} = 26$ milligram of nicotine. It is known that $\sigma = 8$ milligrams of nicotine. With degree of confidence $1-\alpha = 0.99$, we find that the average nicotine content μ for the brand of cigarettes satisfies. The valvue of $Z_{1/2}$ is 2.58.

$$26 - 2.58 * \frac{8}{\sqrt{100}} \leq \mu \leq 26 + 2.58 * \frac{8}{\sqrt{100}}$$

which is equivalent to $23.94 \leq \mu \leq 28.06$

3.3.1.9.1.2 *Example (Quirin 1978)*

A sample of 40 exams scores taken from a large population of exam scores is found to have an average of $\bar{x} = 77.125$. The standard deviation for this type of exam scores is known to be $\sigma = 13.36$. With a degree of confidence 0.95, we estimate that the average score μ for this exam lies in the interval.

$77.125 - 1.96 * 13.36/ \sqrt{40} \leq \mu \leq 77.125 + 1.96 * 13.36 / \sqrt{40}$, which is equivalent to $72.985 \leq \mu \leq 81.265$. Of course, there is a 5% chance that our sample will mislead us, and the true population mean μ will not lie in this interval.

3.3.1.9.1.3 *Differences Between Two Population Mean or Proportions*

One of the most important and frequently occurring problems in statistics is to determine whether or not two populations have a common parameter. This parameter could be a mean and we might be asking, for example, if two similar products have the same life span, or that the parameter could be a proportion, and our problem might be to determine if the citizens of two cities feel the same way about a certain political issue. Perhaps the most popular variation of this type of problem is the determination of the impact of change. For example, one might wish to determine if a new drug is more effective than a drug presently in use, or if a new product is, in some way, better than an older product.

The approach to such problems will be to take a sample from each population and then determine the sample parameter (perhaps a mean or a proportion) for each sample. Needless to say, our sample parameters will probably differ. We will then have to determine if this difference is statistically significant and reflects a true difference in the corresponding population parameters, or if the difference is small and can reasonably be attributed to chance.

Since we have already seen that sample proportions can be looked upon as a particular case of sample means, we shall set up the machinery for studying the differences between sample means, and then obtain the results to study the differences between sample proportions as corollaries.

Let us suppose that we have two populations, the first with mean μ_1 and standard deviation σ_1, and the second with mean μ_2 and standard deviation σ_2. Suppose that we take a sample of size n_1 from the first population and a sample of size n_2 from the second population, and obtain sample means of $\bar{x}_1$ and $\bar{x}_2$, respectively. This is equivalent to having set up n_1 independent, identically distributed random variables $x_1, x_2, \ldots\ldots x_n$, for sampling from the first population, and n_2 independent, identically distributed random variables $y_1, y_2, \ldots, y_n$ for sampling from the second population. Now,

$E(\bar{x}_i) = \mu_1$ and var $(x_i) = \sigma^2_1$ for $i = 1, 2 \ldots\ldots, n_1$, and

$E(\bar{y}_i) = \mu_2$ and var $(y_i) = \sigma^2_2$ for $i = 1, 2 \ldots\ldots, n_2$,

by definition, the two sample means equal

$$\bar{x} = \left(\sum_{i=1}^{n} x_i \right) /n_1 \text{ and } \bar{y} = \left(\sum_{i=1}^{n} y_i \right) /n_2$$

under these conditions, we can prove :

(a) $\bar{x} - \bar{y}$ is approximately normal

(b) $E(\bar{x} - \bar{y}) = \mu_1 - \mu_2$

(c) var $(\bar{x} - \bar{y}) = (\sigma^2_1/n_1) + (\sigma^2_2/n_2)$

Proof :

(a) Both $\bar{x}$ and $\bar{y}$ are normal and, consequently, their difference $\bar{x} - \bar{y}$ is normal

(b) $E(\bar{x} - \bar{y}) = E\left[\sum_{i=1}^{n_1} x_i/n - \sum_{i=1}^{n_2} y_i/n_2 \right]$

$= 1/n_1 \sum_{i=1}^{n_1} E(x_i) - 1/n_2 \sum_{i=1}^{n_j} E(y_i)$

$= (n_1 \mu_1/n_1) - (n_2 \mu_2/n_2)$

$= \mu_1 - \mu_2$

(c) Using the fact that $x_1, x_2, ..., x_n$ are independent, that $y_1, y_2, ..., y_n$ are independent, we have

$$\text{var}\,(\bar{x} - \bar{y}) = \text{var}\left(\sum_{i=1}^{n_1} x_i/n_1 - \sum_{i=1}^{n_2} y_i/n_1\right)$$

$$= \left(\frac{1}{n_1}\right)^2 \sum_{i=1}^{n_1} \text{var}\,(x_i) + \left(\frac{1}{n_2}\right)^2 \sum_{i=1}^{n_2} \text{var}\,(y_i)$$

$$= \frac{n_1 \sigma_1^2}{n_1^2} + \frac{n_2 \sigma_2^2}{n_2^2}$$

$$= \frac{\sigma_1^2}{n_1} + \frac{\sigma_2^2}{n_2}$$

Now let us apply these results to the case where the parameters being studied are the proportion, rather than the means. If $\bar{A}_1$ represents the proportion of items with a some characteristic in a sample taken from the first population, and $\bar{A}_2$ the proportion of items with this characteristic in a sample taken from the second population, then it is seen that both $\bar{A}_1$ and $\bar{A}_2$ can be looked upon as sample means. Since, in this case, one has $E\,(x_i) = A_1$ (the true proportion of items with the given characteristic in the first population) and var $(x_i) = A_1 - (1 - A_1)$, for $i = 1, 2, ..., n_1$ and $E\,(y_i) = A_2$ (the true proportion for the second population) and var $(y_i) = A_2\,(1 - A_2)$, for $i = 1, 2, ..., n_2$, one has consequences.

(a) $\bar{A}_1 - \bar{A}_2$ is approximately normal

(b) $E\,(\bar{A}_1 - \bar{A}_2) = A_1 - A_2$

(c) var $(\bar{A}_1 - \bar{A}_2) = A_1\,(1 - A_1) / n_1 + A_2\,(1 - A_2) / n_2$

The results just derived are usually put to use testing the hypothesis that the two populations do share a common parameter. For example, if we are attempting to determine if the two populations means μ_1 and μ_2 are, in fact, the same, we would test the null hypothesis.

$$H_0 : \mu_1 = \mu_2$$

against one of the alternative hypothesis $\mu_1 \neq \mu_2$, $\mu_1 > \mu_2$ or $\mu_1 < \mu_2$ at a given level of significance. We would take samples of size, n_1 and n_2 from the two populations, calculate the sample means $\bar{x}_1$ and $\bar{x}_2$, and then consider the standard statistic.

$$z = \frac{(\bar{x}_1 - \bar{x}_2) - E(\bar{x}_1 - \bar{x}_2)}{\sigma_{\bar{x}_1 - \bar{x}_2}}$$

$$= \frac{(\bar{x}_1 - \bar{x}_2) - (\mu_1 - \mu_2)}{\sqrt{\sigma_1^2 / n_1 + \sigma_2^2 / n_2}}$$

Under the assumption that the null hypothesis is true, this statistic becomes

$$= \frac{\bar{x}_1 - \bar{x}_2}{\sqrt{\sigma_1^2 / n_1 + \sigma_2^2 / n_2}}$$

We then either reject, or do not reject the null hypothesis depending on whether the statistic z does or does not fall in the critical region.

Example (1) : A farmer must decide whether to use the variety of wheat A or B, so a test is made of 100 plants of each variety. Suppose that the 100 plants of A had an average yield of 985 gm, while the 100 plants of B had an average yield of 1004 gm. We then assume that the standard deviation of yield of each variety is 80 gm. We wish to determine if the difference in yield for the two samples is significant at the five percent level of significance, so we test the null hypothesis.

$$H_0 : \mu_A = \mu_B$$

against the alternative

$$H_1 : \mu_A \neq \mu_B$$

where μ_A and μ_B are the actual yield for variety A and B. At the five percent level of significance, we have the two-sided critical region [Z] > 1.96, and if we evaluate

$$z = \frac{985 - 1004}{\sqrt{(80)^2 / 100 + (80)^2 / 100}} = \frac{-19}{\sqrt{128}} = -1.68$$

we find that our sample statistic does not fall in the critical region. Therefore, we cannot conclude that the difference between the two sample means reflects a true difference in the two population means.

Example (2): Suppose we wish to test whether the proposed new rotary engines for tractors are cleaner, with respect to sulphur dioxide emission, than the piston engines currently in use. Suppose that a sample of 50 rotary engines showed an emission level of 26 parts per million, while a sample of 50 piston engines showed an emission level of 29 parts per million. Also suppose that we can assume a standard deviation of 10 parts per million. We test the hypothesis

$$H_0 : \mu_1 = \mu_2$$

that the average emission levels for the two types of engines are equal against the alternative

$$H_1 : \mu_1 < \mu_2$$

that the rotary engines are cleaner, at a 1% level of significance. We have the one-sided critical region $Z < -1.64$, and the sample statistic.

$$z = \frac{26 - 29}{\sqrt{(10)^2 / 50 + (10)^2 / 50}} = -1.50$$

does not fall in the critical region. Therefore, on the basis of evidence contained in these two small samples, we can not conclude that the new rotary engine is cleaner than the old piston engine.

Proportion Parameter

We test the null hypothesis

$$H_0 : A_1 = A_2$$

Determine whether or not the statistic

$$z = \frac{(\overline{A}_1 - \overline{A}_2) - (A_1 - A_2)}{\sqrt{A_1 q_1 / n_1 + A_2 q_2 / n_2}}$$

lie in the appropriate critical region. Under the assumption that the null hypothesis is true, i.e. $A_1 = A_2$, this statistic reduces to

$$z = \frac{(\overline{A}_1 - \overline{A}_2)}{\sqrt{A_1 q_1 / n_1 + A_2 q_2 / n_2}}$$

One problem is that the values of the unknown parameters A_1 and A_2 appear in the formula for z. To overcome this problem, we caculate the pooled proportion.

$$\overline{A} = \frac{n_1 A_1 + n_2 A_2}{n_1 + n_2}$$

the proportion in the merged sample of items with the given characteristic. Under the assumption that the null hypothesis is true, that the two populations have the same percentage of items with the given characteristics, then $\overline{A}$ is the best guess we can make concerning the true value of $A_1 = A_2$, based on the information contained in the two samples. Using $\overline{A}$, as a substitute for both A_1 and A_2 our statistics Z becomes.

$$z = \frac{(\bar{A}_1 - \bar{A}_2)}{\sqrt{\bar{A}\bar{q}/n_1 + \bar{A}\bar{q}/n_2}}$$

Example (3): We wish to determine whether the male and female voters in a certain city are of the same mind with respect to an issue that Sh. A.B. Vajpayee is the best prime minister so far. A sample of 144 men contained 70 or 48.6 percent in favor of the issue, while a sample of 100 women contained 55 or 55 percent in favor. We wish to determine if the sample proportions 0.486 and 0.550 indicate a substantial difference of opinion or if the difference between these two samples can reasonably be attributed to chance. For this purpose, we evaluate first the pooled proportion

$$\bar{A} = \frac{144(0.486) + 100(0.55)}{144 + 100}$$

$$\bar{A} = \frac{125}{244} = 0.512$$

and then the statistic

$$z = \frac{0.468 - 0.550}{\sqrt{0.512 * 0.488 * [(1/144) + (1/100)]}}$$

$$= \frac{-0.64}{0.065}$$

$$= -0.985$$

which clearly is not large enough to fall in any one percent or five percent critical region. Hence, on the basis of the sample alone, we cannot say, for example, that the women feel more strongly about this issue than the men.

Example (4): Suppose that we wish to test two different types of herbicides to kill weeds. Suppose that in a group of 100 plants, using the first type, 70 plants of weed species are killed within 24 hours, while in a group of 100 plants using the second type, only 60 plants were killed within 24 hours. Is this sufficient proof that the first type of herbicide killed faster than the second type? To answer this question, we test the null hypothesis

$$H_0 : A_1 = A_2$$

that the two types of herbicides are equally effective, i.e. the same percentage of plants will be killed within 24 hours using either brand of herbicides against the alternative

$$H_1 = A_1 > A_2$$

that the first type will kill a higher percentage. At a one percent level of significance, we have the one-sided critical region $Z > 2.33$, and since our sample statistic

$$z = \frac{0.70 - 0.60}{\sqrt{0.65 * 0.35 * (2/100)}} = 1.48$$

does not fall in the critical region, we cannot conclude, on the basis of these samples, that one herbicide is more effective than the other.

3.3.1.9.1.4 Interval Estimation

Instead of approximating a population parameter with a single value, with no method of determining how accurate that estimate might be, it would be much more informative if we could determine a range of values, where, we could say with a certain prescribed amount of confidence, that the parameter will lie. The objective of this section is to describe the method for determining such a range for both population means and population proportions. Quirin (1978) has given an excellent description of interval estimation.

Given any fraction α between 0 and 1, i.e. $\alpha \in (0, 1)$, we denote by $z_{\alpha/2}$ the point on the x axis in such a way that the area under the standard normal curve between $-z_{\alpha/2}$ and $+z_{\alpha/2}$ equals $1-\alpha$ (Fig. 3.21). Hence,

$$\frac{1}{\sqrt{2\pi}} \int_{-Z_{\alpha/2}}^{+Z_{\alpha/2}} e^{-(1/2)x^2} dx = 1-\alpha$$

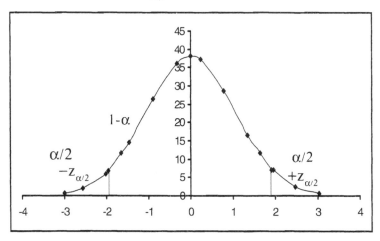

Fig. 3.21 The area under the standard normal curve

Since the standard normal curve is symmetric about the line $x = 0$, the remaining area, which equals α, is divided equating between the two tails of the curve. In other words, the area under the curve to the right of $z_{\alpha/2}$ equals $\alpha/2$, and the area under the curve to the left of $-z_{\alpha/2}$ also equals $\alpha/2$ (Quirin 1978). Table 3.21 gives the values of $z_{\alpha/2}$ corresponding to frequently used value of α (Quirin 1978).

Therefore, the probability that a random variable with standard normal distribution assumes a value between -1.96 and $+1.96$ equals 0.95. Recalling that the mean and standard deviation of the standard normal distribution are 0 and 1, respectively, we see that the probability of such a random variable assuming a value within two deviations of its mean is slightly higher than 0.95. This result can be extended to normally distributed random variables in general. Suppose that x is such a random variable with parameter μ and σ. Under the change of variables $Z = (x - \mu)/\sigma$, we find that

Table 3.21: The values of $z_{\alpha/2}$ corresponding to frequently used value of α

α	$1-\alpha$	$z_{\alpha/2}$
0.10	0.90	1.64
0.05	0.95	1.96
0.02	0.98	2.33
0.01	0.99	2.58

$$\frac{1}{\sigma\sqrt{2\pi}} \int_{\mu-z_{\alpha/2}\sigma}^{\mu+z_{\alpha/2}\sigma} exp\left[-\frac{1}{2}\left(\frac{x-\mu}{\sigma}\right)^2\right] dx$$

$$= \frac{1}{\sqrt{2\pi}} \int_{-z_{\alpha/2}}^{z_{\alpha/2}} e^{-(1/2)Z^2} dz$$

Therefore, the probability that x assumes a value within 1.96 deviations σ of its mean μ equals 0.95. Compare this with the information obtained from Chebyshev's inequality; the probability that *any* random variable assumes a value within two deviations of its mean is at least $1-(1/2)^2 = 3/4$. (Quirin 1978).

Now, suppose we wish to estimate the mean μ of a population. We will assume for the moment that the standard deviation σ of the population is known. We take a sample of size n, with replacement, and find that our sample has mean $\bar{x}$. Our sample's average is but one value of the sample mean, a random variable which itself has mean μ and, for samples of size n, deviation $\sigma/\sqrt{n}$. According to the central limit theorem, the distribution of the sample mean can be closely approximated by a normal curve (with parameter μ and $\sigma/\sqrt{n}$), and therefore, we can say that

$$p\left[\mid \bar{x} - \mu \mid \leq z_{\alpha/2}\,(\sigma/\sqrt{n}\,)\right] = 1 - \alpha$$

That is we can assert with probability $1-\alpha$ that the value of our sample's average (the value assumed by the sample mean in this case) will differ from the population's average (which is also the average of the sample mean) by no more than $z_{\alpha/2}$ (sample mean) deviations. Since

$$\mid \bar{x} - \mu \mid \,\leq z_{\alpha/2}\,(\sigma/\sqrt{n}\,)$$

is equivalent to

$$-z_{\alpha/2}\,(\sigma/\sqrt{n}\,) \leq\, \bar{x} - \mu \leq z_{\alpha/2}\sigma/\sqrt{n}\ \ \text{or}$$

$$\mu - z_{\alpha/2}\sigma/\sqrt{n}\, \leq\, \bar{x}\, \leq \mu + \,z_{\alpha/2}\,\ \sigma/\sqrt{n}$$

we can assert with probability $1-\alpha$ that our sample average falls in the interval

$$(\mu - z_{\alpha/2}\ \sigma/\sqrt{n}\,,\, \mu + z_{\alpha/2}\ \sigma/\sqrt{n}\,)$$

Unfortunately, μ is unknown, and so it is impossible to determine the end points of this interval. But the interval is important in the sense that it does contain a certain percentage of the sample averages. That is what we mean when we say 'we can assert with probability $1-\alpha$'. We are saying that a certain proportion $1-\alpha$ of all possible sample averages will fall in the above interval. A small percentage α of the samples will be so unrepresentative of the population that they will produce sample averages that will fall outside this interval.

However, we are not interested in what happens to the majority of the sample averages but only in the one sample average $\bar{x}$ that we have observed, and what this sample average allows us to conclude about the population average μ. Now the inequality

$$\mid \bar{x} - \mu \mid\, \leq z_{\alpha/2}\ \sigma/\sqrt{n}$$

can be restated as

$$\mid \mu - \bar{x} \mid\, \leq z_{\alpha/2}\ \sigma/\sqrt{n}$$

which is equivalent to

$$- z_{\alpha/2}\sigma/\sqrt{n}\, \leq \mu \leq\, \bar{x}\, +\, z_{\alpha/2}\sigma/\sqrt{n}\ \ \text{or}$$

$$\bar{x} - z_{\alpha/2}\sigma/\sqrt{n} \leq \mu \leq \bar{x} + z_{\alpha/2}\,\sigma/\sqrt{n}$$

Therefore, we can assert with probability $1-\alpha$ that the population average μ falls in the interval.

$$(\bar{x} - z_{\alpha/2}\,\sigma/\sqrt{n}\,,\;\;\bar{x} + z_{\alpha/2}\,\sigma/\sqrt{n}\,)$$

This interval is called a confidence interval for μ, while the endpoints of the interval $\bar{x} \pm z_{\alpha/2}(\sigma/\sqrt{n})$, are called the confidence limits for μ. The value (probability) $1-\alpha$ is called the degree of confidence.

Notice that the confidence interval is extremely dependent on the mean $\bar{x}$ of the sample taken. The value $\bar{x}$ is the midpoint of the confidence interval. Different samples, which probably will have different means, will lead to different confidence intervals for μ. However, we can assert with degree of confidence $1-\alpha$ that the true value of μ will fall somewhere in the confidence interval. What this means is that most [$(1-\alpha)$ percent] of the possible samples of size n will have samples average close enough to the true value of μ that the confidence interval determined by that sample average will contain μ. Only α percentage of such samples have a mean $\bar{x}$ so different from μ that the confidence interval about $\bar{x}$ fails to include μ (Quirin 1978).

The length of the confidence interval is controlled by three different factors, as is suggested by the formula for the length $2z_{\alpha/2}(\sigma/\sqrt{n}\,)$;

(1) One factor is the degree of confidence $1-\alpha$, which determines the size of $z_{\alpha/2}$. A higher degree of confidence corresponds to a larger value of $z_{\alpha/2}$, and therefore, to a wider confidence interval. Hence, the more confidence we desire, the less we will have to be confident about. A larger confidence interval gives less information about the true value of μ. A smaller confidence interval pinpoints the value of μ much better than does a larger confidence interval.

(2) A second factor is the standard deviation of the population. A population with a large deviation will give rise to samples with equally large deviations and widely varying means. On the other hand, a population with a small deviation will give rise to samples with small deviations and quite similar means. In other words, the larger the value of σ, more we would expect the sample averages to differ from the true population average which they are supposed to approximate. For small values of σ, we would expect the sample averages to concentrate around μ. Thus, we would expect the length of the confidence interval to be directly proportional to the size of σ.

(3)   The final factor that is of importance is the size of the samples. We will naturally expect that the average of a large sample gives us a better approximation to the population average than would the average of a smaller sample. Consequently, we will expect that the length of the confidence interval to be inversely proportional to the size of the sample.

It is customary to require that the sample size n should be at least 30 in order for the normal approximation to the sample mean to be sufficiently accurate. In the case $n \geq 30$, the value of the sample deviation s can be substituted in the formula for the confidence interval for the population deviation σ, should the latter be unknown (Quirin 1978).

Example (1) A random sample of 100 seeds of a variety yields an average of $\bar{x} = 26$ milligrams of protein. It is known that $\sigma = 8$ milligrams of protein. With degree of confidence $1-\alpha = 0.99$, we find that the average protein content μ for the variety of seeds satisfies

$$26 - 2.58 * 8/\sqrt{100} \leq \mu \leq 26 + 2.58 * 8/\sqrt{100}$$

which is equivalent to

$$23.94 \leq \mu \leq 28.06$$

Example (2) A sample of 40 seasonal yields of irrigated wheat crop from a large population of seasonal yields is found to have an average yield of $\bar{x} = 77.125$ quintal/ha. The standard deviation for this type of seasonal yield is known to be $\sigma = 13.36$ quintal/ha. With the degree of confidence 0.95; we estimate that the average yield μ for this location and crop lies in the interval

$$77.125 - 1.96 * \frac{13.36}{\sqrt{40}} \leq \mu \leq 77.125 + 1.96 * \frac{13.36}{\sqrt{40}}$$

which is equivalent to $72.985 \leq \mu \leq 81.265$. Of course, there is a five percent chance that our sample will mislead us, and the true population mean μ will not lie in this interval.

If we were to use our sample mean $\bar{x}$ as a point estimate for the population mean μ, then we could say with degree of confidence $1-\alpha$ that the error we make will not exceed

$$E = z_{\alpha/2} \frac{\sigma}{\sqrt{n}}$$

This is because the true population mean will fall, with probability $1-\alpha$ in an interval of length $2E$ centred about $\bar{x}$. Should we wish to decrease the size of this error term, while simultaneously holding the degree of confidence constant (and working with a known σ), our only choice would be to increase the sample size n. If, for example, we wish to have $E \leq E_0$, for some specific value of E_0, then solving the equation

$$z_{\alpha/2} \frac{\sigma}{\sqrt{n}} \leq E_0$$

for n, we find that we must choose n so that

$$n \geq \left(\frac{z_{\alpha/2}\sigma}{E_0} \right)^2$$

Example (1) (continued) : Suppose that we wish to reduce the length of the confidence interval to 2. That is, we want $E = 1$. We must choose a sample size.

$$n \geq \left(\frac{2.58 * 8}{1} \right)^2 = 424.8$$

Hence, a sample of size 425 will suffice. This means that, regardless of which sample of size 425 we take, the confidence interval will extend one unit above and one unit below the samples average, and 99 percent of the time, this interval will contain the true population mean.

Example (2) (continued): If we wish to reduce our error to at most 2 with 95 percent degree of confidence, then we must choose our sample size

$$n \geq \left(\frac{1.96 * 13.36}{2} \right)^2 = 171.42$$

That is, a sample of size 172 will suffice.

Now, suppose that we wish to estimate the proportion A of the elements in a population that has a certain characteristic. We take a sample of size n, with replacement, and find that the proportion of elements with the given characteristic in the sample is $\overline{A}$. Now our particular sample proportion is just one particular value assumed by the random variable known as the sample proportion. We have seen that this random variable has mean A and standard deviation $\sqrt{Aq/n}$, and that its distribution is approximately that of the normal distribution with the same mean and standard deviation. As a result, we can say that

$$P\left(|\overline{A} - A| \le z_{\alpha/2} \sqrt{\frac{Aq}{n}}\right) = 1 - \alpha$$

and consequently, we can assert with probability $1-\alpha$ that

$$\overline{A} - z_{\alpha/2} \sqrt{\frac{Aq}{n}} \le A \le \overline{A} + z_{\alpha/2} \sqrt{\frac{Aq}{n}}$$

We have, therefore, determined the form of the confidence interval for the unknown parameter A.

In practice, the unknown A and q in the expression for the confidence limits can be replaced by $\overline{A}$ and $\overline{q} = 1 - \overline{A}$. Thus, in practice, the $1-\alpha$ percent degree of confidence interval for A is

$$\overline{A} - z_{\alpha/2} \sqrt{\frac{\overline{A}\overline{q}}{n}} \le A \le \overline{A} + z_{\alpha/2} \sqrt{\frac{\overline{A}\overline{q}}{n}}$$

Example (3): A researcher wishes to estimate the percentage of the plants of a cultivar of drought-resistant wheat. He takes a sample of 100 plants and finds that 55 of them are drought tolerant. What can we conclude about the variety chance of being drought resistant? Since $n = 100$, $\overline{A} = 0.55$ and $\overline{q} = 0.45$, the 95 percent confidence interval is

$$0.55 - 1.96 \sqrt{\frac{0.55 * 0.45}{100}} \le A \le 0.55 + 1.96 \sqrt{\frac{0.55 * 0.45}{100}}$$

which is equivalent to $0.45 \le A \le 0.65$. Hence, although 55 percent of the plant sample is drought resistant, the researcher cannot conclude—at a 95 percent level of confidence—any more than that the true percentage of plants of drought resistance lies in the interval between 45 and 65 percent. Most importantly, the researcher cannot conclude that a majority of the plants are drought resistant.

Example (4) : A large pot contains true and false variety seeds. In order to estimate the proportion **A** of true variety seed in the pot, a sample of 49 seeds is drawn with replacement and found to contain 35 true variety seeds. As a result, with 95 percent confidence, we can say that A lies in the following confidence interval about $\overline{A} = 35/49 = 0.714$

$$0.714 - 1.96 \sqrt{\frac{0.714 * 0.286}{49}} \le A \le 0.714 + 1.96 \sqrt{\frac{0.714 * 0.286}{49}}$$

which is equivalent to $0.588 \leq A \leq 0.841$. Once again, we have a rather large confidence interval; in this case, one that covers all of 25 percent. Obviously, we must take large samples if we wish to obtain more precise confidence intervals. If, for example, we had taken a sample of $n = 400$ seeds, and found that 289 were true to the type, we would have obtained the following confidence interval about $\overline{A} = 284/400 = 0.7225$

$$0.7225 - 1.96 \sqrt{\frac{0.7225 * 0.2775}{400}} \leq A \leq 0.7225 + 1.96 \sqrt{\frac{0.7225 * 0.2775}{400}}$$

which is equivalent to $0.6786 \leq A \leq 0.7664$

If we were to use the sample proportion as a point estimate for the population proportion, we would make an error that would not exceed.

$$E = z_{\alpha/2} \sqrt{\frac{Aq}{n}}$$

Should we wish to control this error term, we would be faced with a problem that did not arise in connection with the error term for sample means. In the case of sample proportions, the error term contains the unknown population proportion A. To alleviate this problem, we note that the function $x(1 - x)$ assumes its maximum over the interval $[0, 1]$ at the point $x = 1/2$, and so $x(1-x) \leq 1/4$ over the interval $[0,1]$. Consequently, $0 \leq Aq \leq 1/4$, and so

$$E = z_{\alpha/2} \sqrt{\frac{Aq}{n}} \leq \frac{z_{\alpha/2}}{\sqrt{4n}}$$

should we wish to make $E \leq E_0$, for some specific value of E_0, it would suffice to make

$$\frac{z_{\alpha/2}}{\sqrt{4n}} \leq E_0$$

or $$n \geq \left(\frac{z_{\alpha/2}}{2E_0} \right)^2$$

Example (3) (Continued): Should the researcher wish to obtain a sample proportion that he could be 95% certain was within 2 percent of the true population proportion, i.e., should he desire an error term of not more than 2 percent or a confidence interval of length 4 percent, then he would have to take a sample of size

$$n \geq \left(\frac{1.96}{2*0.02} \right)^2 = 49^2 = 2401$$

Roughly speaking, a sample of size 2400 would be necessary.

3.3.1.10 Parameter Estimation of Samples and the Universe of Discourse

The mean of an infinite number of measurements will concide exactly with the 'true value' (Lewis, 1971).

In principle, since there is a finite number of normal people on this earth, we may be able to measure the hemoglobin concentration in everyone on this earth; we can also measure the hemoglobin concentration in all of them and arrive at an average or arithmetic mean. Since this is impractical, we can obtain an *estimate* of this mean. We tend to assume that the more people we measure, the closer will our estimate approach the universal mean. In order to specify which set of a theoretically infinite number of measurements are being considered, statisticians use the expression *universe of discourse* (Lewis, 1971).

We recognize that the parameters of any universe of discourse can never be known exactly. However close our estimates may come to 'true' values, the mean and standard deviation of any universe of discourse represent an ideal but generally unknowable value. We distinguish between the ideal value and the measured estimate in our algebraic notation by using Greek letters for the ideal values and Roman letters for estimates. Thus, the ideal mean for a universe is usually symbolized by the letter μ and the corresponding standard deviation by σ. The measured estimates use $\bar{x}$ or m for the mean and s for the standard deviation (Lewis, 1971).

Although we may be unable to attain the ideal mean of an infinite number of measurements, we can arrive at a satisfactory evaluation of the multitude of error in an estimate by considering the effect of sample size on the mean. In Figure 3.22, the curve for $n = 1$ represents the normal distribution curve of a population whose mean is μ and standard deviation is σ. Suppose that in estimating the characteristics of a population, instead of making individual measurements, we make the measurements in pairs and treat the means of the two measurements as a variate. Intuitively, we can see that this derived population of means will have a distinctly narrower scatter, but obviously the mean of these pairs will be identical to the universe mean of the parent population. This narrower, steeper distribution is shown superimposed on the parent population in Figure 3.22.

If we repeat this process using groups of four in each sample and plotting the means of these subgroups, the mean of these sample means will be identical to the universe mean. This curve, however, is even narrower and steeper than the curve for the means of sample size of two. Thus, as we increase the size of the samples making up of the subgroups, the means remain unchanged but the standard deviation of the distributions of distributions of sample means diminishes. It can be shown that if σ is the standard deviation of a universe, the standard deviation of means for samples of n variates is $\sigma/\sqrt{n}$. This value is usually called the standard error of the mean.

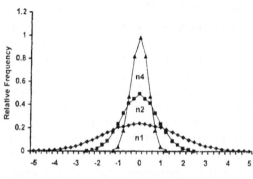

Fig. 3.22 Effect of sample size on distribution of sample means.

Each of the increasingly narrow distributions described above is still a normal distribution. It follows then that we can utilize the tabulated values of proportionate area expressed as a function of the standardized deviate, U. Thus, if we have a mean, $\bar{x}$, and estimated standard deviation, s, we can write :

$$U = \pm \frac{(\mu - \bar{x})}{s/\sqrt{n}} \qquad (3.36)$$

Note that in this expression s is used as if it were identical or at least not appreciably different from σ. By simple algebraic transformation, we may rewrite this as:

$$\bar{x} = \mu \pm Us/\sqrt{n} \qquad (3.37)$$

Written this way, the equation states that the estimate of the mean using a sample size n, lies within a certain distance of the universe mean, μ, if we can select an appropriate value of U.

Suppose, using the table of normal distribution (double-tails), proportion of area lying outside the ordinates $U = \pm (\mu - x)/\sigma$, we select a value of U for which the excluded or shaded areas constitute 5% of the total area. This means that 95% of the estimates, $\bar{x}$'s, will lie no further from μ than $Us/\sqrt{n}$. This value would be found to be 1.96. In other words, 95% of the time the estimate, $\bar{x}$, would be found somewhere between $\mu + 1.96\,s/\sqrt{n}$ and $\mu - 1.96\ s/\sqrt{n}$; or in terms of probability, we could say that the probability being 0.95, there would be 20:1 odds that our estimate would lie within these limits.

If, for example, the mean weight, μ, of all the women in the world were 120 pounds and the standard deviation were 5 pounds, 95% of the time the mean weight of samples of 100 women selected at random would fall within the limits described by the equation.

$$\bar{x} = 120 \pm 1.96\ (5/\sqrt{100}) \qquad (3.38)$$

Thus, we could give 20:1 odds that $\bar{x}$ would lie somewhere between 119 and 121 pounds.

The difficulty at this point is that we do not know that μ is 120. Suppose, however, that our sample mean, $\bar{x}$, on a sample of 100 women turned out to be 119 pounds. Since $s = 5$, we can see that this is consistent with all that has been said so far. However, if μ were really 118 mislead of 120, the observed value for $\bar{x}$, 119, could happen with the same degree of probability. Thus :

$$\bar{x} = \mu + U\,s/\sqrt{n} \qquad (3.39)$$

$$119 = 118 + 1.96\ (5/\sqrt{100})$$

We see then that if these were two populations so that in one $\mu_1 = 120$, and in the other $\mu_2 = 118$, as long as both had a standard deviation $s = 5$, 119 could be obtained as a sample mean from either population with the same degree of probability. It follows from here that if $\bar{x}$ is 119, it could have come from a population with a universe mean of 120 or even from a different population with a universe mean of 118, but we have no way of knowing which one is the parent population. However, we can reverse the reasoning and say that of $\bar{x} = 119$ and $s = 5$, the odds are $20:1$ that μ lies between 118 and 120. Thus, we can write

$$\mu = \bar{x} \pm 1.96\ s/\sqrt{n} \qquad (3.40)$$

Even with appreciable variability, large samples do tend to yield reliable estimates of μ and σ. With large samples, it is assumed that s was so close to σ that we were able to use it in establishing the fiducial limits. With small samples, we cannot do so safely without making some allowance for the possible unreliability of s as a measure of σ. Essentially, to make this allowance, we multiply U by a factor g when setting of fiducial limits of the mean. Thus, we would have

$$\mu = \bar{x} \pm (gU)s$$

where the factor g would depend on the size of the sample. For small samples, g must be larger than 1.0, but as the sample size increases, g must approach 1.0 as its limiting value. This correction factor was first solved by W.S. Gosset and published in 1908 under the pseudonym 'Student'. The value of (gU) for various probabilities as a function of sample size were tabulated under the heading t, and use of the corrected value is still commonly called student's t test. For samples of 30 or more, t and U are practically equal; therefore, when samples contain less than 30, we must use t in place of U. Thus :

$$\mu = \bar{x} \pm ts$$

Note carefully that values of t are tabulated as a function of the number of degrees of freedom rather than as a function of the number n of items in the sample. The value of t are used in other applications besides establishing confidence or fiducial limits around a mean. A degree of freedom or $n - 1$ was used instead of n for samples of less than 30 in computing s. In any event, although the t values were designed to set fiducial limits for the means obtained with small samples, its widest application is found in statistical comparisons of two means, such as those obtained for an experimental group and a control group of data.

For example, the mean hemoglobin concentration was reported as 14.3 g/100 ml of blood. Suppose this was obtained on a sample of 20 subjects. The standard deviation was reported as 1.2. In practical applications, we generally use the value 2.00 instead of 1.96 as this simplifies calculation, and the error is about 2%. From this information, we can state that there is a probability of 0.95 that the universe mean is somewhere between the fiducial limits of 14.3 $\pm$ t (1.2). The t value for (20–1) or 19 degree of freedom = 2.09.

$$\mu = 14.3 \pm 2.09 \ (1.2)$$

$$= 14.3 \pm 2.508$$

Note that the amount to be added or subtracted from 14.3 is 2.508, so the fiducial limits are 16.808 to 11.792.

Problem:

1. Calculate the mean and standard deviation for the following data:

   Serum calcium levels

10.5	
11.4	
11.4	mean = 10.61
11.4	s = 0.6488
9.7	
10.2	$(n - 1)$ or $df =$
10.8	at 9 $df\ t_{0.05} = 2.26$
9.7	
10.6	$t_{0.01}$ at 9 $df = 3.25$
10.4	

Using the mean and standard deviation, estimate the 95% confidence limits of the mean. What would be the 99% confidence limit?

$$\mu = \bar{x} \pm ts$$

$$= 10.61 \pm 2.26 * 0.65$$

$$= 10.61 \pm 1.5$$

μ will be situated between 12.11 to 9.11

Confidence limit at 5% level of confidence

$$= 12.11 \leq 10.61 \geq 9.11$$

at 1% confidence limit =

$$10.61 \pm 3.25 \times 0.65$$

$$10.61 \pm 2.11$$

$$12.72 \leq 10.61 \leq 8.5$$

It will be situated between = 12.72 to 8.5

3.3.1.11 Parameter Estimation and Hypothesis Testing

Suppose again that we have a population whose distribution is determined by an unknown parameter π, and suppose that we have good reason to believe that $\pi = \pi_0$. Rather than using information obtained from a sample to specify either a point estimate or a confidence interval estimate for π we shall, instead, use the information contained in the sample to either support or deny the hypothesis that $\pi = \pi_0$. In other words, we shall use the sample information as a basis for deciding between two possible courses of action. Either we will accept the hypothesis that $\pi = \pi_0$, or atleast subject it to further testing, or reject it completely and look for a more reasonable estimate for the parameter π.

3.3.1.11.1 Example (1)

A manufacturer wishes to improve his productivity which claims that 90 percent of the items produced by the machine are non-defective. The advertisement also states that prospective buyers may observe the machine in operation and inspect its produce (Quirin, 1978).

Our friend, realizing that such a machine would be of great value to his business to observe the production of 25 items and, on the basis of this sample, has to decide whether or not to purchase the machine. He must, therefore, decide whether or not to accept the hypothesis that $p = 0.90$, where p is the probability that an item produced by the machine is non-defective. All he will have to base his decision on is the number of non-defectives in the sample of size 25 that he observed—that, and a little probabilistic reasoning.

If the hypothesis that $p = 0.90$ is true, and if k denotes the number of non-defectives in the observed sample, then

$$P(k \leq 19) = \sum_{k=0}^{19} b(k; 25, 0.90) = 0.0334$$

Thus, it is unlikely that a machine producing 90 percent non-defectives will produce only 19 or fewer non-defectives in a sample of size 25, although it nevertheless is possible. (We point out that the value 19 was determined by trial and error. The probabilities $P(k \leq c)$ were calculated, for $c = 25, 24, 23 \ldots$, until the value for $P(k \leq c$ became exceptionally small.)

Our manufacturer decides upon a rule for making his decision. He will accept (believe) the advertiser's claim that $p = 0.90$, and purchase the machine, provided that the observed value of k is at least 20. We emphasize that this rule depends only on the value of k, and not at all (except as it effects the value of k) on the actual value of p. Regardless of what the actual value of p is, the manufacturer will follow the above rule because he does not know what the value of p really is. Remember, all that the manufacturer knows is the observed value of k.

Now, there is a chance that the manufacturer will make an error. It is possible that the hypothesis $p = 0.90$ is true, and still the machine produces $k \leq 19$ non-defectives in a sample of 25. But such an observed value of k will lead the manufacturer to reject the true hypothesis that $p = 0.90$, and not buy a machine he probably should buy. Such an error, rejecting a true hypothesis, is called a type 1 error. The probability of making such an error, which we shall denote by α, is called the level of significance of the test (rule). In our example, $\alpha = 0.0334$. The set of outcomes that lead to the rejection of the hypothesis is called the critical region of the test. The critical region for the test above is $k \leq 19$.

If the advertiser's claim is, in fact, a lie, the manufacturer would hope that his rule would lead him to reject the hypothesis that $p = 0.90$, and not buy the machine. Suppose that, in fact, the true value of p is 0.80. Then the probabilities that the hypothesis $p = 0.90$ will be rejected is

$$P\ (k \leq 19)\ \sum_{k=0}^{19}\ b\ (k; 25, 0.80 = 0.3833\ )$$

Notice that in calculating $P\ (k \leq 19)$ here, the value $p = 0.80$ is used rather than the value $p = 0.90$. Therefore, the probability of the false hypothesis $p = 0.90$ being accepted is 0.6167. An error of this type accepting a false hypothesis and is called a type 2 error. The probability of making such an error depends on the actual true value of the parameter p. We define the power of the test to be the function which gives the probability of rejecting the hypothesis $p = 0.90$ for all possible values $0 \leq p \leq 1$. If, for example, $p = 0.90$, the power is 0.0334, the level of significance of the test.

For any other value of p, the power $P\ (k \leq 19)$ can be determined from a table of binomial probabilities. Several values of the power function for the test above are given in Table 3.22 below.

Table 3.22 The value of p (probability) and P (Power) value of p

Value of p	Power
0.80	0.3833
0.75	0.6217
0.70	0.8065
0.60	0.9264
0.50	0.9979

The graph of this power function is shown in Fig. 3.23.

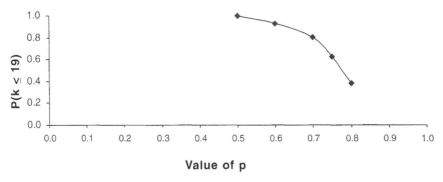

Fig. 3.23 The relationship between value of p and power (P).

The example above is typical of many problems in statistics, although the procedure followed above is not the usual order. The general problem involved is whether or not to reject a hypothesis concerning some population parameter. Since most of our parameters are closely related to the mean, we now discuss a general procedure to be followed in deciding whether or not to reject the hypothesis that $\mu = \mu_0$. The hypothesis $\mu = \mu_0$ is commonly known as the null hypothesis and is denoted H_0. One reason for the negative aspect of the name is that it is common practice to set up such a hypothesis in the hope of rejecting it.

If the null hypothesis were false, we would want to accept one of the alternatives $\mu \neq \mu$, $\mu < \mu_0$, or $\mu > \mu_0$. Whichever alternative hypothesis fits the particular problem shall be denoted H_1. If H_1 is $\mu = \mu_0$, it is called a *two-sided alternative*; if H_1 is either $\mu < \mu_0$ or $\mu > \mu_0$, it is called a *one-sided alternative*.

After having decided upon the null hypothesis, we must then decide upon the level of significance of the test. The deciding factor here should be the seriousness of making a type 1 error. We might also wish to control the power function of the test, i.e., control the chances of making a type 2 error. Thus, we might seriously wish to keep α near zero and the value of the power function near one. The normal way to do this is by means of a wise choice of sample size, although this at times might require such a large sample as to become economically not feasible. At times a compromise between the level of significance, the value of the power function, and the size of the sample becomes necessary.

We then take a sample size n with replacement, calculate the sample average $\bar{x}$, and consider the statistic

$$z = \frac{\bar{x} - \mu_0}{\sigma / \sqrt{n}}$$

If the standard deviation σ of the population is unknown, it may be replaced by the standard deviations s of the sample, provided that the sample size $n \geq 30$. Notice that z is the standardized form of the sample mean $\bar{x}$, under the assumption that $\mu = \mu_0$, i.e., under the assumption that our null hypothesis is true. Consequently, should the null hypothesis be true, the statistic z would have a standard normal distribution.

The values of $\bar{x}$ will, naturally, vary from being fairly close to μ_0 to being either well above or well below μ_0, provided, of course, that μ_0 is the true population mean. Should the logical alternative hypothesis be one sided, these extreme values of $\bar{x}$ would tend to be on just the one side of μ_0. The problem facing us is how to interpret the value of $\bar{x}$ (or z), particularly the extreme values. This is done by means of the critical region of the test, which is determined by the level of significance α and the nature of the alternative hypothesis H_1.

Table 3.23 will prove helpful in determining critical regions in the examples that follow :

Table 3.23 showing level of gnisicance (α), one sides critical region (z_α) and two sided critical region ($z_{\alpha/2}$)

α	One sided z_α	Two sided $z_{\alpha/2}$
0.05	1.64	1.96
0.01	2.33	2.58

To explain what these numbers mean, suppose that we consider the case $\alpha = 0.05$. The values ± 1.96, are such that 95 percent of the area under the standard normal curve lies in the interval between them, with the remaining 5 percent of the area equally distributed in the two tails of the curve, as indicated in Fig. 3.24. Thus, these tails contain those values of z that correspond to the most unlikely 5 percent of all possible sample averages.

Table 3.24 The values of abscissas and ordinates

abscissas	ordinates
2.5	0.002
1.96	0.024
1.64	0.069
1.00	0.234
0.50	0.410
0.00	0.492
−0.50	0.410
−1.00	0.234
−1.64	0.069
−1.96	0.024
−2.50	0.002

On the other hand, the value 1.64 is such that 95 percent of the area under the standard normal curve lies to its left, with the remaining 5 percent to its right, as indicated in Fig. 3.25. In this case, the most unlikely 5 percent of all possible sample averages correspond to the tail to the right. Reverting the role of right and left, similar remarks would be made concerning the value −1.64.

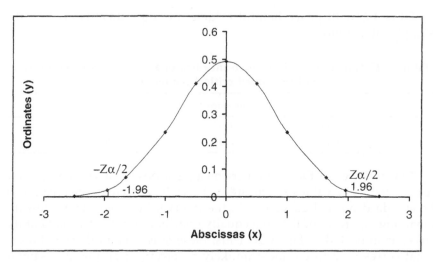

Fig. 3.24 The area lying between −1.96 and + 1.96 (95%) and 5% in the two tails

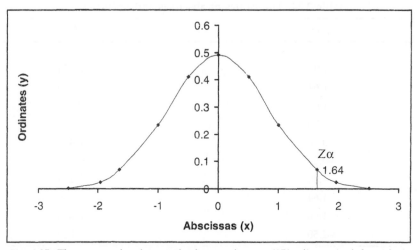

Fig. 3.25 The area under the standard normal curve (95%) lies to its left, with the remaining 5% to its right.

In general, then, the critical region for our test is defined to be

$$|Z| > Z_{\alpha/2} \quad \text{if } H_1 \text{ is two sided}$$

$$Z > Z_{\alpha} \quad \text{if } H_1 \text{ is } \mu > \mu_0$$

$$Z < -Z_{\alpha} \quad \text{if } H_1 \text{ is } \mu < \mu_0$$

The critical region is simply the tail (or tails) corresponding to the extreme values of $\bar{x}$, the most unlikely α percent of all possible sample averages.

We then set up the following criterion for rejecting or not rejecting the null hypothesis H_0.

'If the statistic z, defined above, falls in the critical region, reject the null hypothesis; otherwise, either accept H_0 or reserve judgement on H_0.'

The logic behind our criterion is as follows: If we are working at level of significance $\alpha = 0.05$, for example, then the critical region is defined in such a manner that five percent of our possible samples will have means which fall in the critical region, provided that H_0 is true, i.e. $\mu = \mu_0$. If the sample that we take has a mean which falls in the critical region, then we are faced with the dilemma of either accepting the fact that our sample is that one bad sample in 20 whose mean lies in the critical region (How could we have such bad luck?), or trying to explain why this unusual sample has occurred. The obvious explanation, of course, is that the null hypothesis is not true after all, that the true value of the population mean is some value other than μ_0, and that in the light of the true value of the population mean, the occurrence of the 'unusual' sample would no longer seem so unlikely. Indeed, we shall adopt this latter approach.

We could assert with probability $1-\alpha$ (in our case, $1-\alpha = 0.95$) that our sample average lies in the interval

$$\mu - z_{\alpha/2}\, \frac{\sigma}{\sqrt{n}} \le \bar{x} \le \mu + z_{\alpha/2}\, \frac{\sigma}{\sqrt{n}}$$

centered around the true population mean μ. If our sample value $\bar{x}$ does happen to lie in the two-sided critical region at level of significance $\alpha = 0.05$, then $\bar{x}$ does not lie in the interval

$$\mu_0 - z_{\alpha/2}\, \frac{\sigma}{\sqrt{n}} \le \bar{x} \le \mu_0 + z_{\alpha/2}\, \frac{\sigma}{\sqrt{n}}$$

centered around the suspected mean μ_0. Our only possible reaction is that since 95% of all possible averages should lie in the former interval, and since our sample average did not fall in the latter interval, the 'only' possible conclusion is that these two intervals are not the same. That is, we must conclude that $\mu = \mu_0$.

3.3.1.11.2 Example (2)

When operating properly, a leaf area meter will measure an area of a standard solid circle 2 cm^2, with a deviation of 0.05 cm^2. To insure that the machine remains in proper operating condition, at regular intervals, samples of size 49 are checked. Suppose that such a sample produced an average area of 1.98 cm^2. What can be concluded about the conditions of the machine? The statistical hypothesis that we wish to test here is:

$$H_0: \mu = 2 \text{ cm}^2$$

Hopefully, we will not be able to reject this hypothesis, because to do so would mean incurring the expense of having the machine repaired. If the machine were operating improperly, the area it measures may be either too small in area or too large. (It could also happen that the machine is measuring the area of a standard metallic thin solid circle too large or too small.) Consequently, we must use the two sided alternative.

$$H_1: \mu <> 2 \text{ cm}^2$$

A type 1 error here would mean having a properly functioning machine sent in for repairs, while a type 2 error would mean not having an improperly functioning machine sent in for repairs. Since a type 1 error would mean nothing more than an unnecessary expense, we will set a rather high level of significance $\alpha = 0.05$. This gives the critical region $|z| \geq 1.96$. Since our statistic

$$z = \frac{1.98 - 2.00}{.05/\sqrt{49}} = -2.8$$

does fall in this critical region, we must reject our null hypothesis that μ equals 2 cm^2, and have the machine repaired.

3.3.1.11.3 Example (3)

Suppose that it is common knowledge that if a brand of cigarettes averages 30 milligrams (or more) of nicotine per cigarette, then lung cancer is almost certain to develop in the user. On the other hand, if the average nicotine content is less than 30 mg per cigarette, then it is relatively safe to continue smoking this particular brand of cigarettes. A random sample of 100 brand x cigarettes yields an average of 27 mg of nicotine per cigarette. Can a person safely continue to smoke this brand of cigarettes? What we wish to do here is test the hypothesis.

$$H_0: \mu = 30 \text{ mg}$$

against the one-sided alternative

$$H_1: \mu < 30 \text{ mg}$$

Actually, our null hypothesis is that cigarettes are dangerous ($\mu \geq 30$), while the alternative (hypothesis) is that they are not dangerous. We wish to test our hypothesis at a very low level of significance, since making a type 1 error means continuing to smoke a potential dangerous brand of cigarettes, one which has an average 30 mg (or more) of nicotine per cigarette. Thus, making a type 1 error would certainly lead to the development of lung cancer. We choose $\alpha = 0.01$ to simplify our calculations, although it may be desirable to choose α much smaller. Our critical region is, therefore, z < − 2.33. Our statistic if σ = 8 mg is

$$z = \frac{27 - 30}{8 / \sqrt{100}} = -3.75$$

which certainly falls in our critical region. Therefore, we must reject the null hypothesis (that the cigarettes are potentially dangerous), and may safely continue to smoke this brand of cigarette.

3.3.1.11.4 Example (4)

The contents of medium-sized box of Kellog's cornflakes are supposed to weigh 15 ounces, according to the statement on the package. In order to avoid problems with the government, the people at Kellogs must be careful that the content of these boxes do weigh 15 ounces. To ensure that this is, in fact, the case, they might proceed as follows. At regular intervals, the contents of 50 boxes are weighed, and the sample average is used to test the null hypothesis.

$$H_0: \mu = 15 \text{ ounces}$$

against the one-sided alternative

$$H_1: \mu < 15 \text{ ounces}$$

In this case, making a type 1 error means concluding that the contents of the boxes are deficient when in fact, they are not. On the other hand, making a type 2 error means incorrectly concluding that the boxes weigh the required 15 ounces when, in fact, they weigh less. Obviously, a type 2 error is more critical, since it is this type of error that will get Kellogs in trouble with the government. Therefore, we use the moderately high level of significance α = 0.05, which gives the critical region z < −1.64. Suppose a supply of 50 boxes revealed a sample average of 14.9 ounces per box. If we accept the fact that such measurements produce a standard deviation of σ = 0.5 ounces per box, then our statistic

$$z = \frac{14.9 - 15.0}{0.5 / \sqrt{50}} = -1.43$$

does not fall in the critical region. Therefore, unless other tests prove differentially, Kellogs has no cause for concern, at least not on the basis of this particular sample. Since making a type 2 error is very important in this example, let us evaluate the power of this test at μ = 14.9 ounces. That is, let us calculate the probability that the hypothesis μ = 15 ounces will be rejected if, in fact, μ = 14.9 ounces. By definition, the power is the probability

$$P\left(\frac{\bar{x}-15}{0.5/\sqrt{50}}<-1.64\right)$$

However, the term $(\bar{x}-15)/0.5/\sqrt{50}$ is no longer standard normal, since 14.9, rather than 15, is the true population mean. Therefore, the term $(\bar{x}-14.9/0.5/\sqrt{50})$ is the standard normal, and our calculation of the power of the test at $\mu = 14.9$ proceeds as follows:

$$=P\left(\frac{\bar{x}-14.9}{0.5/\sqrt{50}}+\frac{14.9-15}{0.5/\sqrt{50}}<-1.64\right)$$

$$=P\left(\frac{\bar{x}-14.9}{0.5/\sqrt{50}}-1.43<-1.64\right)$$

$$=P\left(\frac{\bar{x}-14.9}{0.5/\sqrt{50}}<-0.21\right)=0.4168$$

(from the table of Normal Distribution (single tail proportion of area lying to right of ordinate through $z = \pm (x - \mu)/\sigma$).

So, the probability of rejecting $\mu = 15$, if, in fact, $\mu = 14.9$, equals 0.4168. In a similar fashion, we find the power of $\mu = 14.8$ to be 0.8888 and the power at $\mu = 14.7$ to be 0.9960. Now suppose that we wanted both $\alpha = 0.05$ and the power of the test of $\mu = 14.9$ to be 0.90. This could be achieved by taking the sample size n large enough so that

$$P\left(\frac{\bar{x}-15}{0.5/\sqrt{n}}<-1.64\right)=0.90$$

when, in fact, $\mu = 14.9$. Notice that the level of significance $\alpha = 0.05$ appears in this expression in the value −1.64 on the right hand side of the inequality. In case $\mu = 14.9$, the term $(\bar{x} - 14.9)/0.5/\sqrt{n})$ is standard normal, and so

$$0.90=P\left(\frac{\bar{x}-15}{0.5/\sqrt{n}}<-1.64\right)$$

$$=P\left(\frac{\bar{x}-14.9}{0.5/\sqrt{n}}+\frac{14.9-15}{0.5/\sqrt{n}}<-1.64\right)$$

$$=P\left(\frac{\bar{x}-14.9}{0.5/\sqrt{n}}-\frac{\sqrt{n}}{5}<-1.64\right)$$

$$= P\left(\frac{\bar{x} - 14.9}{0.5/\sqrt{n}} < \frac{\sqrt{n}}{5} - 1.64\right)$$

since $(\bar{x} - 14.9/(0.5/\sqrt{n})$ is standard normal, we have

$$\frac{\sqrt{n}}{5} - 1.64 = 1.281$$

(since , if y is a standard normal, $P\,(y < c) = 0.90$ implies that $c = 1.281$, which is equivalent to $n = 213$. Therefore, a sample of size 214 would be sufficient at level of significance 0.05 to guarantee that the power of the test at $\mu = 14.9$, is 0.90.

The null hypothesis concerning a population proportion would be in the form

$$H_0 = A = A_0$$

with either a one-sided or a two-sided alternative. The statistic z tested would be the standardized form $\bar{A}$, the sample proportion, standardized under the assumption that the null hypothesis $A = A_0$ is true, that is

$$z = \frac{\bar{A} - A_0}{\sqrt{A_0 q_0 / n}}$$

Example 3.3.1.11.1 (continued) : Suppose we wish to test the hypothesis H_0: $A = 0.90$ that the advertiser's claim is legitimate at the five percent level of significance. The alternative hypothesis here would be H_1 : $A < 0.90$ since values of $A > 0.90$ would be just as acceptable as the value $A = 0.90$. The one-sided critical region, together with the five percent level of significance, define the critical region $Z < -1.64$. If, in the observed sample of 25 items, we found that 19 were non-defective, the $\bar{A} = 0.76$, and

$$z = \frac{0.76 - 0.90}{\sqrt{.90 * 0.10 / 25}} = \frac{-0.14}{.06} = -2.33$$

which is certainly critical. If, however, we have found that 20 of the 25 items observed were non-defective, we would have $\bar{A} = 0.80$, and

$$z = \frac{0.80 - 0.90}{\sqrt{0.90 * .1 / 25}} = -1.67$$

This value of z lies just slightly over the border line in the critical region. Thus, at the five percent level of significance, at least 21 of the 25 items observed must be non-defective in order to support the advertiser's claim. Recall that our rule previously had been to accept this claim unless the observed number of non-defective was 19 or fewer but this was at level of significance $\alpha = 0.0334$. The one-sided critical region corresponding to this value of would be $z < -1.83$, which would not contain the statistic $z = 1.67$ corresponding to $K = 20$ non-defective in the sample observed.

3.3.1.11.5 Example (5)

A production process is regarded as being under control if it produces five percent or fewer defects. Suppose that a sample of 100 such items contained eight defective ones. Is this cause for concern that the process is no longer under control? To answer this question, we test the hypothesis.

$$H_0 : A = 0.05$$

against the one-sided alternative.

$$H_1 : A > 0.05$$

at the five percent level of significance (Do you understand why we chose a one-sided critical region?) the critical region is, therefore, equal to $z > 1.64$. Since the simple proportion $\overline{A} = 0.08$, the statistic

$$z = \frac{0.08 - 0.05}{\sqrt{0.05 * 0.95 / 100}} = 1.38$$

does not fall in the critical region and so have no reason, based on this test performed on this sample, to believe that the process is no longer under control.

3.3.1.11.6 Example (6)

It is known that approximately 1 out of every 10 people use a certain product. After a concentrated advertising campaign in a certain area, a sample of 400 people contained 60 who were using the product. We wish to determine if the advertising campaign has had a significant effect on consumer preference, so we test the null hypothesis

$$H_0 : A = 0.10$$

that the advertising campaign has had no effect (this is one reason for the term null hypothesis), hoping that the evidence available in the sample will allow us to reject H_0. The alternative hypothesis would be

$$H_1 : A > 0.10$$

and at level of significance $\alpha = 0.01$, we have the critical region $z > 2.33$. Our sample proportion was found to be $\overline{A} = 60/400 = 0.15$, and so our sample statistic

$$z = \frac{0.15 - 0.10}{\sqrt{0.1 * 0.90 / 400}} = \frac{10}{3}$$

which is resoundingly a critical value. Hence, we may reject the null hypothesis, and feel confident that the advertising campaign has significantly raised the percentage of people using this product.

3.3.1.11.7 *Example (7)*

Suppose that we wish to estimate the number of fish in a certain lake. Rather than catching all the fish and counting them (and then eating them), we adopt the following plan. On one day, we catch 1000 fish, mark each with a red dot, and then release them. A few days later, we again catch 1000 fish (both days catching them from many different parts of the lake), and record the number of fish we have just caught that bear our red mark. This number, which we shall denote k, will be critical in our estimation of the size of the fish population of the lake. For suppose that these are N fish in the lake. Then, on the second day, we go out to catch fish, the probability that a fish caught bears our red mark is $A = 1000/N$. But the proportion of fish bearing a red mark in our second sample is $\overline{A} = k/1000$. Our problem is to determine which values of N make this sample proportion seem not unreasonable. Therefore, at level of significance $\alpha = 0.05$, we test the hypothesis

$$H_0 : A = 1000/N$$

against a two-sided alternative we wish to determine for which values of N the sample proportion $\overline{A}$ is not critical, at a five percent level of significance. The sample statistic, we must test is

$$z = \frac{k/1000 - 1000/N}{\sqrt{(1000/N) \ [(N - 1000)/N]/1\,000}}$$

$$z = \frac{kN - 10^6}{1000 \sqrt{N - 1000}}$$

A computer program was written to determine the smallest and largest values of N corresponding to any given value of k, for which the sample statistic z is not critical. In other words, we have calculated the values of N that correspond to the left- and right-hand boundaries of the critical region. The results can be found in Table 3.25.

Table 3.25 Estimation of the size of the fish population

Value of k	Smallest possible N	Largest possible N
10	54616	183801
20	35584	77021
30	23528	47409
40	18516	33886
50	15311	26227
60	13077	21324
70	11427	17929
80	10156	15445
90	9147	13552
100	8323	12062
110	7639	10860
120	7063	9873
130	6568	9045
140	6139	8343
150	5765	7740
160	5435	7216
170	5141	6757
180	4878	6353
190	4642	5993
200	4427	5670

Calculations (Example 3.3.1.11.7)

$$H_0 : A = \frac{1000}{N}$$

$$Z = \frac{k/1000 - 1000/N}{\sqrt{(1000/N)\,[(N-1000)/N]/1000}}$$

Numerator

$$= \frac{k}{1000} - \frac{1000}{N} = \frac{kN - 10^6}{1000\,N}$$

Denominator

$$= \sqrt{\left(\frac{1000}{N}\right)\left(\frac{N-1000}{N}\right)/1000}$$

$$= \sqrt{\left(\frac{1000\,N - (10)^6}{N^2}\right) \bigg/ 1000}$$

$$= \sqrt{\frac{1000\,N - (10)^6}{N^2 \times 1000}}$$

$$= \sqrt{\frac{N - (10)^3}{N^2}}$$

$$= \frac{\sqrt{N - 1000}}{N}$$

$$\frac{\text{Numerator}}{\text{Denominator}} = \frac{kN - (10)^6}{1000\,N} \bigg/ \frac{\sqrt{N - 1000}}{N}$$

$$= \frac{kN - (10)^6}{1000\,N} \times \frac{N}{\sqrt{N - 1000}}$$

$$= \frac{kN - (10)^6}{1000\,\sqrt{N - 1000}}$$

$$z = \frac{kN - (10)^6}{1000\,\sqrt{N - 1000}}$$

At 5% level of significance, the value of $z = \pm 1.96$.

$$\text{Let } k = 100$$

$$1.96 = \frac{100\,N - 10^6}{1000\,\sqrt{N - 1000}}$$

$$1.96 \times 1000 = \sqrt{N - 1000} = 100N - 10^6$$

$$1960\sqrt{N - 1000} = 100N - 10^6$$

by squaring both the sides

$$N^2 - 20384\,N + 100384160 = 0$$

$$N = \frac{-b \pm \sqrt{b^2 - 4ac}}{2a}$$

$$N = \frac{20384 \pm \sqrt{(20384)^2 - 4(100384160)}}{2}$$

$$N = \frac{20384 \pm \sqrt{415507460 - 401536640}}{2}$$

$$N = \frac{20384 \pm \sqrt{13970820}}{2}$$

$$N = \frac{20384 \pm 3737}{2}$$

Largest possible N

$$= \frac{20384 \pm 3737}{2} = \frac{24121}{2} = 12060$$

Smallest possible N

$$= \frac{20384 - 3737}{2} = \frac{16647}{2} = 8323$$

Computer program to solve the equation with optimum seeking design approach (3.4.1.7)

$$1.96 = \frac{kN - (10)^6}{1000\sqrt{N - 1000}}$$

5 REM to find the smallest and largest N values is Z = –1.96 and +1.96 respectively.

10   K = 10

15   For N = 4000 to 200000

20     $Z = \dfrac{kN - (10)^6}{1000\sqrt{N - 1000}}$

30	Next *N*
40	Print, *N*, *Z*
50	END

3.3.1.12 Crop Performance Indices

Fridgen et al., (2004) developed two crop performance indices: (1) Fuzziness performance index (FPI); and (2) normalized classification entropy (NCE) to aid in deciding how many clusters are most appropriate to create management zones.

3.4 NON-STATISTICAL PROCEDURE FOR ESTIMATING THE PARAMETERS (PHYSICAL APPROACH)

Non-statistical procedures deal with the explanatory model, whereas statistical procedures deal with the correlative model. An explanatory model also has to reflect the observed relationships between variables. But, in addition, the structure of the model must reflect some concept of the causal mechanism that underlies the relation. The purpose is not so much to describe the observed relation as to explain it.

Gold (1977) has depicted the difference between the correlative method of parameter estimation and explanatory method of parameter estimation.

3.4.1 Non-statistical Procedure of Parameter Estimation

Poole (1974) also described the non-statistical procedure of parameter estimation as deductive modeling.

Regression models are inductive. Either an experiment is run or observations are made; the data are plotted, and a regression equation is fit to the data depending on the functional relationship suggested by the data. Models may also be derived deductively. In a deductive model, we make logical hypotheses about the process being studied, formulate the hypotheses as a mathematical model, and then try to fit the model to the data. Sometimes, a model may be derived by a combination of deductive and inductive methods. A deductive model is fit to the data and, based on the observed residuals of the model to the data, the model is modified.

The derivation of deductive models usually involves a rather advanced calculus. However, the general principle involved can be illustrated by a few simple examples. Suppose that lead from cars is being added to the roadside at a constant rate *a*. The instantaneous change in the amount of lead in the soil with time can be represented by the differential equation

$$\frac{dP}{dt} = a \qquad (3.41)$$

The differential dP/dt is an instantaneous rate, and is the slope of the line tangent to the curve representing the amount of lead present at time t. In this case, dP/dt is a constant a, and so the slope of the curve is constant at any point t. Because the slope is constant, the functional response between the amount of lead and time is a straight line. Applying the methods of calculus, the differential equation is solved for an explicit value of P at time t, P_t. However, there is an infinite number of lines with slope a, and therefore, the initial value of P at time $t = 0$ must be stipulated to arrive at a unique solution of the differential equation. The solution is $P_t = P_0 + at$, the equation of a straight line. In exponential population growth, the rate of change in the density of population is assumed to be proportional to the density of the population. The proportionality constant is r, and the differential equation is $dN/dt = rN$. Again, to derive a unique solution of the differential equation, the initial value of N must be stipulated. Solving the equation by calculus, we find $N_t = N_0 e^{rt}$. In logistic growth, the rate of increase of the population is not constant, but depends on the density of individuals. We assume that the decrease in r is directly proportional to the density of the population, i.e. rate of increase = $r–aN$, rather than a constant r. Therefore, the differential equation for limited population growth is $dN/dt = N(r - aN)$, rather than $dN/dt = Nr$. If a variable such as larval survival is a function of several variables, the formation of deductive models can become complex.

Watt (1968) used the deductive model for the growth of the world's human population. One means of determining how big the human population will be at various times in the future is to find the population growth equation that most accurately describes growth to the present, then use the equation to extrapolate into the future. To explain the basis of the equation, it is necessary to describe the logical rational underlying mathematical descriptions of population growth.

If a population is introduced into an environment where it is not already found, population growth occurs initially at a rate dependent only on the size of the population present at any instant in time. Where N represents the population size, t is time, b is the instantaneous birth rate, d is the instantaneous death rate, and r, the intrinsic rate of natural increase, is defined as $b - d$, we have

$$\frac{dN}{dt} = (b - d) N$$

$$\text{or} \quad \frac{dN}{dt} = rN \qquad (3.42)$$

However, after a population has been in a new environment for some time, and the density of the population has increased to a high level, competition for food and other resources becomes severe, and this is reflected in dropping rates of birth and survival. We say that the population is being limited by environmental resistance. For any given population of animals or plants and any given environment, there is a certain maximum number of members of the population which the environment is able to support. We shall call this maximum number K. As the population steadily increases, the growth rate becomes smaller and smaller as N approaches K. Hence, beginning with the first small group of immigrants, and following the population to the time at which N equal K, the form of population growth will be described reasonably well by the differential equation

$$\frac{dN}{dt} = rN(K - N) \tag{3.43}$$

A population growing in accord with Equation (3.43) will have its maximum rate of growth when $d^2N/dt^2 = 0$, or when $N = K/2$. However, when a population is just beginning to expand in a new environment, N is very small in relation to K, and, to a very close approximation, changes in the term $K-N$ occur so slowly that we can treat Equation (3.43) as

$$\frac{dN}{dt} = rKN \tag{3.44}$$

We can determine the importance of the term $K-N$ for any population as follows. Equation (3.44) integrates to yield

$$\ln N_t = \ln N_0 + rKt$$

$$\text{or } N_t = N_0\, e^{rkt} \tag{3.45}$$

We can determine if Equation (3.45) describes the growth of a particular population, against t, on semilog graph paper. If $K - N$ has become important for the population in question, the plotted points will not fall on a straight line, but bend more and more downward from the left to the right side of the graph. Watt (1968) stated that the plotted values of N_t against the year show the growing world total human population according to some law different from either the logistic (Eq. 3.43) or the exponential (Eq. 3.42), because exponential growth is the fastest a logistics population will ever show.

Forester et al., (1960 as cited in Watt; 1968) proposed a new law to describe the world population growth of humans, which they derived as follows. As a result of advances in medicine, and the speed with which these advances can be communicated from one country to another, death rates are dropping faster than birth rates, and hence r is not constant, but steadily rising. Furthermore, an analysis of the demographic data shows that the rate of rise in r is itself a function of N, expressed by the relation

$$r = aN^{1/k} \tag{3.46}$$

The support for this particular formula is particular, striking the present data for countries such as India where a rapidly dropping death rate and slowly dropping birth rate together are defined by Eq. 3.46. Substituting for r in Eq. 3.42 yields the new growth law.

$$\frac{dN}{dt} = (aN^{1/k}) \, N = aN^{1 - 1/k}$$

$$\int_{t_1}^{t} \frac{dN}{N^{1+1/k}} = a \int_{t_1}^{t} dt$$

from which we obtain

$$\frac{1}{-(1/k) \, N_t^{1/k}} - \frac{1}{-(1/k) \, N_1^{1/k}} = a \, (t - t_1)$$

(or)
$$\frac{k}{N_t^{1/k}} = a \, (t_1 - t) + \frac{k}{N_1^{1/k}}$$

$$= \frac{aN_1^{1/k} \, (t_1 - t) + k}{N_1^{1/k}}$$

From this, we see that

$$N_t = N_1 \left[\frac{k}{aN_1^{1/k} \, (t_1 - t) + k} \right]^k \tag{3.47}$$

Multiplying the numerator and denominator of the expression in square brackets by $N_1^{-1/k}/a$, and adding and subtracting t_1 in the numerator, gives

$$N_t = N_1 \left[\frac{(k/a) \, N_1^{-1/k} + t_1 - t_1}{(k/a) \, N_1^{-1/k} + t_1 - t} \right]^k \tag{3.48}$$

Forester and his associates, as quoted in Watt (1968), perceptively observed that the expression $\frac{k}{a} N_i^{-1/k} + t_i$ is a measure of time, and the values are all constants, describing the system. Therefore, the time expressed is itself a constant, t_E. Thus, Eq. (3.48) reduces to

$$N_t = N_1 \left(\frac{t_E - t_1}{t_E - t} \right)^k \tag{3.49}$$

in which t_E is the end point for the system. As the difference $t_E - t$ approaches zero, N_t expands very rapidly, approaching infinity at $t = t_E$. For this reason, Forester et al. (as quoted in Watt, 1968), have labeled the time remaining, or $t_E - t$, as "time to doom". Note that in Eq. (3.49), N_t, t_E, t_1, and k are all constants, so that defining two new symbols

$$K = N_1 (t_E - t_1)^k \text{ and } T = t_E - t$$

simplifies Eq. (3.49) still further

$$N_t = KT^{-k} \tag{3.50}$$

$$\text{or } \ln N_t = \ln K - k \ln T \tag{3.51}$$

We can determine if this equation describes the growth of the world human population by plotting the population size against year on a log-log graph paper, with the years plotted backward (i.e., with the earliest year plotted at the right-hand side of the graph and the latest at the left). Forester and his colleagues (1960, as quoted in Watt, 1968) used the least square analysis to obtain the values

$$K = 1.79 \times 10^2$$

$$k = 0.990$$

$$\text{and } t_E = 2026.87 \text{ AD}$$

Watt (1968) suggested that the reader may determine for himself or herself if these parameter values are still realistic by calculating N_t for each year from

$$N_t = K (t_E - t)^{-k} \tag{3.52}$$

and plotting the observed and calculated values of N against the year on a semi-log graph paper. When Watt's students, as quoted in Watt (1968), perform this exercise, they discover to their surprise that the world population is growing faster than the rate predicted by Eq. (3.52) in 1960. If this equation is correct, the world population of humans will become infinite and, therefore, squeeze itself to death, as Forester et al. as quoted in Watt (1968), describe our doom, some time prior to 2026 AD.

However, as might be suspected, there is a flaw in the reasoning underlying Eq. (3.52). Returning to Eqs. (3.42) and (3.46), we see that

$$r = b - d \text{ and also } r = aN^{1/k}$$

$$\text{hence } b - d = aN^{1/k} \tag{3.53}$$

Clearly, there is an absolute lower limit below which the death rate cannot drop and an absolute upper limit above which the birth rate cannot rise. Therefore, $aN^{1/k}$ cannot increase forever, but ultimately will reach an absolute upper limit, which will occur when the difference $b - d$ is $b_{max} - d_{min}$. The difficulty is that because of changing medical procedures, we do not know exactly how low d can sink before it reaches d_{min}, and because of changing social values, we do not know how high b_{max} will be. All we do know is that $a\,N^{1/k}$ can rise considerably above its present level, and that (Eq. 3.52) will, therefore, apply for some time into the future, during which the average international standard of living will drop to a really appalling (shocking) level.

Our aim is not to formulate a model for predicting human population, but it is one example of formulating a deductive model. After a model has been formulated either inductively (statistically) or deductively (conceptually), the values of parameters should be estimated either by methods of statistics or by methods of 'guestimate'.

Dent and Blackie (1979) referred to the option of subjectively estimating data as a new concept in our discussion. Since it is used with reasonable frequency, it should be examined carefully. We should note that the option is also available to the modeler who has data in an inappropriate format. The decision to estimate values subjectively is not such a wild procedure at this stage in our model development as may be thought and it could save from working on that an experimental program with all the expenses incurred in time, facilities and resources this implies. Assume that the difficulty lies with a single piece of information—a coefficient representing a rate-of-change constant. It will also be assumed that one has only the broadest notion of the value of this coefficient. The procedure would be as follows:

1. Determine the best 'guestimate' for the coefficient.
2. Proceed with the development of the rest of the model and complete its structure and implementation on the computer.
3. Establish appropriate parameters to monitor the performance of the entire model.
4. 'Run' the model with the decided value for the unknown coefficient—note the values of the performance parameters.
5. 'Run' the model several times with different, widely spaced values of the unknown parameter each time without changing any other aspect of the model, noting the values of the performance parameters.

6. Compare the values of the performance parameters for each run.

7. In case the different assumed values of the unknown coefficient do not markedly influence the performance parameters, it means that the model is not sensitive to the selected value for the particular coefficient. Our judgment might then be that the 'guestimate' is acceptable and may, without loss of value to the model as a whole, be incorporated in its structure.

8. In case where markedly different output parametrs are consequent on the different assumed values of the unknown coefficient, it means that the model is sensitive to this particular coefficient. The 'guestimate' should now be used with caution and it would be advisable to extend the data search, perhaps by way of a new research program. Alternatively, it may be possible to restructure the model in such a way as to reduce the sensitivity of the coefficient.

3.4.1.1 *'Guestimate' of the Intrinsic Rate of Increase*

Poole (1974) stated that the method of calculation of the parameter r, the intrinsic rate of increase, is based only on the females of a population, and it is assumed that there are enough males to go around. The first step in the calculation of the intrinsic rate of increase is the formulation of a life table for the population. Birch (1948 as quoted in Poole, 1974) has estimated l_x and m_x statistics for a cohort of the weevil *Colandra oryzae* in an unlimited environment of wheat at 20°C and 14 percent relative humidity (Table 3.26).

Table 3.26   A partial life table of the weevil *C. oryzae*

Age in weeks. X	l_x	m_x	$l_x m_x$
4.5	0.87	20.0	17.400
5.5	0.83	23.0	19.090
6.5	0.81	15.0	12.150
7.5	0.80	12.5	10.000
8.5	0.79	12.5	9.875
9.5	0.77	14.0	10.780
10.5	0.74	12.5	9.250
11.5	0.66	14.5	9.570
12.5	0.59	11.0	6.490
13.5	0.52	9.5	4.940
14.5	0.45	2.5	1.125
15.5	0.36	2.5	.900
16.5	0.29	2.5	.800
17.5	0.25	4.0	1.000
18.5	0.19	1.0	.190

R_0 = 113.560.
Source: Birch L.C., 1948.

l_x is the percent of the cohort alive at the mid-point of the time interval, while m_x is the total number of individuals produced per female per generation (fecundity of the female). A female in the age group 8 and 9 weeks could produce 12.5 offspring (Table 3.26) but has only a 79 percent chance of reaching that age. Therefore, the total number of offspring produced per female during the interval of time is

$$l_x \, m_{x \ (8-9)} = (0.79) \, (12.5)$$

$$= 9.88 \text{ offspring}$$

The total number of female offspring produced per female during a single generation is equal to the sum of the $l_x \, m_x$ products, or

$$R_0 = \sum_{x=0}^{n} l_x m_x$$

The sign Σ (sigma) stands for the sum of a series of numbers. The numbers summed are indicated by the limits of summation found above and below the sign. The first number to be summed, the first age group, is indicated by $x = 0$, 17.400 in Table 3.26. The x refers to the subscript of the numbers being added and can be thought of as a counter taking values from $x = 0$ to n, the last age group. The upper limit is n, sometimes written more completely as $x = n$, where n refers to the last number summed, 0.190. In the weevil example, $n = 14$ because there are 15 numbers to be added. The limits of summation are not always from 0 to the last number. The limits of summation might be

$$\sum_{x=r}^{n-r}$$

meaning to sum up a series of numbers beginning at some r^{th} number and ending at the $(n-r)^{\text{th}}$ number. If, in some particular example, n equals 20 and r is 5, this is equivalent to

$$\sum_{r=5}^{15}$$

In some situations, $x = 1$ is the first number of a series rather than $x = 0$, depending on convention and the situation. In some cases, if the limits of summation are by obvious, they are left off the summation sign to conserve space.

In the weevil example in Table 3.26, R_0 (termed the net reproduction rate) equals 113.56, or the population is multiplying 113.56 times each generation under the given set of environmental conditions and in an unlimited environment. If the time interval t equals to the generation time T,

$$R_0 = e^{rT}$$

The mean generation time can be roughly estimated by dividing the log to the base e of R_0 by the intrinsic rate of increase r

$$T = \frac{\log_e R_0}{r} \tag{3.54}$$

If λ is close to 1.0, a simple approximation to Eq. (3.54) (Dublin and Lotka, 1925 as quoted in Poole, 1974) is

$$T = \frac{\sum x l_x m_x}{\sum l_x m_x}$$

λ is the multiplication of the population in one interval of time. It is also a constant. If the population at time t consists of 50 individuals and at time $t + 1$ of 75 individuals, then λ equals 1.5. In other words, the population has multiplied by a factor of 1.5 during the time t to $t + 1$. $N_{t+1} / N_t = e^r = \lambda$ or $\log_e \lambda = r$. The intrinsic rate of increase in the example is 0.405. The two statistics r and λ are appropriate only if the population has stable age distribution, or the sum of the $x l_x m_x$ column (x is the midpoint age in weeks for the weevil) divided by R_0. In Birch's example, $T \approx 943.09/113.56 \approx 8.3$ weeks. Given T, a rough estimate of r is

$$r = \frac{\log_e R_0}{T} \tag{3.55}$$

$$\text{or} \quad r = \frac{\log_e 113.56}{8.30}$$

$$= .57$$

This is only a rough, crude approximation but may be fairly accurate if λ is close to 1.0.

If mortality and fertility remains the same and the population has a stable age distribution, the value of r may be estimated from the equation

$$\sum_{x=0}^{n} e^{-rx} l_x m_x = 1 \tag{3.56}$$

To find r, possible values of r are substituted into the equation, possibly beginning with the rough estimate calculated from Eq. (3.55) until the left side of the equation equals 1. This method of computation (estimation) is called iteration or an iterative process or 'guestimate'. The computations in Eq. (3.56) are quick if a calculator with on e^x key is available. This can also be done through a computer by way of the system simulation approach. A general trick in iteration is to chose two numbers, one of which is sure to be larger than the true value of r, and the second sure to be smaller. Instead of 1, a number larger than 1 and a number smaller than 1 will result as the solution of the left hand side of the equation. From an estimation of the deviations of these two numbers from 1 can be chosen. This rough estimate of r can be further improved by successive steps, iteration, until the deviation of the solution from the left-hand side of the equation is as small as is wanted.

In the weevil example, the rough approximate value for the intrinsic rate of increase of *C. oryzae* was 0.57. Beginning with two estimate of r as 0.5 and 0.8.

$$r = 0.5 \sum_{x=0}^{n} e^{-0.5x} \, l_x m_x = 4.0920 \quad \text{deviation from 1, } + 3.092$$

$$r = 0.8 \sum_{x=0}^{n} e^{-0.8x} \, l_x m_x = 0.8215 \quad \text{deviation from 1, } - 0.1785$$

In examining the two deviations, r is seen to be much closer to 0.8 than to 0.5. By taking $r = 0.76$, the deviation is $+ .0104$. The deviation is small and positive, indicating a number slightly larger than 0.76. A final estimate of r is about 0.762, the intrinsic rate of increase of *C. oryzae* reared at 20°C and 14 percent relative humidity.

If the life table is known and the intrinsic rate of increase calculated, the stable age distribution of the population can be computed. In a stable-age distribution population of N individuals, the birth rate β during one interval of time t can be defined as the number of female births during the interval t to $t + 1$, B_t, divided by N_t. For computing the finite birth rate, which has the same relation to b as λ does to r, the following equation should be used:

$$\frac{1}{\beta} = \sum_{i=1}^{n} e^{-r(x+1)}$$

In the weevil example $1/\beta = 0.81167$ and, dividing 1 by 0.81167, $\beta = 1.23067$. The proportion p_x of individuals in the age group x to $x + 1$ in the stable age distribution is given by

$$p_x = \beta \, l_x \, e^{-r(x+1)}$$

Table 3.27 from Birch (1948 as quoted in Poole, 1974) shows the calculation of the stable age distribution of *C. oryzae* at 29°C and 14 percent relative humidity. If the probabilities associated with the fecundity and mortality of each age group remain the same for every generation, the relative proportion of the age groups of the population approach with time, a fixed distribution, is known as the stable age distribution (Lotka, 1925 as quoted in Poole, 1974).

Table 3.27 Calculation of the stable age distribution of the weevil
C. oryzae

Age group (x)	l_x	$e^{-r(x+1)}$	$l_x e^{-r(x+1)}$	Stable age distribution $\beta l_x e^{-r(x+1)}$	
0	0.95	0.4677	0.4443150	54.740	
1	0.90	0.2187	0.1968300	24.249	
2	0.90	0.10228	0.0920520	11.341	Immatures
3	0.90	0.04783	0.0430470	5.304	
4	0.87	0.02237	0.0194619	2.398	
5	0.83	0.01046	0.0086818	1.070	
6	0.81	0.00489	0.0039609	0.488	
7	0.80	0.002243	0.0017944	0.221	
8	0.79	0.001070	0.0008453	0.104	
9	0.77	0.000500	0.0003850	0.047	Adults
10	0.74	0.000239	0.0001769	0.022	
11	0.66	0.000110	0.0000726	0.009	
12	0.59	0.000051	0.0000301	0.004	
13	0.52	0.000024	0.0000125	0.002	
14	0.45	0.000011	0.0000050	0.001	

Source: From Birch L.C. 1948.

The older age groups are not included because they make almost no contribution to the final age distribution. The stable age distribution is heavily loaded with the immature stages of the weevils. Only 4.5 percent of the total number of individuals in the stable age distribution are adults.

The intrinsic rate of increase r is simply the difference between the instantaneous birth rate and the instantaneous death rate: $r = b - d$. If r has been determined, these two rates can be found by first finding b as

$$b = \frac{\beta r}{\lambda - 1}$$

and then determining d from $r = b - d$. In the weevil experiment 0.76 = 0.82 − d, or $d = 0.06$. The finite death rate δ is found from the solution of

$$d = \frac{\delta r}{\lambda - 1}$$

for δ.

3.4.1.2 *Computer Language Programming and Simulation Studies on Large Computer As a Non-statistical Approach for Estimating Parameters and for Sensitivity Analysis.*

Dent and Blackie (1979), while discussing model-construction, mentioned that one of the limitations may be that although data are available for a given area of the model, they may be unsuitable for the specific requirements of the model. This situation is frequently encountered since data are rarely generated with the needs of a particular model in mind. When faced with this situation, the model-builder can either:

1. Use the data available but modify them in some subjective way; or
2. Reject them and redesign the model; or
3. Reject them and setup a research program to generate the data in the required format.

The decision depends on the precision required in the particular component of the model (and the obvious need to balance the precision between the various components), the time factor involved in generating new data in relation to the data set for the completion of the model and finally on the resources available either to make effective subjective modifications to inappropriate data or even to regenerate new data. A final limitation with respect to data that the modeler will encounter is that the suitable data for specific relationships or rates of change are simply not available.

The option of subjectively estimating data is a new concept in our discussion and since it is used with reasonable frequency, it should be examined carefully. (This option is also available to the modeler who has data in an inappropriate format.) The decision to estimate values subjectively is not such a wild procedure at the stage in the model development as might be thought and it could save entering into an experimental program with all the costs in time facilities and resources this implies. Assume that the difficulty lies with a single piece of informations—a coefficient representing a rate-of-change constant. We have only the broadest idea of the value of this coefficient. The procedure could be as follows:

1. Determine the best 'guestimate' for the coefficient.
2. Proceed with the development of the rest of the model and complete its structure and its implementation on the computer.
3. Establish appropriate parameters for monitoring the performance of the whole model.
4. 'Run' the model with the decided value for the unknown coefficient—note the values of the performance parameters.
5. 'Run' the model several times with different, widely-spaced values of the unknown parameter each time and without changing any other aspect of the model noting the values of the performance parameters.
6. Compare the values of the performance parameters for each 'run'.

7. In case different assumed values of the unknown coefficient do not markedly influence the performance parameters, we may say that the model is not sensitive to the selective value for the particular coefficient. Our judgment might then be that the 'guestimate' is acceptable and may, without loss of value to the model as a whole, be incorporated in the structure.

8. In case where markedly different output parameters are consequent on the different assumed values of the unknown coefficient, we would say that the model is sensitive to this particular coefficient. The 'guestimate' should now be used with caution and it would be advisable to extend the data search perhaps by way of a new research program. Alternatively, it may be possible to restructure the model in such a way as to reduce the sensitivity of the coefficient. However, the potential for guiding analytical research in a problem-solving mode by way of simulation model and sensitivity analysis should be noted and underlined: we will return to this theme again in chapter 6 (model application).

3.4.1.3 Non-statistical Approach for Parameter Estimate in Stochastic Models

As mentioned by Dent and Blackie (1979), the output from a stochastic model can be compared with that of real system using conventional statistical methods since both sets of output are random variable. If a model is deterministic, there will be no true 'experimental error' and, hence, no error (or no appropriate measure of error) for computation of F test in analysis of variance or confidence limits in regression. There may be an apparent 'error': in an analysis of variance, this would represent lack of fit of the linear statistical model. It is obviously possible to highlight the mean difference between two sets of output using a deterministic model but note Anderson's (1976, as quoted in Dent and Blackie, 1979) cautionary words on the use of averages for decision making in an uncertain world.

3.4.1.4 Estimation of Binomial Coefficient with Non-statistical Method

Binomial coefficient/parameter:

$$(p + q)^n = \sum_{k=0}^{n} C_k^n \, p^k \, q^{n-k}$$

p is the probability of an event that occurs.

q is the probability of the event that does not occur.

n is the total number of outcomes including both the events.

C_k^n is the total number of ways of dividing n outcomes into two groups, one of size k and the other of size $n - k$ is

$$n!/k!\ n{-}k!$$

The terms C_k^n are called the binomial coefficients.

3.4.1.4.1 *Example from Lewis (1971)*

Consider the population of plants in which the proportion of albinos is 1/4. Suppose that we examine a group of five plants selected at random. Here the sample size, n, is 5; $p = 1/4$, and $q = 3/4$. Suppose now that we wish to predict how many albinos are likely to be in the sample. We can readily see that on an average, we can expect to find 5 (1/4) or 1.25 plants. Of course, the decimal 0.25 is meaningless for any single sample but indicates that if we took many such samples and collected the average number per sample, we would obtain something other than a whole number. The mean gives the number of albinos that we might expect in a sample of five, but it is entirely possible that a random samples of five plants might contain anywhere from zero to five albino specimens. What we must do is to calculate the probabilities of finding 0, 1, 2, and so on, upto five albinos in the group. We do this by writing out the expansion of $(1/4 + 3/4)^5$ or $(p + q)^5$. There is more than one way to accomplish the mathematical chore; the simplest way is to apply the general formula of the terms of the expanded binomial. The following form of the formula is easier to use:

$$P = \frac{n!\,p^k\,q^{n-k}}{k!\,(n-k)!}$$

where n is the number in the sample, p is the proportion of the population that shows the attributes in equation, and q is the proportion that does not; k is the number of items in the sample showing the attributes (k may take any whole number value from 0 to n) and P is the probability that a random sample of size n will contain k such items. Thus, if we seek the probability that a random sample of 5 plants in the above group could contain 3 albinos, we would find our answer as follows:

$$P = \frac{5!\,(1/4)^3\,(3/4)^2}{3!\,2!}$$

$$C_k^n = \frac{5\times4\times3\times2}{3\times2\times2} = 10$$

so 10 is one of the binomial coefficients for 3 albinos.

If we calculate in this way the binomial coefficients for each of the six possible sample types, we could tabulate the results of different binomial coefficients as follows:

Table 3.28 The results of different binomial coefficients

k	c_k^n	$p^k q^{n-k}$	P	*cumulative P*
0	1	0.2373	0.2373	0.2373
1	5	0.0791	0.3955	0.6328
2	10	0.02636	0.2636	0.8969
3	10	0.008789	0.08789	0.98429
4	5	0.002929	0.01464	0.99893
5	1	0.00009765	0.00009765	0.9990276
				$= 1.0$

3.4.1.4.2 Binomial Distribution (Theorem)

If

1. An experiment has two possible outcomes called, say, a_1, and a_2
2. We let $P\{a_1\} = p$, so that $P\{a_2\} = (1-p)$
3. The experiment is repeated n times and each outcome is independent of any combination of other outcomes.

Then,

The probability of getting $\{a_1\}$ exactly k times and $\{a_2\}$ exactly $n - k$ times is

$P[a_1 \ (k \text{ times}), a_2 \ (n - k) \text{ times}] =$

$$\frac{n!}{k!(n-k)!} \ p^k \ (1-p)^{n-k}$$

In the Binomial distribution, the term distribution is used in the following sense: a probability distribution is a pattern according to which a possible and relevant event may be assigned a probability. Various types of probability distributions pertain to different classes of phenomena. Part of the study of probability theory involves the explicit study of commonly encountered probability distribution.

To recapitulate, the above equation was arrived at by the following sequence of steps:

(a) When n experiments are performed, there may be many sequences that give $\{a_1\}$ k times and $\{a_2\}$ $n - k$ times. In the case of the garden pea experiment, $\{a_1\}$ = {pink flowers} and $\{a_2\}$ = {white flowers}; k = 705 and $n - k$ = 224. Many different sequences of 929 plants could give rise to 705 pinks and 224 whites (Gold, 1977).

(b) Any one of the sequences has probability $p^k \ (1-p)^{n-k}$. For the garden pea experiment, the assumption is that $p = 0.75$, so any one sequence with 705 pinks and 224 white would have probability $(0.75)^{705} (0.25)^{224}$

(c) The sequences are mutually exclusive, so their probabilities simply add. Since all the sequences events $\{a_1\}$ exactly k times and $\{a_2\}$ exactly $(n - k)$ times have the same probability, the sum is simply the total number multiplied by the probability of one of them.

(d) The total number of ways of dividing the outcomes into two groups, one of size k and the other of size $n - k$ is $n!/k! \ n - k!$ For the garden pea experiment, the number of sequences with 705 whites and 224 pinks is 929! / 705! 224!

3.4.1.5 *Multinomial Distribution*

If a given trial can result in the k outcomes $E_1, E_2, ...E_k$ in n independent trials, multinomial distribution is $f(x_1, x_2, ..., x, ; p_1, p_2 ..., p_k, n) =$

$$\left(\frac{n!}{x_1!, x_2!, ...x_k!} \right) p_1^{x1} p_2^{x2} ....p_k^{xk}, \text{ with } \sum_{i=1}^{k} x_i = n \text{ and } \sum_{i=1}^{k} p_i = 1$$

The multinomial distribution derives its name from the fact that the terms of multinomial expansion of $(p_1 + p_2 + ... p_k)^n$ correspond to possible values of $f(x_1, x_2, ..., x_k; p_1, p_2 ... p_k, n)$.

3.4.1.5.1 *Example*

If a pair of dice is tossed 6 times, what is the probability of obtaining a total of 7 or 11 twice, a matching pair once, and any other combination three times?

Solution: We list the possible events
E_1: a total of 7 or 11 occurs,
E_2: a matching pair occurs,
E_3: neither a pair nor a total of 7 or 11 occurs.

The corresponding probability for a given trial are $P_1 = 2/9$, $P_2 = 1/6$, and $P_3 = 11/18$. These values remain constant for all 6 trials. Using the multinomial distribution with $x_1 = 2$, $x_2 = 1$, and $x_3 = 3$, we obtain the required probability:

$f(2, 1, 3; 2/9, 1/6, 11/18, 6) =$

$$\left(\frac{6}{2,1,3} \right) \left(\frac{2}{9} \right)^2 \left(\frac{1}{6} \right) \left(\frac{11}{18} \right)^3 = \frac{6!}{2!1!3!} \cdot \frac{2^2}{9^2} \cdot \frac{1}{6} \cdot \frac{11^3}{18^3} = 0.1127$$

(Walpole, 1982)

The multinomial coefficient here is

$$\frac{6!}{2!1!3!} = \frac{6\times5\times4\times3\times2}{2\times3\times2} = 60$$

3.4.1.6 Poisson Distribution

$$P = \frac{e^{-nP}(np)^x}{x!}$$

where e stands for the base of the natural logarithms. Usually the mean, μ, is substituted for np in this equation. The probabilities that an event will happen 0, 1, 2, 3, ... times is given by the series of terms:

$$e^{-\mu}, \; e^{-\mu}.\mu, \; (e^{-\mu}.\mu^2)_{/2!}, \; (e^{-\mu}.\mu^3)/3!, \; ....$$

The poisson coefficient here is $1/x!$.

3.4.1.7 Optimum Seeking Designs As a Non-statistical Approach in Design of Simulation Experiments

Systematic search on computers

The object of computer simulation studies on management strategies is to explore the manner in which values of the function $Y = f(X_1, X_2, ..., X_n)$ change in response to changing values of the independent variables $X_1, X_2, ..., X_n$. For large values of n, there exist an enormous number of combinations of values of the n independent variables for which Y must be calculated. Rather than computing millions or billions of Y values in order to find the set of X_i values that produce the largest or smallest Y, it is desirable to have some systematic procedure that will lead us quickly to the optimal combination of values of the x_i's. (Watt, 1968).

We shall consider the computational procedure before explaining the method. We wish to find the combination of variate values for optimum procedures. This operating procedure maximizes the productivity of natural resources, Y. We must first calculate Y for an arbitrarily chosen array of X. This chosen array of X produced a Y reasonably close to the maximum Y. We compute a new set of Y values, one for each array of X. The new arrays of X produces some value. If this value is greater than the starting array, then this value becomes the starting point for the next step.

The new X values will be obtained by adding to or subtracting from, a vector h of step sizes of some predetermined value.

The yardstick is the number of steps taken to find a value of Y. This value of Y improves at less than a predetermined rate per iteration. There are *n* independent variables. The number of steps taken to find the optimum *Y* is proportional to n^3. The number of possible search directions increases outward from a point of origin of search with increase in the number of dimensions.

Figure 3.26 explains this phenomenon more clearly.

Unidimensional search

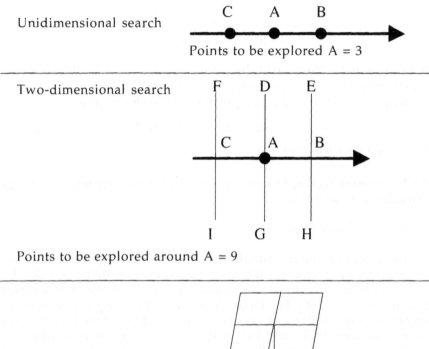

Two-dimensional search

Points to be explored around A = 9

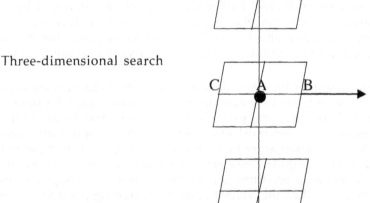

Three-dimensional search

Points to be explored around A = 27

Fig. 3.26 Increase in computing volume created by a search with increasing dimensionality of the surface being searched (Watt, 1968).

There is only one independent variable. We progress along the X axis. We calculate Y values for the given X values to find the optional (maximum or minimum) Y. We move to A in the direction C to B. A is the center of a search area. We can only calculate Y for C, A, and B. Y's are calculated for the center point of the search. Y's are also calculated for h units out from the center. Now there are two independent variables. We calculate Y at C, A, and B. We also calculate Y at D, E, F, G, H, and I. Now there are three dimensions. We have 27 points at which Y should be calculated.

Where there is only one dimension, we find out by working with numbers Y at the center point along with a dot one step-size in front of it and one step-size behind. Increasing the dimensionality by one, we now have 2 dots either side of each of the original 3 at which Y must be calculated. The number of dots to be examined is now $3 + 2 (3) = 9$. Where there are 3 independent variables, we must compute Y at one dot above and one below each of our previous 9 dots or a total of $9 + 2 (9) = 27$. It is evidently dealt with a series of the ensuing type (Table 3.29).

Table 3.29 Series of the ensuing type

Number of dimensions	*Number of dots at which Y must be calculated around each search area*
1	$3 = 3^1$
2	$3 + 2 (3) = 9 = 3^2$
3	$9 + 2 (9) = 27 = 3^3$
4	$27 + 2 (27) = 81 = 3^4$
5	$81 + 2 (81) = 243 = 3^5$
n	3^n

Really, the number of searches needed does not increase quite this steeply, for two reasons. First, we can make use of tricks in experimental design to cut down the required number of examinations. Second, the dots to be searched around a particular search center may already have been examined around the previous center, and the computer program should test to ensure that there is no duplication of effort. The classical minimization techniques, the volume of computation grow as n^3, not as 3^n. The reduction of computation is demonstrated by the ensuing data in Table 3.30.

3.4.1.8. *Fitting Model Equations to Experimental Data*

Model equations of various forms may be derived analytically from theoretical considerations. However, it should be noted that these techniques only determine the form of the equation. Specific values for coefficients and exponents, which are necessary to implement these model equations, must be obtained through the process of curve fitting (Spain, 1982).

Table 3.30 Reduction of computation by the ensuing data

Number of Dimensions	Number of dots at which Y must be calculated around each search area	
	Theoretical	*Actual*
1	$3^1 = 3$	$1 = 1^3$
2	$3^2 = 9$	$8 = 2^3$
3	$3^3 = 27$	$27 = 3^3$
4	$3^4 = 81$	$64 = 4^3$
5	$3^5 = 243$	$125 = 5^3$
6	$3^6 = 729$	$216 = 6^3$
7	$3^7 = 2187$	$343 = 7^3$
8	$3^8 = 6561$	$512 = 8^3$
n	3^n	n^3

Theoretical equations are those which have been derived from analytical considerations or at least have the potential for analytical derivation, perhaps only by way of analogy to known processes. Empirical equations, on the other hand, are those which have neither direct nor implied theoretical origin. This situation often occurs when there is a complex cause and relationship between the variables which the equation is intended to describe. Suppose, for example, one needed an equation to describe the density of water at different temperatures as a component of lake stratification model. There is, of course, a theoretical basis for this temperature effect. However, it is neither pertinent nor essential to the lake stratification model. In such cases, an empirical model equation obtained directly from experimental data may be used to provide the needed information. The most commonly employed empirical equations are in the polynomial class. The example of the polynomial equation:

$$y = A + Bx + Cx^2 + Dx^3$$

The curve fitting procedure would be employed to determine the best values for the coefficients A, B, C, and D. From a mathematical standpoint, theoretical and empirical equations are equally useful to describe the function of a system component. It is generally true, however, that a modeler would prefer to employ equations having a theoretical origin, because they usually contribute more to the understanding of the system being studied (Spain, 1982).

3.4.1.8.1 Selecting Equations for Fitting

Descriptive names and equations are given in Table 3.31.

Equations are selected by entering the equation number. For some equations, it will be necessary to estimate either A or B. These parameters are usually an asymptote or intercept of the approximating curve. One must simply make an educated guess and enter the value. The best value of A or B may be obtained by repeating the curve fitting process several times and accepting the constant which results in the highest value for the F statistic or lowest residual mean square.

Table 3.31 The descriptive name and equation of an analytical model

Number	Descriptive name	Equation
1.	Straight line	$y = Ax + B$
2.	Exponential growth or decay	$y = Ae^{nx}$
3.	Power function	$y = Ax^n$
4.	Hyperbolic	$y = Ax / (B + x)$
5.	Exponential saturation	$y = A (1 - e^{nx})$
6.	Sigmoid (forward or reverse)	$y = A / (1 + Bx^n)$
7.	Exponential Sigmoid	$y = A / (1 + Be^{nx})$
8.	Modified inverse	$y = A / (B + x)$
9.	Modified power function or cut off	$y = Ax^n + B$
10.	Maxima function	$y = Ax . e^{nx})$

3.4.1.8.2 Standard Equation Types

Name, standard form, linear form and parameters defined by slope and intercept are given in Table 3.32.

The data for exponential equation are given in Table 3.33 and shape of the exponential equation is shown is Fig. 3.27. The parameters of the exponential equation are: $A = 1$, $n = 0.05$.

The data for power function equation are given in Table 3.34 and shape of the power function equation is shown in Fig. 3.28. The parameters of such equation are: $A = 1.5$, $n = 0.7$.

The data for hyperbolic saturation equation are given in Table 3.35 and shown in Fig. 3.29. The parameters are: $A = 100$, $B = 50$.

The data for fitting the exponential saturation equation are given in Table 3.36 and shown in Fig. 3.30. The parameters are: $A = 100$, $n = -0.1$

The data for fitting the sigmoid equation are given in Table 3.37 and shown in Fig. 3.31. The parameters are $A = 100$, $n = -3$, $B = 5000$.

Table 3.32 Standard equation types

No.	Name	Standard form	Linear form	Parameters defined by slope and intercept		
				A	B	n
1.	Straight line	$y = Ax + B$	$[y] = A[x]$	S	I	
2.	Exponential	$y = Ae^{xp}$	$[\ln y] = \ln A + n[x]$	Exp (I)		S
3.	Power function	$y = Ax^{n}$	$[\ln y] = \ln A + n[\ln x]$	Exp (I)		S
4.	Hyperbola	$y = Ax/(B+x)$	$[x/y] = B/A + 1/A[x]$	1/S	1/S	
5.	Exponential saturation	$y = A(1-e^{nx})$	$[\ln(A-y)] = \ln A + n[x]$	Exp (I)		S
6.	Sigmoid	$y = A(1 + Bx^{n})$	$[\ln(A/y-1)] = \ln B + n[\ln x]$	Estimate	Exp (I)	S
7.	Exponential sigmoid	$y = A/(1+Be^{rx})$	$[\ln(A/y-1)] = \ln B + n[x]$	Estimate	Exp (I)	S
8.	Modified inverse	$y = A/(B+x)$	$[1/y] = B/A + 1/A[x]$	1/S	1/S	
9.	Modified power function	$y = Ax^{n} + B$	$[\ln(y-B)] = \ln y + n[\ln x]$	Exp (I)	Estimate	S
10.	Maxima function	$y = Ax.\,e^{rx}$	$[\ln(y/x)] = \ln A + n[x]$	Exp (I)		S

S = Slope of fitted line, I = Intercept of fitted line, Exp = Exponential
Transformed values of x and/or y are enclosed in brackets in the linear form of the equation.
Source : Spain, 1982.

Table 3.33 Data for exponential equation

x	y	Exponential equation
10	1.649	$Y = Ae^{nx}$
20	2.718	$A = 1$
30	4.482	$n = 0.05$
40	7.389	
50	12.182	
60	20.085	
70	33.115	
80	54.598	
90	90.017	
100	148.413	

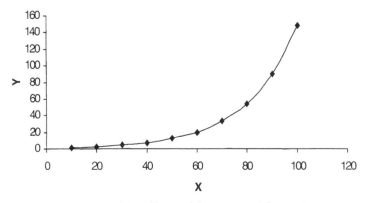

Fig. 3.27 Shape of the exponential equation

Table 3.34 Data for power equation

x	y	Equation
10	7.518	$Y = Ax^n$
20	12.213	$A = 1.5$
30	16.221	$n = 0.7$
40	19.840	
50	23.194	
60	26.351	
70	29.353	
80	32.229	
90	35.000	
100	37.678	
110	40.278	
120	42.807	
130	45.274	
140	47.685	
150	50.044	
160	52.357	

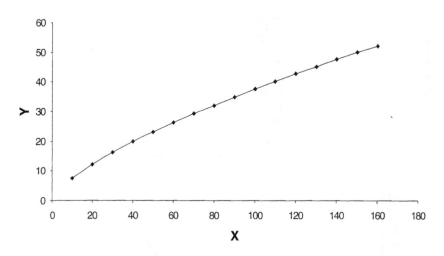

Fig. 3.28 Shape of the power function equation.

Table 3.35 Values of variables of *x* and *y* fitting the hyperbolic saturation equation

x	*y*	Equation Hyperbolic saturation functions
10	16.667	$Y = Ax/(B + x)$
20	28.571	Where $A = 100$
30	37.5	$B = 50$
40	44.444	
50	50.000	
60	54.545	
70	58.333	
80	61.538	
90	64.286	
100	66.667	
110	68.750	
120	70.588	
130	72.222	
140	73.684	
150	75.00	
160	76.190	

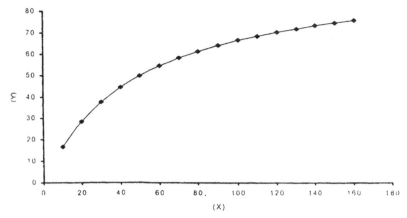

Fig. 3.29 Shape of the hyperbolic saturation equation.

Table 3.36 **The values of variables of x and y fitting the exponential saturation equation. The parameters are $A = 100$, $n = -0.1$**

x(10)	y	Exponential saturation equation
1	63.212	$Y = A(1 - e^{nx})$
2	86.466	Where $A = 100$
3	95.021	$n = -0.1$
4	98.168	
5	99.326	
6	99.752	
7	99.908	
8	99.966	
9	99.988	
10	99.995	

Table 3.37 **The values of variables of x and y fitting the sigmoid equation**

x	y	Sigmoid equation
5	2.4	$Y = A/(1 + Bx^n)$
10	16.6	Where $A = 100$
20	61.5	$n = -3$
30	84.3	$B = 5000$
40	92.7	
50	96.1	
60	97.7	
70	98.5	
80	99.0	
90	99.3	
100	99.5	

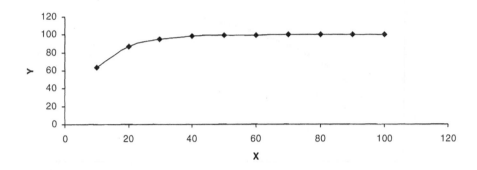

Fig. 3.30 Shape of the exponential saturation equation

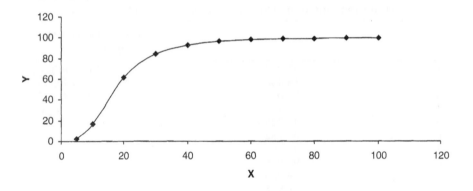

Fig. 3.31 Shape of Sigmoid equation

The data for fitting the exponential sigmoid equation are given in Table 3.38 and shown in Fig. 3.32. The parameters are:

$$A = 100, B = 50, n = -0.2.$$

The data for fitting the modified inverse equation are given in Table 3.39 and shown in Fig. 3.33. The parameters are:

$$A = 2000, B = 0.$$

The data for fitting the modified power function equation are given in Table 3.40 and shown in Fig. 3.34. The parameters are:

$$A = -0.00001, B = 100, n = 3.$$

The data for fitting the maxima function equation are given in Table 3.41 and shown in Fig. 3.35. The parameters are: $A = 15$, $n = -0.06$.

Table 3.38 **Giving the values of independent variable x, dependent variable y, exponential sigmoid equation and parameters**

x	y	Exponential sigmoid equation
10	12.8	$y = A/(1 + Be^n)$
20	52.2	Where $A =100$
30	89.0	$B = 50$
40	98.4	$n = -0.2$
50	99.8	
60	100	
70	100	
80	100	
90	100	
100	100	
110	100	
120	100	
130	100	
140	100	
150	100	
160	100	

Table 3.39 **The values of independent variable, x, dependent variable, y of modified inverse equation and its parameters**

x	y	Modified inverse equation
10	200	$y = A/(B + x)$
20	100	$A = 2000$
30	66.6	$B = 0$
40	50	
50	40	
60	33.3	
70	28.5	
80	25	
90	22.2	
100	20	
110	18.1	
120	16.6	
130	15.3	
140	14.2	
150	13.3	
160	12.5	

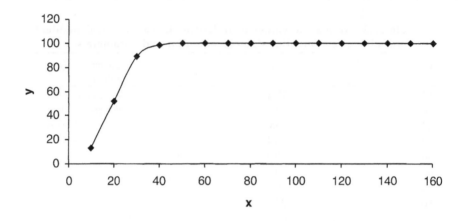

Fig. 3.32 Showing the shape of the exponentialy sigmoid equation.

Table 3.40 **The values of independent variable, x, dependent variable, y of modified power function equation and its parameters**

x	y	Modified power function equation
10	99.9	$Y = Ax^n + B$
20	99.9	$A = -0.00001$
30	99.7	$B = 100$
40	99.3	$n = 3$
50	98.7	
60	97.8	
70	96.5	
80	94.8	
90	92.7	
100	90.0	
110	86.6	
120	82.7	
130	78.0	
140	72.5	
150	66.2	
160	59.0	

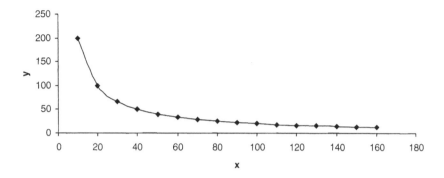

Fig. 3.33 Shape of the modified inverse equation

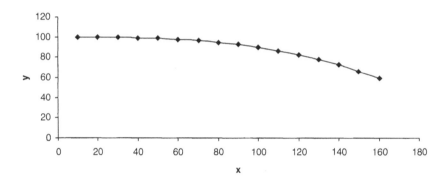

Fig. 3.34 Shape of the modified power function

Table 3.41 **The values of independent variable, *x*, dependent variable, *y* of maxima function equation and its parameters**

x	y	Maxima function equation
0	0	
10	82.3	$Y = Ax \cdot e^{nx}$
20	90.3	$A = 15$
30	74.3	$n = -0.06$
40	54.4	
50	37.3	
60	24.5	
70	15.7	
80	9.8	

contd....

Table 3.41 Contd.

90	6.0
100	3.7
110	2.2
120	1.3
130	0.7
140	0.4
150	0.2
160	0.1

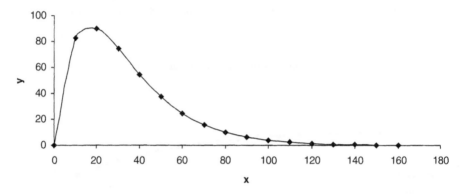

Fig. 3.35 Shape of the maxima function equation

3.4.1.9 *Mathematical formulation for solving the differential equation (Analytical solution)*

Analytical models are classified according to the type of analysis or theoretical derivation on which they are based. The models which are based on the solution of differential equations are called analytical model based on differential equations. Differential equations define a rate of change of some dependent variable with respect to some independent variable, usually time or distance. To show how differential equations are utilized to develop model equations, let us examine the development of the model for unlimited population growth.

The exponential growth model

The advantage of using differential equations results from the ease with which they are usually obtained from a function equation. In this example, we start with the general observation that population growth is a function of population density. From this generalization, we can write a simple function equation for population growth.

$$G = f(N) \qquad (3.57)$$

where G is the growth rate, and N is the population density. If we assume that growth is a direct function of N, i.e. the growth rate is directly proportional to population and no other factors are involved, then the growth rate equation could take the form:

$$G = k . N \qquad (3.58)$$

where k is a proportionality constant. The curve for this equation would be a straight line with slope = k. In other words, with increase in N, there is a proportionate increase in G. The equation has very limited value in this form, since interest is in the population density at any time during the growth process, rather than the growth rate. To convert the growth rate equation to a form which gives this kind of information, it is essential to write it as a differential equation in which growth is redefined as dN/dt. This expression symbolizes the derivative of N with respect to time t, and means the instantaneous rate of change of population density with respect to time. Since this is equivalent to G in Equation 3.58, the equation now has the form

$$\frac{dN}{dt} = kN \qquad (3.59)$$

Now the equation has two variables, N and t. It is now possible to reconstruct the population growth curve equation from this by the process of integration. The equation 3.59 is used for the slope of the growth curve to reconstruct the equation for the growth curve. Mathematicians know how to perform the process of integration only as a result of their experience with the reverse process of differentiation. These experiences have led to a large number of integration rules. A few useful integration rules are presented in Table 3.42.

The first step in the integration process is to collect all the terms dealing with the dependent variable, N, on the left side of the equal sign and the independent variable and associated constants on the right side. In this example, the result is as follows:

$$\int_{N_0}^{N_t} \frac{dN}{N} = \int_{0}^{t} K dt \qquad (3.60)$$

The integration symbol gives instructions to perform an integration according to the rules. In this case, we will be integrating N and t between the limits of $t = 0$ and $t = t$. The result without limits (according to rule 1 and 3 (Table 3.42) is

Table 3.42 Common integration rules

1. $\int a\,dx = ax + c$

2. $\int x^n dx = \dfrac{x^{n+1}}{n+1} + c\ (n <> -1)$

3. $\int \dfrac{dx}{x} = \ln x + c$

4. $\int e^x dx = e^x + c$

5. $\int e^{ax} dx = \dfrac{e^{ax}}{a} + c$

6. $\int \dfrac{dx}{a+bx} = \dfrac{\ln(a+bx)}{b} + c$

Source: Span (1982). [These rules are written for the infinite integral; hence each contains an integration constant.]

$$\ln N = k.t + c \tag{3.61}$$

When integrating between limits, we take the integral at the upper limit minus the integral of the lower limit, first doing the left side of the equation, then doing the right. The result is the following:

$$\ln N_t - \ln N_0 = k.t - k.0 \tag{3.62}$$

Since the difference between logs equals the log of the quotient, we obtain

$$\ln(N_t / N_0) = k.t \tag{3.63}$$

By taking exponents of both sides, we get the form:

$$\dfrac{N_t}{N_0} = e^{kt} \tag{3.64}$$

$$\text{and} \qquad N_t = N_0 e^{kt} \tag{3.65}$$

Thus, a differential equation is solved.

Exercise: Program equation $C_t = C_0 \cdot e^{-kt}$ as a model of radioactive decay. Let C_0, the original specific activity, equal 100μ curies, and the value of $k = 0.1733$/day which is the decay rate constant of I^{124}. Increment the time in days from 0 to 30.

The sample program of exponential decay model is given below:

```
10    REM MODEL OF EXPONENTIAL DECAY
20    REM Ct = C0 . e^{-kt}
30    REM N0 = Initial number of radioactive atoms.
40    REM N = Number of atom remaining at time T
50    REM K = Decay Constant
60    REM F = The fraction of atoms remaining at time T.
70    REM P = Percent of atom remaining
80    READ K, N0
90    FOR T = 0 to 20
100   N = N0 * Exp (-k * T )
110   F = N / N0
120   P = 100 * F
130   Y = INT (P * 50 /100)
135   PRINT T(I), N (I), F (I), P (I) and Y (I)
140   NEXT T
150   DATA .1733, 1000
160   END
```

Table 3.43 gives the data on T (I), N (I), F (I), P (I).

Figure 3.37 shows the relationship betwen t (I) as independent variable and P (I) as the dependent variable in the form of exponential decay.

Table 3.43 Showing T, N, F and P.

T	N	F	P
1	844.88	0.8408	84.08
2	707.08	0.707	70.7
3	594.58	0.594	59.4
4	499.97	0.499	49.9
5	422.42	0.422	42.2
6	353.53	0.353	35.3
7	297.27	0.297	29.7
8	249.97	0.249	24.9
9	210.19	0.210	21.0
10	176.75	0.176	17.6
11	148.58	0.148	14.8
12	124.98	0.124	12.4
13	105.09	0.105	10.5

contd....

Table 3.43 Contd.

T	N	F	P
14	88.37	.088	8.83
15	74.31	.074	7.4
16	62.48	.062	6.2
17	52.54	.052	5.2
18	44.18	.044	4.4
19	37.15	.037	3.7
20	31.24	.031	3.1

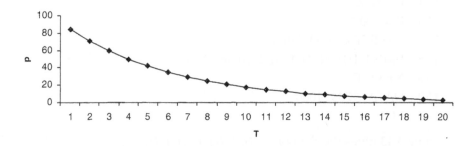

Fig. 3.36 Showing the relationship between τ and P as exponential decay.

3.4.1.10 Mathematical Formulation for Solving the Difference Equation (Numerical Solution)

The analytical solution of rate equation (differential equations) uses the rules of integral calculus. As a result of their nonlinear nature, most differential equations of biological interest are not readily solved by such analytical techniques. Thus, most simulations depend on a numerical solution of rate equations, or more precisely, numerical integration. It is this technique more than any other which makes simulations dependent upon the digital computer.

Numerical integration, in its simplest form, is easily understood. It employs the very straight forward techniques of calculating an incremental change in a variable and adding it to the old value in order to obtain the new value of the variable.

For example, if one is calculating the population of organisms at time $t + \Delta t$, one would add the change in population that occurs during time increment Δt, to the population size at time t. The computer is ideally designed to perform repetitive operations of this nature and to print or plot the result for any number of time intervals. Numerical integration can be a highly accurate technique. The results approach those of the analytical solution to varying degrees, depending upon the method employed. In any case, the small errors involved are not serious, since in simulation, one is more interested in the behavioral modes exhibited by a model than in precise duplication of real data (Spain, 1982).

3.4.1.10.1 *The finite difference approach*

All numerical integration techniques are based upon the fact that for small incremental changes in x (the independent variable), the difference quotient, $\Delta y / \Delta x$, approximates the derivative of the differential equation, dy/dx. The dy/dx is actually defined as follows:

$$\frac{dy}{dx} - \underset{\Delta \to 0}{\text{Lim}} \frac{\Delta y}{\Delta x}$$

Therefore, if

$$\frac{dy}{dx} = f(y)$$

then

$$\frac{\Delta y}{\Delta x} \cong f(y)$$

and

$$\Delta y \cong f(y) \cdot \Delta x \qquad (3.66)$$

If

$$f(y) = ky$$

then

$$\Delta y = ky.\Delta x \qquad (3.67)$$

This leads to the general equation upon which all numerical integration techniques are based.

$$y_{1 \cdot \Delta x} = y_1 + \Delta y \qquad (3.68)$$

Equation (3.67) defines the change in y for a finite change in x, and (3.68) calculates the new value of y after the change has occurred (Spain, 1982).

3.4.1.10.2 The Euler Technique

Equations (3.67) and (3.68) actually define what has been called the Euler Integration Technique. Let us examine the differential equation for exponential growth.

$$\frac{dN}{dt} = kN \tag{3.69}$$

where N is population density and k is the growth rate constant. Like most differential equations of interest to modelers, the independent variable is time t. The differential equation is approximated by the following difference equation:

$$\frac{\Delta N}{\Delta t} = kN \tag{3.70}$$

or

$$\Delta N = kn\ \Delta t \tag{3.71}$$

and the new value of N is obtained by the equation

$$N_{(t - \Delta t)} = N_t + \Delta N \tag{3.72}$$

To carry out an Euler integration, the increment time from one to t_{max} and at each time interval, calculate the change in N_t, and the new value of the population density, $N_{(t-\Delta t)}$. The new value then becomes the N_t for the following time interval.

Although this equation is simple enough to be calculated by a single combined equation, it is better to develop the habit of first calculating the change using an equation like (3.71) and then updating the variable with an equation such as (3.72). This procedure is referred to as the 'two-stage' approach.

The following coding sequence uses subscripted variables to accomplish this for the integration of the exponential decay models.

```
10   REM A sample program of exponential decay
20   REM DN/DT = KN
30   REM DN = KN.dt
40   REM N₀ = Initial Number of radioactive atoms.
50   REM N = Number of atoms remaining at time T.
60   REM K = Decay constant
```

```
70    REM F = The fraction of atoms remaining at time T
80    REM P = Percentage of atom remaining
90    dN/dt = Difference Quotient
100   READ K, N₀
110   FOR DT = 0 to 20
120   DN (I) = K*N (I) * DT (I)
130   N (I) = N (I) + DN (I)
```

$$140 \quad F(I) = \frac{N(I)}{N_o}$$

```
150   P (I) = 100 * F (I)
160   Y (I) = INT (P * 50/100)
170   Print T(I), N (I), F (I), P (I) and Y (I)
180   Next T
190   Data − .1733, 1000
200   End
```

Table 3.44 Showing *T (I)*, *N (I)*, *F (I)*, *P (I)* using *DT = 1*

T(I)	N (I)	F (I)	P (I)
0	1000	1	100
1	826.700	0.826	82.6
2	683.432	0.683	68.3
3	564.993	0.564	56.4
4	467.079	0.467	46.7
5	386.134	0.386	38.6
6	319.216	0.319	31.9

Significance of time increment in numerical integration

The further Δt departs from the limiting value of zero, the farther the results of the finite difference approach depart from the true value of the integral.

The error for a single step is directly proportion to the size of the Δt. In addition, the error is cumulative, since each new value of P is dependent upon the last value rather than the original value, as in the analytical solution.

Table: 3.45 Details of *T (I), N (I). F (I), P (I)* using *DT = 0.1*

T(I)	N (I)	F (I)	P (I)
0	1000	1	100
0.1	982.670	0.982	98.2
0.2	965.640	0.965	96.5
0.3	948.905	0.948	94.8
0.4	932.460	0.932	93.2
0.5	916.300	0.916	91.6
0.6	900.420	0.900	90.0
0.7	884.815	0.884	88.4
0.8	869.481	0.869	86.9
0.9	854.412	0.854	85.4
1.0	839.60	0.839	83.9

3.4.1.10.3. An iterated second order Runge-Kutta method

Since one can improve the estimate of N_t by basing the change, ΔN, on both the value of N_t and an estimate of $N_{t-\Delta t}$, it is logical that the accuracy of $N_{t-\Delta t}$ would be further improved by repeating the process using the improved estimate of $N_{t-\Delta t}$ to calculate the change. Thus, it is possible to use an iterative process to further improve the second order Runge-Kutta integration. The following coding sequence uses subscripted variables to accomplish this to integrate the exponential growth model:

```
10  REM EXPONENTIAL GROWTH MODEL
20  REM INTEGRATED SECOND ORDER RUNGE-KUTTA
SOLUTION
30  N = 2: K = 0.3: DT = 0.1
40  IM = INT (1/DT + 0.1)
50  For T = 1 to 20
60  For I = 1 to IM
65  N(1) = N
70  For J = 1 to 3
80  DN (J) = K * N (J) * DT
90  N (J + 1) = N + (DN (J) + DN (I))/2
100 Next J
110 N = N (4)
120 NEXT I
130 PRINT T  N
140 NEXT T
150 STOP
```

The first iteration through the J-loop calculates the first estimate of $N_{t \cdot \Delta t}$ designated N (2). The next iteration ($J = 2$) uses the value of N (2) and the original value N (1) to make the first improved estimate of $N_{t \cdot \Delta t}$, called N (3). The third iteration calculates the final value for $N_{t \cdot \Delta t}$, based on the average of the DN_t, i.e. $\Delta N(1)$, and the best estimate of $\Delta N_{t \cdot \Delta t}$ i.e. DN (3). Further iterations of this loop do not seem to make any significant improvements.

REFERENCES

Anderson, J.R. (1976). Essential probabilities in modeling agricultural systems. Agr. Systems **1**: 219-232.

Birch, L.C. (1948). The intrinsic rate of natural increase of an insect population. J. Ecol. **17**: 15-26.

Dent, J.B. and Blackie, M.J. (1979). Systems Simulation in Agriculture, 180p. London: Applied Science Publishers Ltd.

Dublin, L.I. and Lotka, A.J. (1925). On the true rate of natural increase. J. Am. Stat. Assoc. **20**: 305-339.

Emigh, T.H. and Goodman, M.M. (1985). Multivariate analysis in nematode taxonomy. Chapter 15, pp.197-204. In: An advanced treatise on Meloidogyne. Vol. II, Methodology, (eds) Barker, K.R., Carter, C.C. and Sasser, J.N. Releigh: North Carolina State University Graphics.

Forester, H.V., Mora, P.M. and Amiot, L.W.(1960). Doomsday, Friday, 13 November, A.D. 2026. Science **132**: 1291-1295.

Fridgen, J.J., Kitchen, N.R., Sudduth, K.A., Drummond, S.T., Wiebold, W.J. and Fraisse, C.W. (2004). Management zone analyst (MZA): Software for subfield management zone delineation. Agron. J. **96** (1): 100-108.

Gold, H.J. (1977). Mathematical modeling of biological system: An introductory guidebook, 357p. New York: John Wiley & Sons.

Gosset, W.S. (1908). As quoted in LEWIS, A.E. (1971). Biostatistics. 227 p. New Delhi: Affiliated East-West Press Pvt. Ltd.

Goulden, C.H. (1962). Methods of Statistical Analysis. 467 p. Mumbai: Asia Publishing House.

Heady, E.O. and Candler, W. (1973). Linear Programming Methods. 597 p. Ames, Iowa, USA: The Iowa State University Press.

Holling, C.S. (1966). The functional response of invertebrate predators to prey density, Mem. Entomol. Soc. Can. no. **48**.

Leslie, P.H. (1945). On the use of matrices in certain population mathematics, Biometrika **33**: 183-212.

Leslie, P.H. (1948). Some further notes on the use of matrices in population mathematics. Biometrika **35**: 213-245.

Lewis, A.E. (1971). Biostatistics. 227 p. New Delhi: Affiliated East-West Press Pvt. Ltd.

Lotka, A.J. (1925). Elements of Physical Biology. Baltimore: Williams and Wilkins.

Manly, B.F. (1994). Multivariate Statistical Methods: A Primer. 2nd edition. 215 p. London: Chapman & Hall.

Pine, E.S. (1975). How to Enjoy Calculus. 128 p. New Jersey: Steinlitz-Hammacher Co.

Poole, R.W. (1974). An Introduction to Quantitative Ecology, 532p. New York: McGraw-Hill.

Quirin, W.L. (1978). Probability and Statistics. 488p. New York: Harper & Row, Publishers.

Rabinovich, J.E. (1970). Vital statistics of Synthesiomyia nudiseta (Diptera: Muscidae). Ann. Entomol. Soc. Am. **73**: 749-752.

Segel, I.H. (1975). Biochemical Calculations. 441p. New York: John Wiley & Sons.

Singh, P., Singh, P. and Singh, D.P. (1978). Optimum allocation of land and water. pp.62-66. In: Agriculture Systems, Theory and Application (eds.) Dent, J.B. and Murthy, V.V.N. Ludhiana (India): Punjab Agricultural University.

Spain, J.D. (1982). *Basic Microcomputer Models in Biology*. 354p. London: Addison-Wesley Publishing Co.

Walpole, R.E. (1982). Introduction to Statistics. 521p. New York: Macmillan Publishing Co.

Watt, K.E.F. (1968). Ecology and Resource Management, 450p. New York: McGraw-Hill Book Co.

4

Computer Implementation

The topic has been dealt with explicitly and in an excellent manner by Dent and Blackie (1979), Watt (1968) and Gold (1977). We move towards the selection of an appropriate computer program. The languages—BASIC, FORTRAN and C⁺⁺ —are applied here. The selection of a specific program depends on the particular modeling exercise. Once the language is selected, the translation from the flow diagram must be made. During the translation, there are chances of making errors which inevitably creep into the translation.

Here, the main objectives are to outline the arguments in the determination of a suitable computer program for a specific modeling project; and to provide guidelines for translating the model into the selected application in such a way as to limit the number of errors.

4.1 MODEL SOFTWARE REQUIREMENT

A large number of computer languages suitable for simulation purposes have been written. In the academic world, a lot of literature is available on this topic. But it is all meant for the specialist computer programr. Eventually, others find them confusing rather than helpful. Here, we will consider computer program in a non-technical way. We will introduce some of the basic ideas behind their design, followed by a discussion on the factors affecting the selection of a suitable language.

Computer operation works on a series of single-operation instructions supplied in the form of a program. Programs written for a computer are broken down into the single operation steps. The computer users make use of the convenient and efficient 'high-level' languages, which are available on modern computers. The 'high-level' languages use simple verbs and standard mathematical notations to represent a series of single-operation instructions. The input of data item requires that the computer is instructed about what form the data item is in; where the data item is to be found; in what position in memory that data item should be stored; and finally, to find the next step in the program to be completed. All this can be replaced by single high-level instructions as in the FORTRAN example below.

Input data on this step   —— | Read (5,10) A | ——

Place variable read-in into this position in memory. This statement defines input format.

Next step to be executed ——➤ | B = A + 2 |

The compilation is referred as the translation from a high-level program into the eventual series of single step instructions. A compiler is a specialist computer program which converts a high level language into single-step instructions. Each high-level language has its own compiler. The requirement of a compiler is that the program follows specified language rules. These rules are referred to as the 'syntax' of the language. Firstly, the program is written in a high level language. Then it is run by a computer. The compiler checks the program for syntax. If the program fails this check, the computer will reject the program.

The two statements below illustrate how the syntax varies between languages. In both statements, the variable 'A' is assigned the value 20. The first statement is in FORTRAN, the second in ALGOL, the third in BASIC and the fourth in C··. Several differences will be noticed between the two statements.

A = 20: FORTRAN assignment statement

A = 20: ALGOL assignment statement

A = 20: BASIC assignment statement

a = 20: C·· assignment statement

The ensuing example shows a FORTRAN statement with an incorrect syntax. The compiler assumes that brackets always occur in pairs. In this example, the right hand bracket is missing

$$A = (20 * B/C + 4$$

It cannot specify whether the programr meant

$$A = (20 * B) /C + 4 \text{ or}$$

$$A = (20* B/C) + 4$$

and, therefore, execution is rejected, because the statement is insufficiently defined. If the programr left out both brackets, execution follows as per certain default rules.

High-level applications make programming easier to understand. Although such languages are designed to ease programming, a larger computer will occupy more of the computer memory and may be more expensive to use and will have a greater likelihood of containing 'bugs' of its own.

A more powerful language may lead the unskilled and unwary modeler into intended errors of logic. Therefore, some compromise has to be arrived at between the power of the language, the computer facilities available and the knowledge and experience of the programr involved. A high grade application may be grouped into two classes—general-purpose and special-purpose languages.

4.1.1 General-purpose Program

The examples of general-purpose programs are BASIC, FORTRAN, ALGOL, C--, PL/1, PASCAL, etc., which are of interest to the simulation modeler. These languages are very general; so they may be used to construct simulation models of any type of system. The modeler has to develop his own input-output routines, set up his own time clock and switches within the model, and write his own special-purpose routines such as normal pseudorandom-number generators. One who uses these applications for modeling must be an expert in programming with a reasonable knowledge of the computer and its associated systems.

4.1.2 Special-purpose Simulation Programs

Special purpose applications have specialized facilities. These languages are convenient for modeling particular types of situation. Here, the objectives is to make it easier for the non-programmer to write his own model without the need of specialist-programmer assistance. Such programs often incorporate automatic time-keeping routines and sophisticated output facilities. Some of these languages are additions of some general-purpose language such as FORTRAN (CSMP is an example). Others are self contained such as GPSS. These languages, constructed to satisfy the requirements of specific problems, differ in the type and range of their possible application. Sometimes, the modeler's requirements do not correspond with the language; there may be considerable problems in representing the real system. Under such conditions, the languages may dictate the structure of the model rather than the real system.

4.1.3 Requirement of General-purpose or Special-purpose Languages

One should use that program which is readily available and supported by the computer center. Languages are software supported by the computer manufacturer, the computer center systems staff or by the language developer. One should not choose the language for which the compiler is not readily available. In such a case, the advice of the computer center should be sought. The modeler will be discouraged if the language compiler is erratic, slow or expensive to use.

Modern computers can provide the facilities of a special-purpose language. The modeler, using a general-purpose language, will have to develop his own input-output routines, data-checking facilities and time counters.

The range of language should be constrained if the model is used at several different centers. One should use a language which is supported at all centers.

Many special purpose languages have high compilation and running time cost. The high compilation cost may be compensated by a possible saving in time for model development. A general-purpose language may have considerable long-term savings if a model is to be repeatedly used.

Either a general-purpose language or a language such as CSMP (superset of a general-purpose language (FORTRAN), may be recommended, because such programs are well maintained and supported by computer-center staff. Also, the models using these languages are more easily transferred to other centres.

4.1.4 Requirement of Special-purpose Languages

Most of the special-purpose languages are designed for military or industrial processes. These programs may not be suitable for biological or agricultural application. Biological or agricultural systems may be classified into two categories: (1) a system continuously changing through time (growth of a plant); and (2) a system with a sequence of operations, each occurring at a discrete point in time.

The plant growth system faces a number of independent, continuously changing relationships responsible for the development of the entire plant through time (water balance, photosynthesis, respiration, nutrient balance). These relationships are not only interlinked, but they also exhibit feedback conditions. For example, the rate of photosynthesis influences the rate of tissue development and this rate of tissue development affects the leaf area, which again influences the rate of photosynthesis. The rates of processes can be integrated over a period of time to give a level of major plant components which are the state variables in the simulation model. To represent such continuous systems, the key characteristics of such languages is the capacity to integrate sets of differential (rate) equations. These are the languages which may be described as 'integrating' in nature. A number of languages are available at many computer centres, e.g. CSMP and DYNAMO. CSMP has a wide choice of approximate integration procedures. These integration procedures can be selected by the modeler depending on the nature of his model. DYNAMO can only cope with simple rectangular integrations. This is illustrated in the DYNAMO level type of equation.

$$\text{LEVEL}_t = \text{LEVEL}_{t-1} + \text{DT (RATE IN} \pm \text{RATE OUT)}$$

Such an integration can be included in CSMP as one of the options.

The grain-harvesting system is not concerned with integrating rates over time. This system is a sequential one. The grain-harvesting system consists of the following stages: the ripe crop is harvested with a combine harvester, the grain is transported in sacks or in bulk trucks to the farmstead where it may be dried and cleaned before transfer to storage bins. The reason behind studying this type of system is to alleviate bottlenecks of delay in harvesting, transporting, unloading, etc. The delaying factors relate to harvesting capacity, grain transport, unloading capacity, etc. For such a system, it is necessary to advance through time, recording the occurrence of events and executing decisions in the appropriate chronological order. Examples of specialized languages of a non-integrating type are GPSS and SIMSCRIPT. SIMSCRIPT is the more flexible; it may be more preferred in representing sequential systems. It also permits the inclusion of FORTRAN statements between its special statements. CSMP serves both the purposes—intergrating as well as non-integrating due to the inclusion of FORTRAN statements between its special statements. Similarly, SIMSCRIPT can deal with sequential and queueing aspects, while FORTRAN provides a facility for integration.

Many systems may fall into both categories. They may have some aspects of differential equations and some of sequential and queueing processes. Hybrid simulation languages exist to solve such systems, but they are not widely available.

Jones and Luyten (1998) mentioned that the choice of language is not a major issue today. Modules developed in different languages may be linked together, if the models use modular structure and adhere to well-defined data standards. However, each programming language has its own advantages and disadvantages. FORTRAN, BASIC and PASCAL are procedural languages used for biological models. Most existing crop models are programmed in FORTRAN. Procedural languages have excellent capabilities for scientific computations. Biological systems are very complex, having non-linearity in their approach. The procedural languages can easily handle the complex biological systems. These languages can be used on many computer platforms with little or no modification. The procedural languages cannot handle modern user interface programming. There is one of the modern visual programming tools such as Visual Basic that can handle the user interface and data manipulations. New applications of these models are being developed in which the model is treated as a module in an overall software package. There are certain object-oriented programming languages such as SMALLTALK and C++. These new languages have potential for improving the modularity of biological models.

There are a number of new specialized computer simulation languages. The model developers with little or no computer programming skills can implement the models on a computer. The systems of continuous differential equations may be solved by continuous simulation languages.

SLAM systems and FSE packages provide modules that perform most of the computer simulation tasks. Users only have to 'program' their specific model equations and provide parameters and other inputs in order to obtain a simulated behavior of a system. Users can build conceptual models on computer screens using icons (symbols) such as Forrester symbols. System software allows uses to build a conceptual model. Two widely used packages are Stella (High performance systems) and VisSim (Visual solution). Both these packages are available for use on personal computers. These packages guide the users in linking together different icons to depict a system with its storage compartments and flow paths, then ask for needed parameter and other inputs to run the model. These packages will then simulate the system for users to give graph or print results. No programming is needed with these packages. These packages have a distinct advantage. They can evaluate the variations in model structure rapidly. They can also test different hypotheses about the system rapidly. Each computer language has its advantages and disadvantages. Factors affecting the best choice for a particular model are: model complexity, intended users of the model, how it will be used, the need to integrate it with other software and databases, and the need to maintain it.

Problem 1: Depict compartment model symbols adapted from Forester.

Solution:

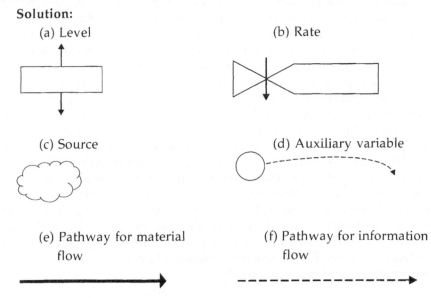

(a) Level

(b) Rate

(c) Source

(d) Auxiliary variable

(e) Pathway for material flow

(f) Pathway for information flow

Problem 2: Find the LEVEL at time *t* with the following conformation: Level at time *t*-1 is 100; DT = 30 minutes; RATE IN = 50; RATE OUT = 30. Calculate using simple rectangular integration technique.

$$LEVEL_t = LEVEL_{t-1} + DT \text{ (RATE IN } - \text{ RATE OUT)}$$

$$= 100 + 30 \ (50 - 30)$$

$$= 100 + 30 \ (20)$$

$$= 100 + 600$$

$$LEVEL_t = 700$$

Problem 3: If dN/dt N = r*N, initial time is zero, intial population is 5, and rate is 0.2, $\Delta t = 0.1$, calculate the population after *t*+1 time and *t*+2 time through Euler (rectangular) integration rule.

$$N \ (t+1) = r*N* \ \Delta t + N_0$$

$$= 0.2 \ {}^*5 * 0.1 + 5$$

$$= 0.1 + 5 = 5.1$$

$$N \ (t+2) = 0.2 * 5.1* \ 0.1 + 5.1$$

$$= 0.102 + 5.1$$

$$= 5.202$$

Problem 4: Create a computer program of a system giving the following characteristics through Euler (rectangular) integration rule:

$$T = 0 \text{ to } 0.9 \text{ Step } 0.1$$

$$\Delta t = 0.1$$

$$N = 5$$

$$\text{Function } dN/dt = rN$$

Solution

Computer Program in BASIC language

$$\varepsilon = \Delta t^2 = (0.1)^2$$

$$\text{For } t = 0 \text{ to } 0.9 \text{ Step } 0.1$$

$$N = t * rN + N + \varepsilon$$

$$\text{Print } N$$

$$\text{Next } t$$

Problem 5: Integrate the population of a living organism through predictor-corrector (trapezoidal) technique. The specifications of the living organism are:

1. Initial population $= N = 5$
2. $r = 0.2$
3. $\Delta t = 0.1$
4. $\dfrac{dN}{dt} = r * N : N_I = N_0 e^{r(I)}$

The analogous equation for exponential growth would be

$$\Delta \hat{N}_I = r * N_I * \Delta t$$
$$= 0.2 * 5 * 0.1$$
$$= 0.1$$

$$\hat{N}_{I \cdot \Delta I} = N_I * \Delta \hat{N}_I$$
$$= 5 + 0.1$$
$$= 5.1$$

$$\Delta \hat{N}_{I \cdot \Delta I} = r * \hat{N}_{I \cdot \Delta I} * \Delta t$$
$$= 0.2 * 5.1 * 0.1$$
$$= 0.102$$

$$N_{I \cdot \Delta I} = N_I + (\Delta \hat{N}_I + \Delta \hat{N}_{I \cdot \Delta I}) / 2$$
$$= 5 + (0.1 + 0.102) / 2$$
$$= 5 + 0.101$$
$$= 5.101$$

So, the Predictor-corrector equation is:

$$N_{I \cdot \Delta I} = N_I + (\Delta \hat{N}_I + \Delta \hat{N}_{I \cdot \Delta I}) / 2$$

$$N_{(I \cdot st)} = N_I + 1/2 (r * N_I * \Delta t + r * N_{I \cdot \Delta I} * \Delta t)$$

$$N_{(I \cdot st)} = N_I + 1/2 (\text{Rate}_I * \Delta t + \text{Rate}_{I + I} * \Delta t)$$

$$N_{(I \cdot \Delta I)} = N_I + 1/2 (\text{Rate}_I + \text{Rate}_{I \cdot \Delta I}) \Delta t$$

Problem 6: Integrate the population of a living organism through Taylor's series expansion. The specifications of the population are as:

1. $N = 5$
2. $\triangle t = 0.1$
3. Integrated relation $= Y_t = Ne^{rt}$
4. $r = 0.2$

$$N_{(t+\triangle t)} = N_t + \triangle t N_t^{(1)} + (\triangle t^2/2) \, N_t^{(2)} + \varepsilon$$

$$= Ne^{rt} + \triangle t \, Nre^{rt} + (\triangle t^2/2) \, Nr^2 \, e^{rt} + \varepsilon$$

$$= Ne^{r(0)} + \triangle t \, Nre^{r(t)} + (\triangle t^2/2) \, Nr^2 \, e^{r(2)} + \varepsilon$$

$$= N + \triangle t \, N*0.2*e^{0.2(0.1)} + (\triangle t^2/2) \, N * .2^2 \, e^{0.2(0.2)} + \varepsilon$$

$$= 5 + 0.1 * 5* 0.2 * 1.02 + 0.01/2 * 5* 0.2^2 * 1.04 + \varepsilon$$

$$= 5 + 0.102 + 0.005 * 5 * 0.04 * 1.04$$

$$= 5 + 0.102 + .005 * 0.208$$

$$= 5.103$$

Since $N_t^{(2)}$ can be approximated by

$$(N_{t+st}^{(1)} - N_t^{(1)})/ \triangle t$$

So,

$$N_{t+\triangle t} = N_t + \triangle t \, N_t^{(1)} + (\triangle t^2/2)* (N_{t+\triangle t}^{(1)} - N_t^{(1)})/ \triangle t + \varepsilon$$

$$= 5 + 0.1 * N* 0.2 \, e^{0.2(0.1)} + 0.005 * (N*0.2*e^{.2(.2)} - N* 0.2 \, e^{.2(.1)})/0.1$$

$$= 5 + 0.102 + 0.005 * (N*0.2*e^{.2(0.2)} - N* 0.2 \, e^{.2(.1)})/0.1$$

$$= 5 + 0.102 + 0.005 * (1.0408 - 1.0202)/0.1$$

$$= 5 + .102 + 0.005* 0.206$$

$$= 5.103$$

Note:
$$Y_t = Ne^{rt}$$

$$Y_{0.1}^{(1)} = Nre^{r(t)}$$

$$= N * 0.2e^{0.2(0.1)}$$

$$Y_{0.2}^{(1)} = N*0.2*e^{0.2^*.2}$$

$$Y_{0.2}^2 = N*(0.2)^2 e^{.2*.2}$$

$$Y_{0.1}^2 = N*(0.2)^2 e^{0.2(0.1)}$$

$$Y_0^0 = N e^{r(0)}$$

$$Y_t^{(1)} = N r e^{rt}$$

Problem 7: Make a computer program for iterative second order Runge-Kutta (Predictor-corrector or trapezoidal method/function/Technique.

```
100   REM EXPONENTIAL GROWTH MODEL
110   REM ITERATIVE SECOND ORDER RUNGE-KUTTA FUNCTION
120   N = 5: K = 0.2: DT = 0.1
130   IM = INT (1/DT) + 0.1))
140   FOR T = 0.1 to 2.1 step 0.1
150   FOR I = 1 TO IM
160   N (1) = N
170   FOR J = 1 TO 3
180   DN (J) = K* N (J)* DT
190   N (J+1) = N + (DN (J) * DN (1)) /2
200   NEXT J
210   N = N(4)
220   NEXT I
230   PRINT T, N
240   NEXT T
250   STOP
```

RUNNH

0.1. 5.05	0.6. 5.31	1.1. 5.61	1.6. 5.95
0.2. 5.10	0.7. 5.37	1.2. 5.68	1.7. 6.02
0.3. 5.15	0.8. 5.43	1.3. 5.74	1.8. 6.09
0.4. 5.20	0.9. 5.49	1.4. 5.81	1.9. 6.17
0.5. 5.26	1.0. 5.55	1.5. 5.88	2.0. 6.25

Problem 8: Create a computer program for Taylor series expansion as an alternative for Runge-Kutta function.

```
5.    REM EXPONENTIAL GROWTH MODEL
```

```
10    REM INTEGRATION THROUGH TAYLOR SERIES EXPANSION
15    DIM N(20)
20    N = 5: R = 0.2: DELT = 0.1: T = 0
25    T = T + 0.1
40    NT = N * (Exp (R *T) ): DNT = NT *R: DNTDT = N*R*EXP
      (R*(T+DELT))
45    N (T+DELT) = NT + DELT * DNT + (DELT²/2) * (DNTDT-DNT)/
      DELT
50    PRINT NT , N (T+DELT), T
52    1F T = > 2 then STOP
55    GO TO 25
```

RUNNH

NT	N(T+DELT)	T	NT	N(T+DELT)	T
5.101	5.204	0.1	6.230	6.356	1.1
5.204	5.309	0.2	6.356	6.484	1.2
5.309	5.416	0.3	6.484	6.615	1.3
5.416	5.525	0.4	6.615	6.749	1.4
5.525	5.637	0.5	6.749	6.885	1.5
5.637	5.751	0.6	6.885	7.024	1.6
5.751	5.867	0.7	7.024	7.166	1.7
5.867	5.986	0.8	7.166	7.311	1.8
5.986	6.107	0.9	7.311	7.459	1.9
6.107	6.230	1	7.459	-	2.0

4.1.5 Recent Softwares Developed

Fridgen et al., (2004) developed MZA (Management Zone Analyst) software for sub-field management zone delineation. Liebig et al., (2004) developed AEPAT (Agro Ecosystem Performance Assessment Tool) software for assessing agronomic and environmental performance.

4.2. GENERALIZED MODEL

Elmasri and Navathe (2000) define the term generalization as the process of defining a generalized entity type from the given entity types. For example, consider the entity types car and truck. They can be generalized into the entity type VEHICLE. Both CAR and TRUCK are now subclasses of the generalized superclass VEHICLE.

Specialization is the process of defining a set of subclasses of an entity type called the superclass of the specialization. The set of subclasses that forms a specialization is defined on the basis of certain distinguishing characteristics of the entities in the superclass. For example, the set of subclasses (wheat, maize, pearlmillet, chickpea, etc.) is a specialization of the superclass CROP that distinguishes among CROP entities based on the taxonomy of each entity. We may have several specialization of the same entity type based on different distinguishing characteristics. For example, another specialization of the crop entity type may yield the set of subclasses (high-yielding crops and low-yielding crops). This specialization distinguishes among crops based on the yielding capacity of the crop (Figure 4.1).

While dealing with Newton's law of cooling, the equation $T = F + Ce^{kt}$ is a generalized model and the equation $T = 76 + -39\ e^{-0.298t}$ is a specialized model.

DBMS Languages

A database is a collection of related data. By data, we mean known facts that can be recorded and that have implicit meaning. A database represents some aspect of the real world, sometimes called the mini-world or Universe of Discourse (UoD). A database is a logically coherent collection of data with some inherent meaning. A random assortment of data cannot be correctly referred to as a database. A database is designed, built, and populated with data for a specific purpose.

Databases and database systems have become an essential component of everyday life in modern society. In the course of a day, most of us encounter several activities that involve some interaction with a database. For example, if we go to the bank to deposit or withdraw funds, if we make a hotel or airline reservation, if we access a computerized library catalog to search for a bibliographic item, or even if we order a magazine subscription from a publisher, chances are that our activities will involve someone accessing a database. Even purchasing an item from a supermarket nowadays in many cases involves an automatic update of the database that keeps the inventory of supermarket items (Elmasri and Navathe, 2000).

Once the design of a database is completed and a DBMS is chosen to implement it, the first order of the day is to specify a conceptual and internal scheme for the database and any mapping between the two. In many DBMSs, where no strict separation of levels is maintained, one language—called the data definition language (DDL)—is used by both the DBA and by database designers to define both schema. The DBMS will have a DDL compiler whose function is to process DDL statements in order to identify descriptions of the schema constructs and to store the schema description in the DBMS catalog.

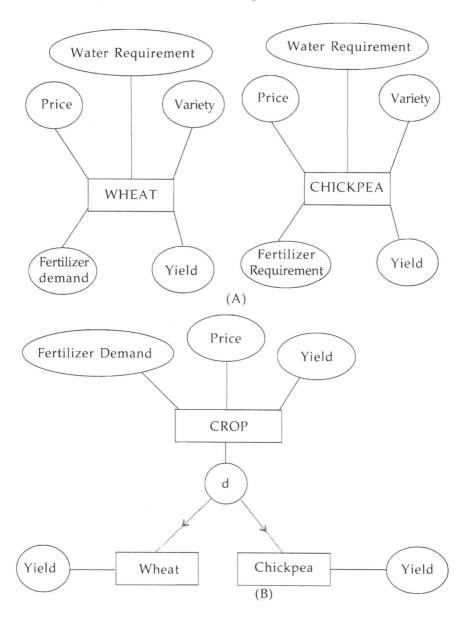

Fig. 4.1   Example of generalization.

(A) Two entity types—wheat and chickpea.

(B) Generalizing wheat and chickpea into crops.

In DBMSs where a clear separation is maintained between the conceptual and internal levels, the DDL is used to specify the conceptual schema only. Another language, the storage definition language (SDL), is used to specify the internal schema. The mappings between the two schema may be specified in either one of these languages. For a true three-schema architecture, we would need a third language, **the view definition language** (VDL) to specify user views and their mappings to the conceptual schema, but in most DBMSs, the DDL is used to define both conceptual and external schemas.

Once the database schemas are compiled and the database is populated with data, the users must have some means to manipulate the database. Typical manipulations include retrieval, insertion, deletion, and modification of the data. The DBMS provides a data manipulation language (DML) for these purposes.

In current DBMSs, the preceding types of languages are usually not considered distinct languages; rather, a comprehensive integrated language is used that includes constructs for conceptual schema definition, view definition, and data manipulation. Storage definition is typically kept separate, since it is used for defining physical storage structures to finetune the performance of the database system, and it is usually utilized by the DBA staff. A typical example of a comprehensive database language is the SQL relational database language, which represents a combination of DDL, VDL, and DML, as well as statements for constraints specification and schema evolution. The SDL was a component in an earlier version of SQL but has been removed from the language to keep it at the conceptual and external level only.

There are two main types of DMLs. A high-level or non-procedural DML can be used on its own to specify complex database operations in a concise manner. Many DBMSs allow high-level DML statements either to be entered interactively from a terminal (or monitor) or to be embedded in a general-purpose programming language. In the latter case, DML statements must be identified within the program so that they can be extracted by a precompiler and processed by the DBMS. A low-level or procedural DML must be embedded in a general-purpose programming language. This type of DML typically retrieves individual records or objects from the data base and processes each separately. Hence, it needs to use programming language constructs, such a looping, to retrieve and process each record from a set of records. For this reason, low-level DML are also called record at a time DMLs. High-level DMLs, such as SQL, can specify and retrieve many records in a single DML statement and hence are called set-at-a-time or set-oriented DMLS. A query in a high-level DML often specifies which data to retrieve rather than how to retrieve it; hence, such languages are also called declarative ones.

DML commands—high-level or low level—are embedded in a general-purpose programming language called the host language and the DML is called the data sublanguage. In object databases, the host and data sub-languages typically form one integrated language—for example, C⁻⁻ with some extensions to 'support data base functionally. Some relational systems also provide integrated languages—for example ORACLE's PL/SQL. On the other hand, a high-level DML used in a stand-alone interactive manner is called a query language (used to describe only retrievals, not updates). In general, both retrieval and update commands of a high-level DML may be used interactively and are, hence, considered part of the query language.

Casual end users typically use a high-level query language to specify their requests, whereas programrs use the DML in its embedded form. For naive and parametric users, usually there are user-friendly interfaces for interacting with the database; these can also be used by casual users or others who do not want to learn the details of a high-level query language (Elmasri and Navathe, 2000, p. 30).

DBMS Interfaces

Menu-based interfaces for browsing

These interfaces present the user with lists of options, called menu, that lead the user through the formulation of a request. Menus do away with the need to memorise the specific commands and syntax of a query language; rather, the query is composed step by step by picking options from a menu that is displayed by the systems. Pull-down menus are becoming a very popular technique in window-based user interfaces. They are often used in browsing interfaces, which allows a user to look through the contents of a database in an exploratory and unstructured manner.

Graphical user interfaces

A Graphical User Interface (GUI) typically displays a schema to the user in diagrammatic form. The user then specifies a query by manipulating the diagram. In many cases, GUIs utilize both menus and forms. Most GUIs use a pointing device, such as a mouse, to pick certain parts of the displayed schema diagram. (Elmasri and Navthe, 2000, pp. 31 and 32).

4.2.1 Specialization and Generalization

Specialization is the process of defining a set of subclasses of an entity type; this entity type is called the superclass of the specialization. The set of subclasses that forms a specialization is defined on the basis of some distinguishing characteristic of the entities in the superclass. For example, the set of subclasses (laboratory, engineer, technician) is a specialization of superclass EMPLOYEE that distinguishes among EMPLOY entities based on the job type of each entity. We may have several specializations of the same entity type based on different distinguishing characteristics. For example, another specialization of the EMPLOYEE entity type may yield the set of subclasses (SALARIED-EMPLOYEE, HOURLY-EMPLOYEE). This specialization distinguishes among employees based on the *method of pay*.

Figure 4.2 shows the manner in which we represent a specialization diagrammatically in an EER diagram. The subclasses that define a specialization are attached by lines to a circle (d=disjoint), which is connected to the superclass. The *subset symbol* on each line connecting a subclass to the circle indicates the direction of the superclass/subclass relationship. Attributes that apply only to entities of a particular subclass— such as Typing Speed of SECRETARY—are attached to the rectangle representing that subclass. These are called **specific attributes** or **local attributes**) of the subclass. Similarly, a subclass can participate in **specific relationship types** such as the HOURLY _ EMPLOYEE subclass participating in the BELONGS _TO relationship in Figure 4.2.

Figure 4.3 shows a few entity instances that belong to subclasses of the (SECRETARY, ENGINEER, TECHNICIAN) specialization. Again, notice that an entity that belongs to a subclass represents *the same real world entity* as the entity connected to it in the EMPLOYEE superclass, even though the same entity is shown twice; for example, e_1, is shown in both EMPLOYEE and SECRETARY in Figure 4.3. As suggested in this figure, a superclass/subclass relationship such as EMPLOYEE/ SECRETARY somewhat resembles a 1:1 relationship at the instance level (Fig. 4.4). The main difference is that in a 1:1 relationship two distinct entities are related, whereas in a superclass/subclass relationships the entity in the subclass is the same real-world entity as the entity in the superclass but playing a specialized role; for example, an EMPLOYEE specialized in the role of SECRETARY, or an EMPLOYEE specialized in the role of TECHNICIAN.

There are two main reasons for including class/subclass relationships and specializations in a data model. The first is that certain attributes may apply to some but not all entities of the superclass. A subclass is defined in order to group the entities to which these attributes apply. The members of the subclass may still share the majority of their attributes with the other members of the superclass. For example, the SECRETARY subclass may have an attribute Typing Speed, whereas the ENGINEER subclass may have an attribute Engineer Type, but SECRETARY and ENGINEER

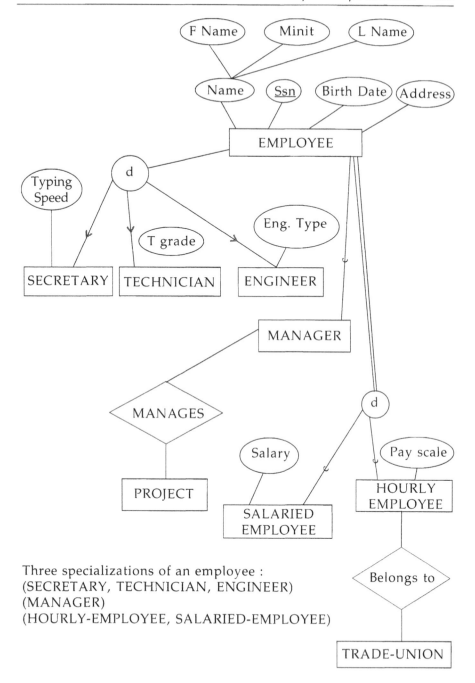

Fig. 4.2 Enhanced Entity Relation (EER) diagram notation for representing specialization and subclasses (From Elmasri and Navathe, 2000).

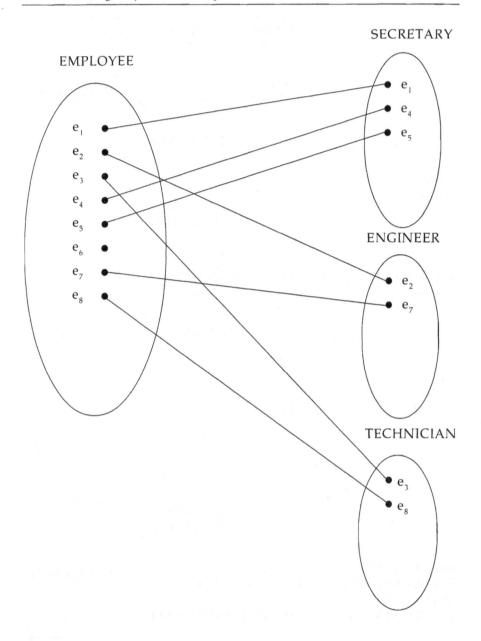

Fig. 4.3 Some instances of the specialization of EMPLOY into the (SECRETARY, ENGINEER, TECHNICIAN set of subclasses).

share their other attributes as members of the EMPLOYEE entity type.

The second reason for using a subclass is that some relationship types may be participated in only by entities that are members of the specific subclass. For example, only HOURLY-EMPLOYEE of EMPLOYEE relating the subclass to an entity type TRADE-UNION via the BELONGS-TO relationship type, as illustrated in Figure 4.2.

In summary, the specialization process allows us to do the following:

* Define a set of subclasses of an entity type.
* Establish additional specific attributes with each subclass.
* Establish additional specific relationship types between each subclass and other entity types of other subclass.

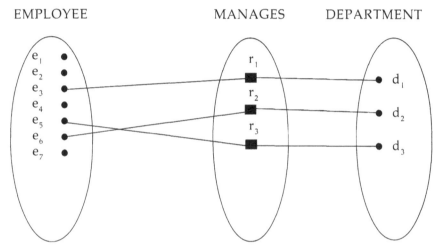

Fig. 4.4 The 1:1 relationship MANAGES with partial participation of EMPLOYEE and total participation of DEPARTMENT.

Generalization

We can think of a reverse process of abstraction where we suppress the differences among several entity types, identify their common features and generalize them into a single superclass of which the original entity types are special subclasses. For example, consider the entity types CAR and TRUCK shown in Figure 4.5a. They can be generalized into entity type VEHICLE, as shown in Figure 4.5b. Both CAR and TRUCK are now subclasses of the generalized superclass VEHICLE. We use the term generalization to refer to the process of defining a generalized entity type from the given entity types.

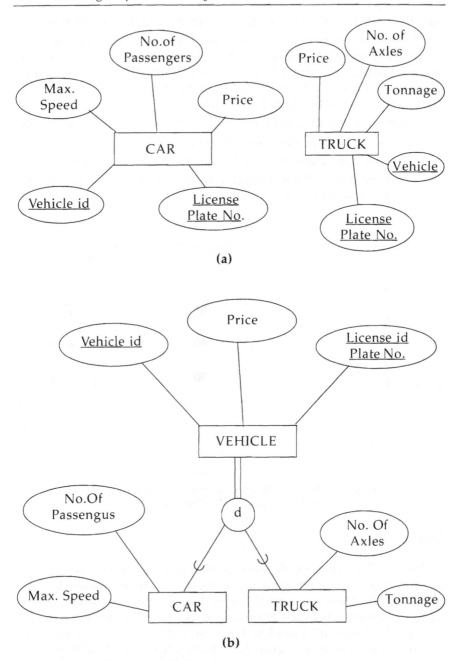

Fig. 4.5 Examples of generalization.
(a) Two entity types, CAR and TRUCK.
(b) Generalizing CAR and TRUCK into VEHICLE.

Notice that the generalization process can be viewed as being functionally the inverse of the specialization process. Hence, in Figure 4.5, we can view CAR and TRUCK as a specialization of VEHICLE, rather than viewing VEHICLE as a generalization of CAR and TRUCK. Similarly, in Figure 4.2, we can view EMPLOYEE as a generalziation of SECRETARY, TECHNICIAN, and ENGINEER.

So far we have introduced the concepts of subclasses and super-classes/subclass relationships, as well as the specialization and generalization processes. In general, a superclass or subclass represents a collection of entities of the same type and, hence, also describes an entity type; that is why superclasses and subclasses are shown in rectangles in EER diagrams (like entity types) (Elmasri and Navathe, 2000, pp 76-79).

4.2.2 Constraints and Characteristics of Specialization and Generalization

Constraints on specialization/generalization. For brevity, our discussion refers only to specialization even though it applies to both specialization and generalization. In general, we may have several specializations defined on the same entity type (or superclass) as shown in Figure 4.2. In such a case, entities may belong to subclasses in each of the specialization. However, a specialization may also consist of a single subclass only, such as the (MANAGER) specialization in Figure 4.2. Here, we do not use the circle notation.

In some specializations, we can determine exactly those entities that will become members of each subclass by placing a condition on the value of some attribute of the superclass. Such subclasses are called predicate-defined (or condition-defined) subclasses. For example, if the EMPLOYEE entity type has an attribute Job Type, as shown in Figure 4.6, we can specify the condition of membership in the SECRETARY subclass by the predicate (Job Type = 'Secretary'), which we call the defining predicate of the subclass. This condition is a constraint specifying that members of the SECRETARY subclass must satisfy the predicate and that all entities of the EMPLOYEE entity type whose attribute value for Job Type is Secretary must belong to the subclass. We display a predicate-defined subclass by writing the predicate condition next to the line that connects the subclass to the specialization circle.

If all subclasses in a specialization have the membership condition on the same attribute of the subclass, the specialization itself is called an attribute-defined specialization, and the attribute is then called the defining attribute of the specialization. We display an attribute-defined specialization, as shown in Figure 4.6, by placing the defining attribute name next to the arc from the circle to the superclass.

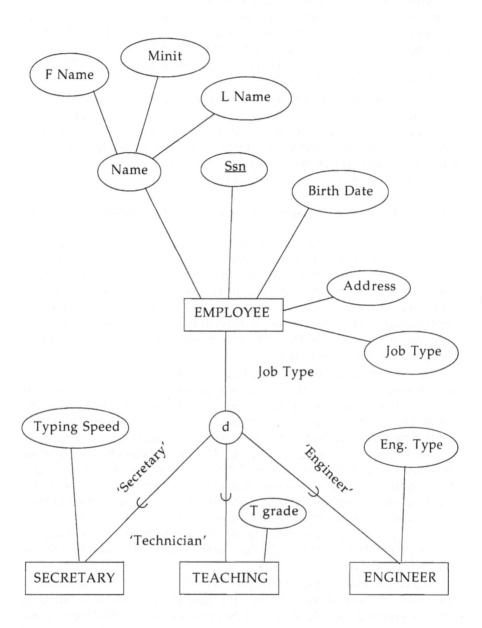

Fig. 4.6 An attribute-defined specialization on the Job Type attribute of EMPLOYEE.

When we do not have a condition of the determining membership in a subclass, the subclass is called user defined. Membership in such a subclass is determined by the database users when they apply the operation to add the entity subclass; hence, membership is specified individually for each entity by the user, not by any condition that may be evaluated automatically.

Two other constraints may apply to a specialization. The first is the disjointness constraints, which specify that the subclasses of the specialization must be disjointed. This means that an entity can be a member of at most one of the subclasses of the specialization. An attribute-defined specialization implies the disjoint constraint if the attribute used to define the membership predicate is single valued. Figure 4.6 illustrates this case, where the 'd' in the circle stands for disjoint. We also use the 'd' notation to specify the constraint that user-defined subclass of a specialization must be disjoint, as illustrated by the specialization (HOURLY-EMPLOYEE, SALARIED-EMPLOYEE) in Figure 4.2. If the subclasses are not constrained to be disjoint, their sets of entities may overlap; that is, the same (real-world) entity may be a member of more than one subclass of the specialization. This case, which is the default, is displayed by placing an 'O' in the circle, as shown in Figure 4.7.

The second constraint on specialization is called the completeness constraint, which may be total or partial. A total specialization constraint specifies that every entity in the superclass must be a member of some subclass in the specialization. For example, if every EMPLOYEE must be either an HOURLY-EMPLOYEE or a SALARIED-EMPLOYEE, then the specializations (HOURLY-EMPLOYEE, SALARIED-EMPLOYEE) of Figure 4.2 is a total specialization of EMPLOYEE, this is shown in EER diagrams by using a double line to connect the superclass to the circle. A single line is used to display a partial specialization, which allows an entity not to belong to any of the subclasses. For example, if some EMPLOYEE entities do not belong to any of the subclass (SECRETARY, ENGINEER, TECHNICIAN) of Figures 4.2 and 4.6 then that specialization is partial. Notice that the disjointness and completeness constraints are independent. Hence, we have the following four possible constraints on specialization:

*   Disjoint, total
*   Disjoint, partial
*   Overlapping, total
*   Overlapping, partial

Of course, the correct constraint is determined from the real world meaning that applies to each specialization. However, a superclass that was identified through the generalization process usually is total, because the superclass is derived from the subclasses and hence contains only the entities that are in the subclasses.

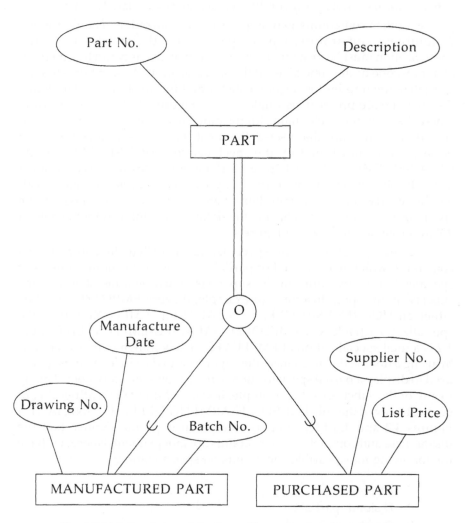

Fig. 4.7 Notation for specialization with overlapping (non-disjoint).

Certain insertion and deletion rules apply to specialization (and generalization) as a consequences of the constraints specified earlier. Some of these rules are as follows:

* Deleting an entity from a superclass implies that it is automatically deleted from all the subclasses to which it belongs.

* Inserting an entity in a superclass implies that the entry is mandatorily inserted in all *predicate-defined (or attribute defined)* subclasses for which the entity satisfies the defining predicate.

* Inserting an entity in a superclass of a total specialization implies that the entity is mandatorily inserted in atleast one of the subclasses of the specialization.

Specialization/generalization hierarchies and lattices

A subclass itself may have further subclasses, forming a hierarchy or lattice of specializations. For example, in Figure 4.8, ENGINEER is a subclass of EMPLOYEE and also a superclass of ENGINEERING-MANAGER. This represents the real-world constraint that every engineering-manager is required to be an engineer. A specialization hierarchy has the constraint that every subclass participates as a subclass in only one class/subclass relationship. In contrast, for a specialization lattice, a subclass can be a subclass in more than one class/subclass relationship. Hence, Figure 4.8 is a lattice.

Figure 4.9 shows another specialization lattice of more than one level. This may be part of a conceptual schema for a university database. Notice that this arrangement would have been a hierarchial except for the STUDENT-ASSISTANT subclass, which is a subclass in two distinct class/subclass relationships. In Figure 4.10, all person entities represented in the database are members of the person entity type, which is specialized into the subclasses (EMPLOYEE, ALUMNUS, STUDENT). This specialization is overlapping; for example, an alumnus may also be an employee as well as a student pursuing an advanced degree. The subclass STUDENT is superclass for the specialization (GRADUATE STUDENT, UNDERGRADUATE), while EMPLOYEE is superclass for the specialization [STUDENT ASSISTANT, FACULTY, STAFF]. Notice that STUDENT ASSISTANT is also a subclass of student. Finally, STUDENT ASSISTANT is superclass for the specialization into [RESEARCH ASSISTANT, TEACHING ASSISTANT].

In such a specialization lattice or hierarchy, a subclass inherits the attributes not only of its direct superclass but also of all its predecessor superclasses *all the way to the root* of the hierarchy, or lattice. For example, an entity in GRADUATE STUDENT inherits all the attributes of that entity as a student and as a person. Notice that an entity may exist in several

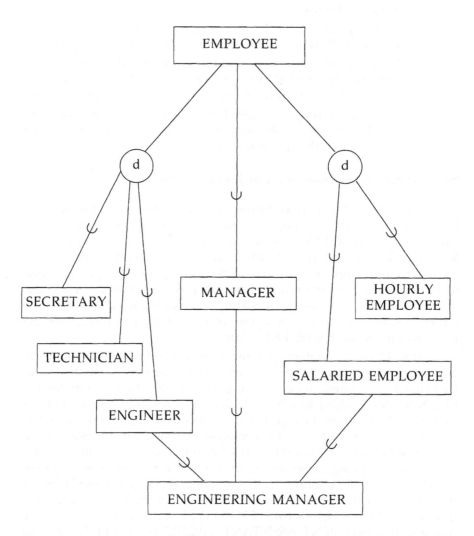

Fig. 4.8 A specialization lattice with the shared subclass ENGINEERING-MANAGER
(*Source*-Elmasri and Navathe, 2000).

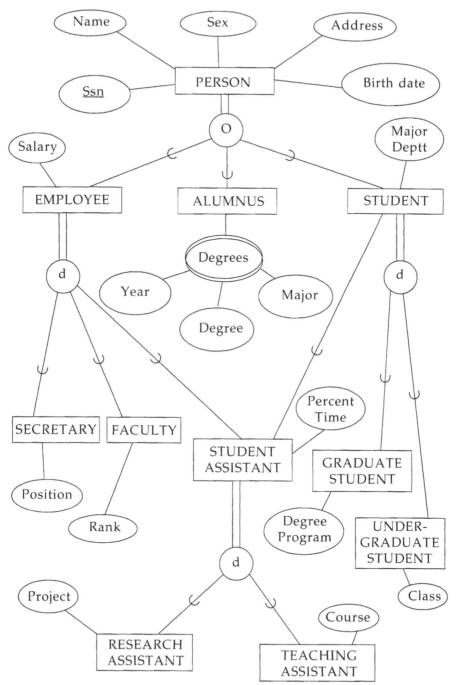

Fig. 4.9 A specialization lattice (with multiple inheritance) for a university database
(*Source :* Elmasri and Navathe, 2000).

leaf nodes of the hierarchy, where a leaf node is a class that has no subclass of its own. For example, a member of GRADUATE STUDENT may also be a member of RESEARCH ASSISTANT.

A subclass with more than one superclass is called a shared subclass. For example, if every ENGINEERING MANAGER must be an ENGINEER—but must also be SALARIED EMPLOYEE and a MANAGER—then the ENGINEERING MANAGER should be a shared subclass of all three superclasses (Figure 4.8). This leads to the concept known as multiple inheritance, since the shared subclass ENGINEERING MANAGER directly inherits the attributes and relationships from multiple classes. Notice that the existence of at least one shared subclass leads to a lattice (and hence to multiple inheritance); if no shared subclass existed, we would have a hierarchy rather than a lattice. An important rule related to multiple inheritance can be illustrated by the example of the shared subclass STUDENT ASSISTANT in Figure 4.9, which inherits attributes from both EMPLOYEE and STUDENT. Here, both EMPLOYEE and STUDENT inherit the same attributes from a person. The rule states that if an attribute (or relationship) originating in the same superclass (person) is inherited more than once via different paths (EMPLOYEE and STUDENT) in the lattice, then it should be included only once in the shared subclass (STUDENT ASSISTANT). Hence, the attributes of person are inherited only once in the STUDENT ASSISTANT subclass of Figure 4.9.

It is important to note here that some inheritance mechanisms do not allow multiple inheritance (shared subclasses). In such a model, it is necessary to create additional subclasses to cover all possible combinations of classes that may have the same entity will belong to all these classes simultaneously. Hence, any overlapping specialization would require multiple additional subclasses. For example, in the overlapping specialization of PERSON into [EMPLOYEE, ALUMNUS, STUDENT] (or (E, A, S) for short), it would be necessary to create seven subclasses of PERSON: E, A, S, E-A, E-S, A-S and E-A-S in order to cover all possible type of entities. Obviously, this can lead to extra complexity.

It is also important to note that some inheritance mechanisms that allow multiple inheritance do not allow an entity to have multiple types, and hence an entity can be a member of only one class. In such a model, it is necessary to create additional shared subclass as leaf nodes to cover all possible combinations of classes that may have some entity belonging to all these classes simultaneously. Hence, we would require the same seven subclasses of person.

Although we have used specialization to illustrate our discussion, similar concepts apply equally to generalization. Hence, we can also speak of generalization hierarchies and generalization lattices.

Utilizing specialization and generalization in conceptual data modeling

We now elaborate on the differences between the specialization and generalization processes during the conceptual database design. In the specialization process, we typically start with an entity type and then define subclasses of the entity type by successive specialization, i.e., we repeatedly define more specific groupings of the entity types. For example, when designing the specialization lattice in Figure 4.9, we may first specify an entity type PERSON for a university database. Then we discover that three type of persons will be represented in the database: university employees, alumni, and students. We create the specialization (EMPLOYEE, ALUMNI, STUDENT) for this purpose and choose the overlapping constraint because a person may belong to more than one of the subclasses. We then specialize EMPLOYEE further into (STAFF, FACULTY, STUDENT ASSISTANT), and specialize STUDENT into (GRADUATE STUDENT, UNDERGRADUATE STUDENT). Finally, we specialize STUDENT ASSISTANT into RESEARCH ASSISTANT, TEACHING ASSISTANT. This successive specialization corresponds to a **top-down conceptual refinement process** during conceptual scheme design. So far, we have a hierarchy, we then realize that STUDENT ASSISTANT is a shared subclass since it is also a subclass of student leading to the lattice.

It is possible to arrive at the same hierarchy or lattice from the other direction. In such a case, the process involves generalization rather than specialization and corresponds to a **bottom up conceptual synthesis.** In this case, designers may first discover entity types such as STAFF, FACULTY, ALUMNUS, GRADUATE STUDENT, UNDERGRADUATE STUDENT, RESEARCH ASSISTANT, TEACHING ASSISTANT and so on, then they generalize (GRADUATE STUDENT, UNDERGRADUATE STUDENT) into STUDENT, they generalize (RESEARCH ASSISTANT, TEACHING ASSISTANT) into STUDENT ASSISTANT; then they generalize (STAFF, FACULTY, STUDENT ASSISTANT into EMPLOYEE); and finally they generalize (EMPLOYEE, ALUMNUS, STUDENT) into person (Figure 4.10).

In structural terms, hierarchies or lattices resulting from either processes may be identical; the only difference relates to the manner or order in which the scheme superclasses and subclasses were specified. In practice, it is likely that neither the generalization process nor the specialization process is followed strictly, but a combination of the two processes is employed. In this case, new classes are continually incorporated into a hierarchy or lattice as they become apparent to users and designers. Notice that the notion of representing data and knowledge by using superclass/subclass hierarchies and lattices is quite common in

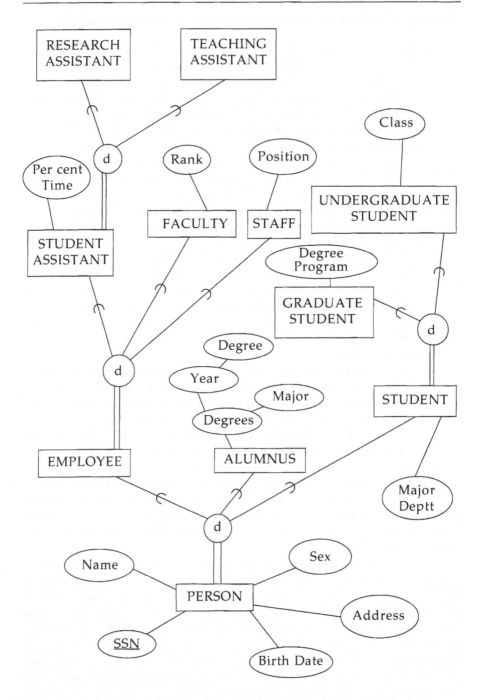

Fig. 4.10 A generalization lattice (with multiple inheritance) for university database.

knowledge-based systems and expert systems, which combine database technology with artificial intelligence technology. For example, frame-based knowledge representation schemes closely resemble class hierarchies. Specialization is also common in software engineering design methodologies based on the object-oriented paradigm.

Generalized/specialized model: Top-down design is a specialization process. Reverse of it is the generalization processes. In a specialization process it means top-down design. Figure 4.11 is a standard form of the specialized model. A top-down design simply involves breaking a large problem into smaller subproblems that can be dealt with individually.

Figure 4.12 is a standard form of the generalized model, also known as bottom-up design. Initialization module, input module, simulation module and output module are the subclasses of the superclass of model (Also see Figure 4.13).

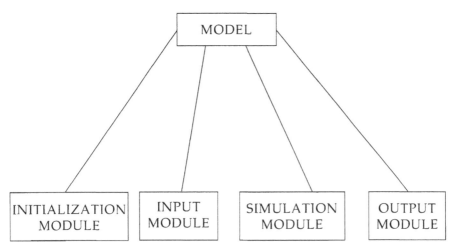

Fig. 4.11 A specialization hierarchy (with single inheritance) for a standard simulation - model design (*Source*: Dent and Blackie (1979).

Note that we are not involved at this stage with computer programs, but are constructing a design which can then be readily programed.

Some guidelines may prove to be useful (Dent and Blackie, 1979).

1. A systematic approach is necessary. First, specify the inputs, outputs and functions of each subproblem or module. Remember, the model is to be a hierarchial structure and lateral communication should only be possible through a higher-level module.

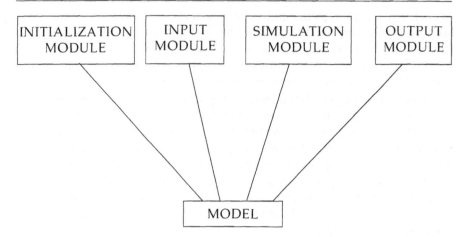

Fig. 4.12 A generalized hierarchy (with single inheritance) for a standard simulation model design.

2. Design the model from the highest-level module down. It is often useful to produce a basic model design and then develop in detail a design for one module or sub module. The module user then programs and tests the design and, if satisfied, moves to designing in detail a further part of the model. Just as it is easier to understand a problem by slowly breaking it down into smaller pieces, so also is it simple to build a model by building the broad outlines (or high-level modules) and filling in the details (or low-level modules) as experience and knowledge of the system develop. The modifications to the structure of the model are easier in a top-down design than if the model has been built from the lowest model upwards. Robustness and flexibility are more likely in models built using the top-down concept.

4.3 SOFTWARE SPECIFICATION

4.3.1 Command Language

4.3.1.1 Data Manipulating Language for the Hierarchial Model

Now we shall discuss the Hierarchical Data Manipulation Language (HDML).This is a record at-a-time language. It introduces the concepts of a hierarchial database manipulation language. A general-purpose programming language (host language) imbeds the commands of the language.

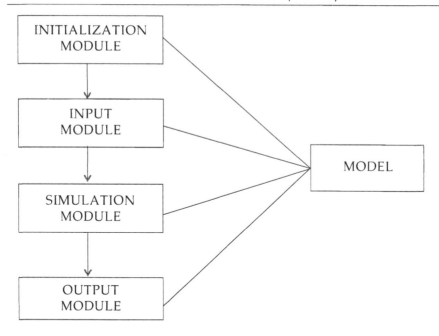

Fig. 4.13 A generalized model showing the four subclasses and one superclass.

The concept of hierarchial sequence is the basis of the HDML. The current database record is the last record accessed by the command. A pointer to the current record is maintained by the DBMS. Subsequent database commands proceed from these current record and may define a new current record, depending on the type of command.

4.3.1.1.1 The GET Command

The HDML command for retrieving a record is the GET command. There are many variations of GET; the structure of two of these variations is as follows, with optional parts enclosed in brackets [   ].

    *   GET FIRST < record type name > [WHERE < Condition>]

    *   GET NEXT < <u>record type name</u> > [WHERE < Condition>]

The simplest variation is the GET FIRST Command, which always starts by searching the database from the beginning of the hierarchial sequence until it finds the first record occurrence of < record type name > that satisfies < condition >. This record also becomes the current of database, current of hierarchy, and current of record type and is retrieved into the corresponding variable. For example, to retrieve the "first" EMPLOYEE record in the hierarchial sequence whose name is John Smith, we write Ex1:

EX 1: $GET FIRST EMPLOYEE WHERE FNAME =
'JOHN' AND LNAME = 'SMITH';

The DBMS uses the condition following WHERE to search for the first record in the hierarchial sequence that satisfies the condition and is of the specified record type. If more than one record in the database satisfies the WHERE condition and we want to retrieve all of them, we must write a looping construct in the host program and use the GET NEXT command. We assume that the GET NEXT begins its search from the current record of the record type specified in GET NEXT, and it searches forward in the hierarchial sequence to find another record of the specified type satisfying the WHERE condition. For example, to retrieve records of all EMPLOYEES whose salary is less then $20,000 and obtain a printout of their names, we can write the program segment shown in Ex 2:

EX 2: $GET FIRST EMPLOYEE WHERE SALARY < '20000.00';
While DB STATUS = 0 do
begin
writeln (P - EMPLOYEE. F NAME, P-EMPLOYEE .LNAME);
$ GET NEXT EMPLOYEE WHERE SALARY < '20000'
end;

In Ex 2, the while loop continues until no more EMPLOYEE records in the database satisfy the WHERE condition; hence the search goes through to the last record in the database (hierarchial sequence). When no more records are found, DB-STATUS becomes none zero, with a code indicating "end of database reached", and the while loop terminates.

4.3.1.1.2 The GET PATH and GET NEXT WITHIN PARENT Retrieval Commands

So far, we have considered retrieving single records by using the GET command. But when we have to locate a record deep in the hierarchy, the retrieval may be based on a series of conditions on records along the entire hierarchial path. To accommodate this, we introduce the GET-PATH command.

GET (FIRST |NEXT) PATH < hierarchial path > [WHERE < condition>]

Here, <hierarchial path> is a list of record types that starts from the root along a path in the hierarchial schema, and <condition> is a Boolean expression specifying conditions on the individual record types along the path. As several record types may be specified, the field names are prefixed by the record type names in <condition>. For example, consider the following query "List the Last name and birth dates of all employ-dependent pairs, where both have the first name John". This is shown in Ex 3:

Ex 3: $get first path employee, Dependent
WHERE EMPLOYEE. FNAME = 'John' and DEPENDENT
DEP NAME = 'John';
while DB - STATUS = 0 do
begin
Writeln (P - EMPLOYEE, FNAME, P - EMPLOYEE. BDATE,
P - DEPENDENT. BIRTHDATE);
$ GET NEXT PATH EMPLOYEE, DEPENDENT
WHERE EMPLOYEE. FNAME = 'John'
AND DEPENDENT. DEPNAME = 'John'
end;

We assume that a GET PATH command retrieves all records doing the specified path into the user work variable, and the last record along the path becomes the current database record. In addition, all records along the path become the current records of their respective record types.

Another common type of query is to find all existing records of a given type that have the same parent record. In this case, we need the GET NEXT WITHIN PARENT Command, which can be used to loop through the child records of a parent record and has the following format:

GET NEXT < CHILD RECORD TYPE NAME>

WITHIN [VIRTUAL] PARENT [Parent record type name>]

[WHERE <condition>]

This command retrieves the next record of the child record type by searching forward from the current of the child record type for the next child record owned by the current parent record. If no more child records are found, DB - STATUS is set to nonzero value to indicate that "there is no more record of the specified child record type that have the same parent as the parent current parent record". The <parent record type name> is optional, and the default is the immediate (real) parent record type of <child record type name>. For example, to retrieve the names of all projects controlled by the 'Research' department, we can write the program segment shown in Ex 4:

EX 4: $GET FIRST PATH DEPARTMENT, PROJECT
WHERE DNAME = 'Research';
(*The above establishes the 'Research' DEPARTMENT record
as current parent of type DEPARTMENT, and retrieves the first child project record under that DEPARTMENT record*)
while DB - STATUS = 0 do
begin
Writeln (P - PROJECT .PNAME);
$ GET NEXT PROJECT WITHIN PARENT
end;

4.3.1.1.3 HDML Commands for Update

The HDML commands for updating a hierarchial database are shown in Table 4.1, along with the retrieval command. The INSERT command is used to insert a new record. Before inserting a record of a particular record type, we must place the field values (data item—integer, real, or string) of the new record in the appropriate user work area program variable.

The INSERT command enters a record into the database, its hierarchial schema, and its record type. If it is a root record, it creates a new hierarchial occurrence tree with the new record as root. The record is then inserted in the hierarchial sequence in the order specified by any ORDER by fields in the schema definition.

Table 4.1 Summary of HDML Commands

RETRIEVAL	
GET	Retrieve a record into the corresponding program variable and make it the current record. Variations include GET FIRST, GET NEXT, GET NEXT WITHIN PARENT, and GET PATH.
RECORD UPDATE	
INSERT	Store a new record in the database and make it the current record.
DELETE	Delete the current record (and its subtree) from the database.
REPLACE	Modify some fields of the current record.
CURRENCY RETENTION	
GET HOLD	Retrieve a record and hold it as the current record so it can be subsequently be deleted or replaced.

To insert a child record, we should make its parent, or one of its sibling records, the current record of the hierarchial schema before issuing the INSERT command. We should also set any virtual parent pointers before inserting the record.

To delete a record from the database, we first make it the current record and then issue the DELETE command. The GET HOLD is used to make the record the current record, when the HOLD key word indicates to the DBMS that the program will delete or update the record just retrieved. For example, to delete all male EMPLOYEEs, we can use Ex5, which also lists the deleted employee name <before> deleting their records.

Ex 5. $GET HOLD FIRST EMPLOYEE WHERE SEX = 'M';
while DB - STATUS = 0 then
begin

Writeln (P_EMPLOYEE; LNAME , P_ EMPLOYEE. FNAME);
$ DELETE EMPLOYEE;
$GET HOLD NEXT EMPLOYEE WHERE SEX = 'M';
end;

4.3.1.1.4 IMS—A Hierarchial DBMS

IMS is one of the earliest DBMSs, ranking as the dominant system in the commercial market for support of large-scale accounting and inventory systems. IBM manuals refer to the full products as IMS/VS (Virtual storage), and typically the complete product is installed under the MVS operating system. IMS DB/DC is the term used for installations that utilize the products'own subsystems to support the physical database (DB) and provide data communication (C).

However, other important versions exist that support only the IMS data language—Data language One (DL/1). Such DL/1–only configurations can be implemented under MVS, but they may also use the DOS/VSE operating system. These systems issue their calls to VSAM files and use IBM's Customer Information Control System (CICS) for data communications. The trade-off is a sacrifice of support features for the sake of simplicity and improvement throughout.

A number of versions of IMS have been marketed to work with various IBM operating systems, including (among the recent systems) OS/VSI, OS/VS2, MVS, MVS/XA, and ESA. The system comes with various options. IMS runs under different versions on the IBM 370 and 30XX family of computers. The data definition and manipulation language of IMS is DL/1. Application programs written in COBOL, PL/1, FORTRAN, and BAL (Basic Assembly Language) interface with DL/1 (Elmasri and Navathe, 2000 pp.949-955).

4.3.2 Program

Computers used at each step in the research program: analysis of sampling data, regression analysis and analysis of variance, tests of the suitability of various models and fitting of parameter values, and finally the development of management strategies through simulation and optimization, can solve the realistically complex problems in ecology and resource management. The other approach other than computers prevents the solution of the complexity of the models and the volume of algebra required by statistical procedures (Watt, 1968).

Computers and the languages are of tremendous importance to ecologists and resource managers for two reasons. The first reason (widely understood) originates from the well-publicized and wonderful properties of computer circuits and memory components. Everyone is now aware that the most powerful available computers can do about 4×10^9 additions a second, and recall from fast-access core-memory any one of 512×10^6 numbers. These machines can read and write on a CD at 4.056×10^8 digits a second. The second reason arises from the conceptual nature of the various branches of mathematics. Each of these (the calculus of Newton and Leibniz, the geometry of Descartes, the algebra of Boole, and the set theory of Cantor) was designed to deal with a certain category of problems. Special qualities of certain branches of mathematics bring them into existence particularly appropriate for the occasion to deal with situations where the variables can vary continuously along a range of values; these branches are alluded to as infinitesimal mathematics. Other branches of mathematics are suitable where variables can assume two or more discrete values; they are alluded to collectively as finite mathematics. Complex happenings in agroecology and crop production resource management complicate in both kinds of processes. To mark sufficiently the real rational structure of such questions, a kind of mathematical language is needed that uses the descriptive characteristics of many different types of mathematics. Suitably, the false-algebraic rules and principles applied by scientists to submit cords of instructions to computers have corrected this feature by design. This is true of COBOL, ALGOL, FORTRAN, BASIC, PL/1, C, C``, Visual Basic and all their related host languages.

It is now necessary to describe the importance of pseudoalgebraic codes. For every kind of developed computer circuitry, numerical or 'machine language' instruction rules had to be developed so the programmer could inform the computer to carry out a unit operation, such as 'Reset the arithmatic register to zero and add to it the number stored in the physical location marked clearly by numerical address number XXXX'. Very soon it became clear that it was too time-taking a process, since many true questions can only have their solutions marked out in terms of several thousand equations, which would need five to ten times as many machine-language instructions. The computer manufacturers spent huge amounts of mathematical manpower to create mathematical languages such as FORTRAN and ALGOL, that were question-oriented, or user-oriented, rather than intended to trigger the circuitry of the computers. Successions of instructions written in these languages are entered into the computer where compiler programs change them to successions of machine-language instructions that verily carryout the operation. It was in the interests of computer manufacturers to ensure that the user-understood false algebraic codes were as strong and adaptable mathematical languages as could be thought out or planned. Thus, languages such as FORTRAN had the characteristics of many different types of mathematical languages even then when first appeared, and additional characteristics have been added since. The computer language PL/1 can undertake a variety of mathematical operations, and explain many different kinds of processes, accurately because the pseudoalgebraic rules were planned to be very powerful and adaptable; they help themselves to realistic explanation of difficult biological processes to a larger extent than any other type of mathematical language yet planned.

4.3.2.1 Flowcharting

The simple technique for communicating between people is flowcharting. It is a diagrammatical display of the logic used on computer. In flowcharting, no specific programming language is employed. Flowcharts may be understood by individuals who do not have specific familiarity with a computer or any computer programming language. Flowcharting may be learned since it involves only a few simple rules. It helps the programmer in programing the information handled by computer. It is a graphic technique. It may be glanced at more quickly than any other way of representing a computer program.

Flowcharts are useful at all stages on the development of simulation models. To define the program, a rudimentary flowchart is used first. A rudimentary flow chart is the sketch of basic input, process, output steps. The sequence of steps is established, as the flow of program takes form. The advantage of rough flowcharts is to test the correctness of the logic employed. After sketching the basically correct flowcharting, the flowchart can be translated into a programming language in order to be run on the computer. The flowcharts may be used again in finding and debugging mistakes in programming language. After the operational program has been obtained, flowcharts should be redrawn using a pattern to trace the lines showing the shape or boundary. The final corrected flowchart then becomes the part of the documentation for the simulation model. Flowcharts may be used to explain to others the logic employed. Flowcharts may also be useful to the modeler for modification of the program if required at a later date. The flowcharting technique and rules for their use have been summarized by Spain (1982).

4.3.2.1.1 General Flowcharting Rules

1. Use simple terminology. Be brief and to the point but unambiguous.
2. Keep the flowchart as simple, clean and clear as possible.
3. Keep the general flow pattern from top to bottom, from left to right. Reverse flows should be marked by arrow heads for clarity. Alternatively, connectors may be employed for reverse flow.

4. The start terminal should be at the top left, and the stop terminal should be at the lower right. Entry connectors should project to the left, and exit connectors should project down or to the right.

5. Use connectors to avoid crossing of flowlines.

6. Draw only one entrance flowline to any outline symbol. (Where more than a single flowline is involved, join them prior to the entry point.)

7. Only the decision block and the incrementing block should have more than one exit flowline. On the decision block, these should be labeled with a 'yes' or 'no' answer to the question posed in the decision outline.

8. Processes should be expressed with algebraic symbolism except that the 'Replaced by' ($\blacktriangleleft$) symbol may be employed instead of the equals symbol.

4.3.2.1.2 Flowchart Symbols and Their Use

Terminal blocks are used to indicate where programs start or stop. Start is usually at top-left and stop is usually down or to the right (Charts 1 and 2).

4.3.2.1.3 Examples of Simple Flowcharts

(a)   Population growth flowchart
Draw a flowchart for equation
     $N_t = N_0.e^{kt}$ will permit you to simulate unlimited growth of bacteria. Increment time from 0 to 50 hours. Assume $k = 0.02/hr$ and $N_0 = 2/ml$. Prepare a graph of the population density at various times upto 50 hrs. It involves the programming of the population growth equation.

$$\int_{N_0}^{N_t} \frac{dN}{N} = \int_0^t kdt$$

The flow chart is given in Figure 4.14

Figure 4.14 contains the basic steps important to most flowcharts:

1. A terminal block outline indicates the start.

2. The read block is used to indicate the list of constants or variables which need to be defined in order for the program to operate.

3. The Z connector permits re-entry into the loop from below.

4. The incrementing block shows that time is to be incremented from 0 to t_{max} (50hrs.) with Δt (1 hr) as the increment size.

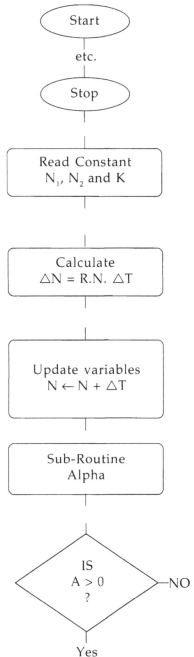

Read blocks are used to indicate the assignment of numerical values to constants and initial values to variables.

Process blocks are used to indicate computational operations. They would contain the equations that normally appear in LET statements of BASIC.

Sub-routine blocks are used for sections of the program which are to be used more than once.

Decision blocks are used for conditional branching of the program. These are analogous to IF.....THEN statements in BASIC.

Chart 1

Incrementing or decrementing blocks are used for looping a specified number of times. These are analogous to FOR -----NEXT statements in BASIC.

(z) ── Increment t from zero to t_{max} stepping Δt ── (STOP)

Preparation blocks are used for initializing a variable or other housekeeping operations.

$D = 0$

Connectors are used to keep the flow chart from being confused with flowlines. They are used mainly for looping back to the top of the page.

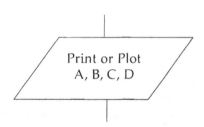

(W) ──── etc.

(W)

Output blocks are used to indicate data output to either a video screen or printer. Output may be a data listing, a graph, or a histogram.

Print or Plot
A, B, C, D

Directions of flow are indicated by flowlines. Arrowheads are not needed if direction of flow is down or to the right.

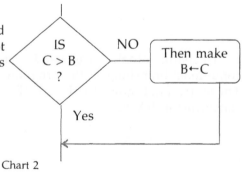

IS C > B ?   NO   Then make B←C

Yes

Chart 2

5. When t_{max} is exceeded, the program exits to the stop terminal.
6. If t_{max} is not exceeded, the program proceeds to the process block and calculates population (P) according to the equation given.
7. The resulting data are printed or plotted and control is returned to point Z by the connector.

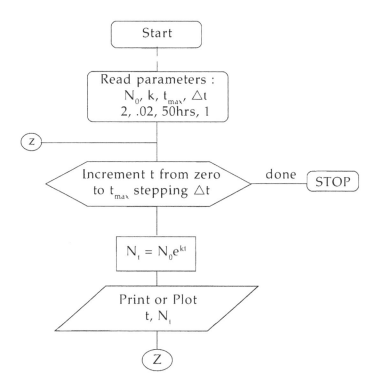

Fig. 4.14 Flowchart of population growth program.

(b) Flowchart of pH/Enzyme activity model

The flowchart in Figure 4.15 of the enzyme-pH model equation

$$\bar{y} = \frac{\alpha}{1 + \dfrac{K_2}{[H]} + \dfrac{[H]}{K_1} \, \alpha \left(1 + \dfrac{K_4}{[H]} + \dfrac{[H]}{K_3} \right)}$$

determines the saturation fraction (essentially the fraction of maximum velocity) for different pH values under two different concentrations of substrate relative to the dissociation constant of the enzyme-substrate compound ($\alpha = 1$ and $\alpha = 10$). $K_1 = 0.0001$; $K_2 = 0.000001$; $K_3 = 0.0005$ and $K_4 = .000001$. Vary the pH from .1 to 14, then convert to [H$\cdot$] and calculate $\bar{y}$ as a function of pH to express the enzyme activity curve. In this example, the same basic input, increment, process, and output format have been employed. In this case, the pH is incremented, and then converted to hydrogen ion concentration for use in the main process block.

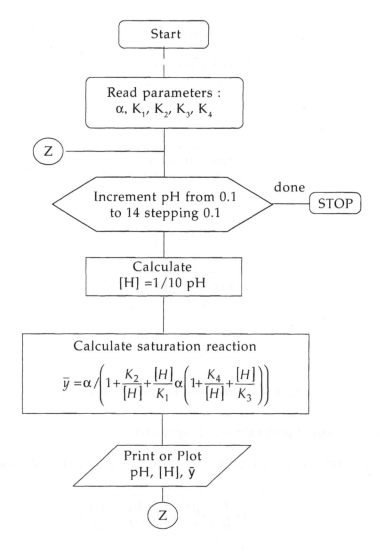

Fig. 4.15 Flowchart for the pH/Enzyme activity model. K_1, K_2, K_3, K_4 are equilibrium constants and α is the substrate concentrations relative to K_5.

The technique of flowcharting has been briefly introduced. It is assumed that as more complicated flow charts are introduced, the students will readily adapt to the system. For further information on the techniques of flowcharting, consult Spain (1982), Boillot, et al., (1974), Edward and Broadwell (1974) and Chapin (1971).

4.3.2.2 Introduction to BASIC Programming

This section describes the BASIC programming language. Here, those aspects are emphasized which are particularly useful for modeling. There are three important books on good general instructional manual dealing with the BASIC computer language: (1) Sack and Meadows (1973), ENTERING BASIC; Science Research Associated, Inc. Chicago, (2) Spencer (1975) A Guide to BASIC PROGRAMMING READING: Addison-Wesley Publishing Company, Massachusetts and (3) Spain (1982), BASIC microcomputer models in biology, Addison-Wesley Publishing Company, Massachusetts.

4.3.2.2.1 BASIC Program

BASIC is an acronym for Beginner's All-purpose Symbolic Instruction Code. A BASIC program consists of a series of lines of instruction, each beginning with a line number followed by a command. Unless directed otherwise, the computer performs the commands in the sequential order. The program commands must be stated precisely, because the computer will execute them exactly as they are written. If a given statement, for instance, statement 110, is not written in proper form, the computer will respond with a SYNTAX ERROR IN LINE 110 message.

4.3.2.2.2 Line Number

Each statement of a BASIC program is assigned a line number which must be an integer between 1 and 63999. These are usually assigned line number 10 for the first statement, 20 for the next, and so on, allowing the modeler to add statements that have been forgotten. It is a good idea to start the actual program in line 100.

4.3.2.2.3 REM

The REMARK statement provides a method for inserting comments into the program. The REM statement may be placed anywhere before the END statement in the program. The form of the statement is

140 REM ANY COMMENTS ABOUT THE PROGRAM

OR:

160 Y = A* X + B: REM EQUATION FOR LINE

Typically, it is used for naming the exercise, telling where it came from, and specifying some of the variables.

4.3.2.2.4 READ and DATA

The READ statement is used to assign numerical values to variables or constants. The READ statement is also designed to work in conjunction with the DATA statement. The form of the READ statement is as follows:

100 READ A, B, N_1, N_2, K_1, K_2, DT

The variables are listed, and separated by commas. The numerical values for variables and constants are listed in the DATA statement in the same order as they were listed in READ. The general form of the DATA statement is:

200 DATA 10, 32 1000, 3000, 2.585, 4.7E8, .1

There may be more data in the DATA statements (or more DATA statements) than constant or variables assigned in the READ statement, but there can never be less. Otherwise the computer responds with OUT OF DATA ERROR IN LINE 100. One may think of the data as being assigned to a data bank from which it is withdrawn in a sequential order. For example, it does not matter whether the above data were entered as a single DATA statement or three DATA statements; it is the sequence of entry that is important. DATA statements are commonly placed together just before the END of the program, although they may be placed anywhere before END. Once data has been read, it cannot be used again unless the READ statement is followed by a RESTORE statement.

4.3.2.2.5 PRINT

The print statement is used to print out the values of variables or explanatory terminology. The general form of the PRINT statement is:

110 PRINT "THE VALUE OF X is; X; " THE VALUE OF Y is"; Y

A typical example for printing column headings is as follows

100 PRINT "TIME", "NO. ORGANISMS", "DENSITY"

The number of characters and the number of print-columns in a line depend on the type of computer and printer being employed. Table 4.2 gives the values for some common equipment.

Table 4.2 Values for some common equipment

Equipment type	Number of Characters	Number of Print zones
Apple II, video display	40	3
Silentype Printer	80	5
TRS -80, video Display	64	4
Contronics Printer	80	5

When using the standard print format, where zones are separated with commas, the size of the print zone is 15 characters, except for the last column which gets the residue. Thus, if you divide the number of characters by 15, you can find the number of print zones. For the Apple II video, $40/15 = 2.67$. Hence, there are two full print zones and one 10-character short zone.

In the standard PRINT format, a comma causes the computer to skip to the next print zone, while a semicolon is interpreted as don't skip. The result is that the next item would be printed immediately after, with no space between. In using quote marks (") note that they must always be in pairs.

To cause the computer to print a space between lines, simply provide a PRINT statement with nothing behind it, as follows:

150 PRINT

The statement required for printing the results of a repetitive calculation usually appears at the end of the calculation loop. For example, the following statement follows the ones listed above after the number of organisms, N, and density, D, have been calculated for each time interval, T:

190 PRINT T, N, D

4.3.2.2.6 LET

Mathematical expressions are usually written as LET statements. The LET statement is used to assign a value to one or more variables. The computer executes the operation designated on the right side of the equals sign and assigns the result to the variable designated on the left side of the equals sign. This causes the value to be computed and used to replace the previous value for that variable. An example of a LET statement follows:

150 LET Y = 3 * x^2/ (2.25 + x) – 3 E9

For most microcomputers, it is not necessary to include the word LET in this statement. The following symbols are used to denote each of the arithmetic operations in BASIC:

Addition +	Division /
Subtraction –	Exponentiation $\wedge$ or $\uparrow$
Multiplication *	Scientific notation, E (4×10^{-3}) is 4 E–3

The example given in statement 150 means:

$$Y = \frac{3x^2}{(2.55 + x)} - 3 \times 10^9$$

Arithmetic expressions are evaluated according to the rules of precedence, in the following order:

1. items enclosed in parentheses
2. exponentiation
3. multiplication and division
4. addition and subtraction

The computer scans from left to right, and evaluate all operations enclosed by parentheses. Next, it evaluates exponentiations, and so on. Since there cannot be too many parentheses, one way to prevent a violation of the rules of precedence is to use parenthesis whenever there is a question. However, make sure that parentheses are always used in pairs; otherwise the computer will respond with a SYNTAX ERROR.

4.3.2.2.7 Variables

Simple variables are designated either by a letter, a letter followed by a single digit, or by two letters. Examples would be A, B, P, Q, N_1, N_2, K_1, K_2, AB or PQ. Letters with double numbers like K_{11} or N_{21} are acceptable variables, but only the first digit is recognized by the computer. Certain double letters may not be used as variables since they have a special meaning to the computer and are called reserved words. These include AT, TO, IF, OR, ON, GR, and FN. Longer variable names may be used, such as TIME or DISTANCE. However, the computer only recognizes the first two letters as being unique. Thus, TIDE cannot be distinguished from TIME by the computer.

4.3.2.2.8 Constants

Values for those constants which are not going to be changed during the use of a model can be written as part of the algebraic expression in which they are required. For example:

110 A = 3.1416* R^2

Another example would be:

110 $Y = 3*x^2 + 5* x - 15$

Those constants which one may wish to change during different runs of a simulation are best included in the algebraic expression as variables. As such, their value would be assigned by the READ and DATA statements. For example, the equation for numerical integration of population growth in a limited environment is

$$N_{t+\Delta t} = N_t + rN_t (1-(N_t/K))$$

In a simulation based on this equation, one is frequently called upon to vary either the intrinsic growth rate, r, or the carrying capacity K, or both. The variables in this equation are the population density, N, and time, t. Assuming time is incremented in units of 1, the value of N is replaced by the new value as follows:

$$N \leftarrow N + rN (1 - (N/K))$$

This would be programmed

100 READ N, K, R

200 LET N = N + R* N (1 – N/K)

300 DATA 20,500, .5

Remember that the computer evaluates the expression to the right of the equals sign, and assigns it to the variable on the left. Thus the old value of N is replaced with the newly calculated value.

The program above would normally include many statements in addition to those listed.

4.3.2.2.9 GO TO

The GO TO statement is used to transfer control to another line of the program. This is sometimes called an unconditional jump statement. It is used to loop back so that a routine may be repeated, or to jump ahead if a segment of program is to be bypassed. A typical example of the use of the GO TO statement follows:

290 GO TO 110

Statement 110 might be a READ statement, in which case the computer would continue to read data from DATA bank until the data were exhausted. The GO TO statement is often combined with the IF .... THEN statement in branching programs.

4.3.2.2.10 STOP

The STOP statement terminates the execution of the program. It is often used to separate the subroutines from the main program.

4.3.2.2.11   IF..............THEN

The IF ....THEN statement is very important to the programr as it allows operations to be executed only if certain conditions are met. The general form of the IF..........THEN statement is as follows:

900 IF (expression) relation (expression) THEN (Operation)

The expression may be an arithmatic expression, a variable, or a constant.

The relation between the expressions should be one of the following:

Relation	Meaning
<	is less than
< = or = <	is less than or equal to
>	is greater than
> = or = >	is greater than or equal to
=	is equal to
≠ or <> or ><	is not equal to

Some examples of simple IF ....THEN statements follow:

```
200    IF N > 20 THEN GO TO 110
210    IF N < A* B THEN N = A*B
220    IF N < 30 THEN N = N* (1–N/K): R = 20: GO TO 100
```

Note that in the last example, all three operations are executed only if the condition $N < 30$ is met.

If $N = > 30$, then the computer moves on to the next statement and no operations to the right of THEN are executed.

The IF .... THEN statement may also be combined with another IF ....THEN statement in a logical manner as follows:

```
230    IF N > 20 OR N < 30 THEN GO TO 300
240    IF X = 5 AND Y > 10 THEN Z = 40: GO TO 100
```

4.3.2.2.12 FOR and NEXT

The best way to cause a routine to be repeated a specified number of times is to employ the FOR and NEXT statements. The FOR statement specifies the number of times that a certain segment of the program is to be repeated. The NEXT statement tells the computer to loop back to the FOR statement and to check if the correct number of repeats has been attained. If it has not, the loop is repeated. Otherwise, control is passed to the statement following the NEXT statement. Time may be incremented or decremented in STEPs of varying size. If the STEPs are in units of 1, the STEP size need not to be specified. Examples follow:

```
100 FOR T = 1 to 30
110 etc
190 NEXT T
```

These statements cause every instruction between 100 and 190 to be repeated 30 times.

STEP size other than 1 is specified as follows:

```
110 FOR N = 0 to 10 STEP 0.5
```

To decrement,

```
115 FOR C = 100 to 0 STEP – 2
```

Every FOR statement must be paired with a NEXT statement. FOR and NEXT statement may be nested, but associated FOR and NEXT statements must either be completely inside or completely outside, any other FOR.........NEXT loops.

Nested FOR ....NEXT loops are often used in programs where time needs to be changed by small, variable increments. In the following example, DT represents a small time increment, Δt, and IM is the number of increments in a major time interval.

```
100     DT = 0.1
110     IM = INT (1/DT + 0.1)
120     FOR T = 1 to 100
130     FOR I = 1 to IM
140     N = N + R* N*DT
150     NEXT I
160     PRINT T, N
170     NEXT T
180     END
```

Note that the inner loop, statements 130 to 150, allows you to vary the size of DT, and still obtain data output only for the major time intervals.

Statement 110 calculates the number of increments, IM, per major time interval, *T*, based on the size of the increment, DT.

4.3.2.2.13 *Numeric Functions*

These functions may be used in any algebraic expression and return a numeric value. Here, X represents any variable or expression.

ABS (X)   This function returns the absolute value of x.

ATN (X)   This function return the Arctangent of x, the angle in radians whose tangent is x.

COS(X)   This function returns the cosine of x when x is expressed in radians.

EXP (X) This function returns a raised to the x power, antilog of x.

EXP (2.302585* X) this function gives the antilog of x if x is a $\log_{10}$.

INT (X) The greatest integer which is less than or equal to x is returned by this function.

LOG (X) This function returns log to the base e of x. To find the log to the base b, use the formula.

$\log_b(x) = \log_e(x)/\log_e (b)$

$\log_{10}(x) = \log (x)/2.303$

RND(1) This function returns a random number uniformly distributed between 0 and 1. Note that the random number will be a decimal fraction.

RND (X) This function produces random integers between 1 and x (TRS -80 only). On the Apple II use R = INT (RND (1)* x +1).

RANDOM This statement normally proceeds the RND (1) function is TRS-80 programs. The RANDOM instruction reseeds the Pseudo random number routine so that a different string of numbers is produced each time.

SGN (x) This function returns 0 if x = 0, −1 if x is negative, and +1 if x is positive.

SIN(X) This function returns the sine of x when x is given in radians.

SQR (X) This function returns the square root of x.

Tan (X) This functions returns the tangent of x when x is given in radians.

It is legitimate to include a mathematical expression involving x in place of x. For example , INT (2.5* $x^\wedge$ 3$/x$ − 2)) would be calculated with no difficulty.

4.3.2.2.14 PRINT Tab

On the Apple II, the print statement listed above could be accomplished by the following statement employing statement HTAB:

```
200     HTAB 3: PRINT "X is "; X;:
        HTAB 30: PRINT"Y is "; Y
```

4.3.2.2.15 PRINT USING (TRS-80 only)

Frequently, the standard four column print format provided by the PRINT statement does not provide enough columns for the amount of information which must be pointed. It is very easy to set up an alternate format with the PRINT USING statement. This statement includes a form of the output and thus tells the computer the number of spaces by using #

symbols, and the position of the decimal. For example, the following would divide the 64 positions of the print line into six columns of 10, each containing 2 decimal places.

220 PRINT USING "#######.##:

A, B, C, D, E, F

In this example A, B, C, ... are variables with less than seven digits to the left of the decimal point. This statement would have been preceded by a statement describing the positions of the column heading, using the PRINT TAB instruction.

       120   PRINT TAB (7); "A", TAB (17); "B";

                TAB (27); "C" TAB (37); "D"

                TAB (47), "E"; TAB (57), "F"

4.3.2.2.16 GOSUB and RETURN

The GOSUB instruction is used to call up a subroutine. A subroutine is a sequence of instructions used repeatedly, or used at several points in the same program. Control is transferred to the statement number that is included in the GOSUB instruction. For example:

       110 GOSUB 240

This statement would transfer control to statement 240 which is presumably the first statement number of the subroutine. The instructions of the subroutine would be followed in order until a RETURN instruction causes control to be transferred back to the statement following GOSUB, normally, statement 120.

SAMPLE PROGRAM (Sack and Meadows, 1973) : Consider, for example, the design of simplified payroll program, WAGES. The program reads DATA statements containing the employee number, the employees' hourly wages, and the hours he worked on each day of the week, Sunday through Saturday.

       DATA 1234, 3.00, 0, 8, 8, 6, 8.5, 10, 3

indicates that employee 1234 earns $3.00 per hour. He has worked eight hours on Monday and Tuesday, six hours on Wednesday, eight and one-half on Thursday, ten on Friday, and three on Saturday.

The program must print the hours worked and the dollars earned by the employee on each day of the week, and total hours worked and dollars earned for the week. The companies pays double time for work on Sunday and time-and-a-half for saturday. If an employee works more than eight hours on any weekday, he gets time-and-a-half for his overtime.

Following is a skeleton flowchart of the logic of the program

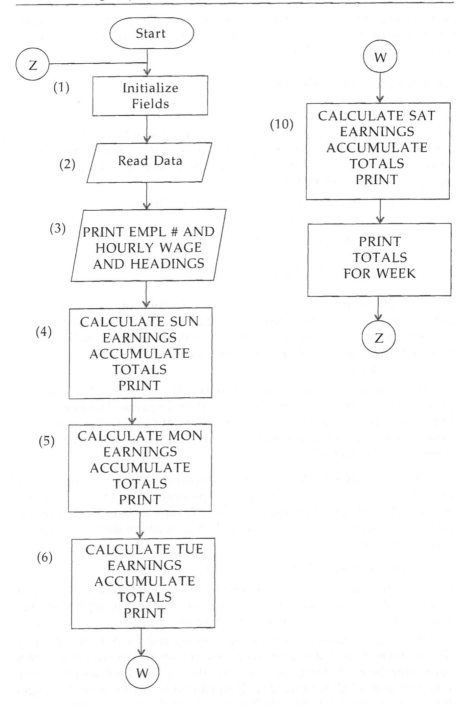

Fig. 4.16 Wages-skeleton Flowchart.

Notice that the logic for boxes 5-9 (Mon-Fri) is almost identical. Instead of coding the logic five times it is coded once as a subroutine at the end of the program (lines 340-420 and is called by GOSUB statements lines 130, 160, 190, 220 and 250).

WAGES

```
10 LET H = 0
20 LET E2 =0
30 READ N, W, D1, D2, D3, D4, D5, D6, D7
40 PRINT "EMPLOYEE" N; "HOURLY WAGES" W
50 PRINT "DAY", "HOURS", "DOLLARS"
60 REM SUNDAY EARNINGS
70 LET E1 = 2* D1*W
80 LET H = H+D1
90 LET E2 = E2 + E1
100 PRINT "SUN", D1, E1
110 X = D2
120 LET D$ = "MON"
130 GOSUB 340
140 LET X = D3
150 LET D$ = "TUES"
160 GOSUB 340
170 LET X = D4
180 LET D$ = "WED"
190 GOSUB 340
200 LET X = D5
210 LET D$ = "THURS"
220 GOSUB 340
230 LET X = D6
240 LET D$ = "FRI"
250 GOSUB 340
260 REM SATURDAY EARNINGS
270 LET E1 = 1.5* D7 *W
280 LET H = H+D7
290 LET E2 = E2 + E1
300 PRINT "SAT", D7, E1
310 REM TOTAL EARNINGS
320 PRINT "TOTAL", H, E2
```

(Contd...)

```
325 PRINT
330 GOTO 10
340 REM WEEKDAY EARNINGS - SUBROUTINE
350 IF X >8 THEN 380
360 LET E1 = X* W
370 GOTO 390
380 LET E1 = 8*W +1.5*W* (X-8)
390 LET H = H + X
400 LET E2 = E2 + E1
410 PRINT D$, X, E1
420 RETURN
500 DATA 1234, 3.00, 0, 8, 8, 6, 8.5, 10, 3
510 DATA 1239, 3.10, 2, 8, 8, 8, 10.5, 0, 7
520 DATA 1339, 3.80, 0, 8, 8, 8, 8, 8, 0
700 END
RUNNH
```

EMPLOYEE 1234	HOURLY WAGES 3	
DAY	HOURS	DOLLARS
SUN	0	0
MON	8	24
TUES	8	24
WED	6	18
THURS	8.5	26.25
FRI	10	33
SAT	3	13.5
TOTAL	**43.5**	**138.75**

EMPLOYEE 1239	HOURLY WAGES 3.1	
DAY	HOURS	DOLLARS
SUN	2	12.4
MON	8	24.8
TUES	8	24.8
WED	8	24.8
THURS	10.5	36.425
FRI	0	0
SAT	7	38.55
TOTAL	**43.5**	**155.775**

(Contd.)

EMPLOYEE 1333	HOURLY WAGES 3.8	
DAY	HOURS	DOLLARS
SUN	0	0
MON	8	30.4
TUES	8	30.4
WED	8	30.4
THURS	10.8	30.4
FRI	8	30.4
SAT	0	0
TOTAL	**40**	**152**

OUT OF DATA IN LINE 30

4.3.2.2.17 GRAPH Subroutine

GRAPH: A Subroutine for Plotting Data
The following description is based on the Apple II Version of the GRAPH
subroutine.

```
30   REM READING HIGH RESOLUTION CHARACTER SET
40   POKE 232, 0: POKE 233, 64: SCALE = 1:
     HCOLOR = 3: ROT = 0
50   PRINT CHR$ (4) " BROAD SMALL CHARACTERS, A $ 4000"
60   REM
100  REM EXAMPLE PROGRAM
110  REM SINE CURVE VS. TIME
120  x$  = "TIME": Y$ = "SINE": YM = 1: YN = –1:
     XM  = 1: GOSUB 3000
125  ZF  = 1: ZG = 0: REM FLAGS FOR LINE PLOT
130  FOR T = 0 TO 1 STEP .01
140  S = SIN (6.28 *T)
150  X = T: Y = S; GOSUB 4000
160  NEXT T
170  STOP
2990 REM
3000 REM AXES AND UNITS FOR GRAPHS REV. SEPT., 2001
3010 REM SUBROUTINE DEVELOPED BY J. SPAIN, MICHIGAN
     TECH UNIV., AND B.J. WINKEL, ROSE-HULMAN INST.
     TECH.
3020 REM X$ = VARIABLE PLOTTED ON X AXIS
```

(Contd.)

```
3030  REM Y$   = VARIABLE PLOTTED ON Y AXIS
3040  REM YM  = MAXIMUM UNITS ON THE Y AXIS
3042  REM YN  = MINIMUM UNITS ON THE Y AXIS
3045  REM XM  = MAXIMUM UNITS ON THE X AXIS
3047  REM XN  = MINIMUM UNITS ON X AXIS
3050  HCOLOR = 3: SCALE = 1: SC = 1 ROT = 0
3055  REM LIST OF RESERVED VARIABLES: X, X0, XM, XN, X$, Y, Y0,
      YM, YN, Y$, Z, ZF, ZG, ZP, Z0, I, L$, SC
3060  HGR: HPLOT 23, 0 TO 23, 149
3070  HPLOT 25, 149 TO 279, 149
3075  FOR I = 9 TO 116 STEP 35
3080  HPLOT 26, I TO 23, I: NEXT I
3085  FOR I = 54 TO 275 STEP 31
3090  HPLOT I, 146 TO I, 149: NEXT I
3093  REM
3095  REM WRITE VARIABLE NAME ON X – AXIS
3100  Z0 = 0: L$ = X$: X0 = 175; Y0 = 150
3110  GOSUB 3500
3120  REM WRITE VARIABLE NAME ON Y – AXIS
3130  Z0 = 1: L $ = Y $: X0 =10: Y0 = 68
3140  GOSUB 3500
3145  REM
3150  REM WRITE UNITS ON THE X – AXIS
3155  Y0 = 150: Z0 = 0
3160  L$ = LEFT$ (STR$ (XM), 3): X0 = 260: GOSUB 3500
3165  L$ = LEFT$ (STR$C (XM – XN)/2 +XN), 3): X0 = 146: GOSUB
      3500
3170  L$ = LEFT$ (STR$ ((XM – XN)/4 + XN), 3): X0 = 82: GOSUB
      3500
3171  L$ = LEFT$ (STR$ (XN), 2): X0 = 23: GOSUB 3500
3172  REM WRITE UNITS ON Y - AXIS
3175  X0 =1
3180  L$ = LEFT$ (STR$ (YM), 3): Y0 = 6: GOSUB 3500
3185  L$ = LEFT$ (STR$ ((YM – YN)/2 +YN), 3): Y0 = 76: GOSUB 3500
3190  L$ = LEFT$ (STR $ ((YM – YN)/4 + YN), 3): Y0 =111:GOSUB
      3500
3192  L$ = LEFT$ (STR$ (YN), 3): Y0 = 147: GOSUB 3500
3200  RETURN
```

(Contd.)

```
3205 REM
3500 REM ALPHANUMERIC CHARACTERS FOR HGR
3510 REM THE FOLLOWING MUST BE DEFINED
3511 REM BEFORE ENTERING THE SUBROUTINE
3512 REM L$ = "CHARACTER STRING"
3513 REM Y0 = THE INITIAL Y POSITION
3514 REM X0 = THE INITIAL X POSITION
3515 REM SET Z0 = 0 IF PRINTING HORIZONTAL
3516 REM SET Z0 = 1 IF PRINTING VERTICALLY
3520 FOR Z = 1 TO LEN (L$) = Z3 = ASC (MID$ (L$, Z,1))
3525 IF X0>275 OR X0 <0 OR Y0 > 153 OR – Y0 < 0 THEN GOTO 3600
3530 IF Z0 <>0 THEN GOTO 3565
3540 IF Z3>64 THEN DRAW Z3 – 64 AT X0 + (Z – 1) *7 *SC, Y0: GOTO
     3600
3550 DRAW Z3 AT X0 + (Z–1)*7*SC, Y0: GOTO 3600
3565 IF Y0 – (Z–1) *7 < 7 THEN GOTO 3600
3570 ROT = 48: IF Z3 > 64 THEN DRAW Z3 – 64 AT X0, Y0 – (Z–1) *7:
     GOTO 3600
3580 DRAW Z3 AT X0, Y0 – (Z – 1)*7
3600 ROT = 0: NEXT Z
3610 RETURN
3615 REM
4000 REM PLOTTING X AND Y VALUES
4005 REM FOR LINE PLOT MAKE ZF = 1: ZG = 0: AT BEGINNING
     OF PLOT
4006 REM TO PLOT (+) SET ZP = 1 BEFORE EACH POINT
4007 REM TO PLOT (0) SET Z0 = 1 BEFORE EACH POINT
4010 X0 = 23 + (X–XN) *252/(XM – XN)
4020 Y0 = 149 – (Y – YN) * 140 /(YM – YN)
4030 IF X0 > 276 OR X0 < 0 OR Y0 > 149 OR Y0 < 4 THEN ZG = 0:
     GOTO 4050
4033 IF ZP = 1 THEN DRAW 43 AT X0 – 2, Y0 – 3: ZP = 0; GOTO 4050
4034 IF Z0 =1 THEN DRAW 79 AT X0 – 2, Y0 – 4: Z0=0: GOTO 4050
4035 IF ZG = 1 THEN HPLOT TO X0, Y0: GOTO 4050
4040 HPLOT X0, Y0: IF ZP = 1 THEN ZG =1
4050 RETURN
4055 END
```

Load the GRAPH subroutine into the computer just before you begin to enter your program. You may then build your program, taking into account that lines 0-21 and 3000 to 4100 are reserved for the GRAPH subroutine. It is recommended that you use line 30-90 for remark statements identifying the exercise number, exercise name, your name, and the date.

The actual program should start at line 100. Typically, it would be written as to plot data as it is calculated although it is possible to store the data in an array and plot it later. Thus, the axes for the graph would be drawn before the main calculation loop is begun. To draw axes, first call for HGR to place the computer in the high resolution graphic mode. Then specify the names of the variables on the *x* and *y* axes, for example, X$ = "time" Y$ = "MONEY". Next specify the maximum units on the *x* axis, such as XM = 100 and the maximum units on the *y* axis, such as YM=900. Note that the maximum number of digits allowed for XM is 3 and the maximum for YM is 4. If you wish to specify minimum values for the *x* or *y* axes, they would be defined as XN and YN. If these minimum are not specified, they will be zero by default. Finally, call for the axes subroutine by GOSUB 3000. The computer set of instructions required to draw and level axes may be placed on a single line as follows:

130 X$ = "TIME": Y$ = "MONEY": XM = 100:
YM = 900; GOSUB 3000

The actual plotting process is executed by specifying the values of X and Y and then calling for the subroutine 4000. The statement is normally placed just before the NEXT T at the end of the time loop. The following example assumes that money, M, is a function of time, T.

150   FOR T = 1 to 100
160   M = 3.95*T
170   X = T; Y = M: GOSUB 4000
180   NEXT T

If you want to plot lines rather than points, the following 'flag' statement should be placed just before the time loop starts.

140 ZF = 1; ZG = 0

If you want points marked with circles or crosses, set flag Z0 = 1 to indicate you want circles and ZP = 1 to indicate you want crosses.

The graph subroutine may also be used to write messages on HGR. Simply specify the message as follows:

L$ = "MESSAGE"

Then specify the *x* and *y* coordinated for the beginning of the message as X0 and Y0 and call for the subroutine by GOSUB 3500. The complete statement would be

135 L$ = "Total AMOUNT": X0 = 50:
Y0 = 100: GOSUB 3500

Note that coordinate positions on HGR go from 0 to 279 on the X-axis and 0 to 159 on the Y-axis. The origin is the top left corner.

4.3.2.2.18 Arrays and Subscripted Variables

Subscripted variables are designated by one or two characters followed by one or more integers or letters enclosed in parentheses. The following are examples of subscripted variables:

A(1), A(2), A (3) ....are all variables in a list or array called A.

B(1), B(2), B (3) .... are all variables in a list of array called B.

On the other hand C (1, 1), C (1, 2), C (1,3), C (2,1) ....are elements of a two-dimensional array called C. Use of these subscripted variables does not preclude the use of regular scalar variables having the same letters. A, B, CC and A1, B3, C0 may all be used in the same program with the above subscripted variables.

BASIC automatically saves space for arrays which involves no more than 10 elements. However, for larger arrays, one must first reserve storage by the use of a dimension of DIM statement. The DIM statement is placed at the beginning of the program, as re-dimensioning will result in an error. The following dimension statement would reserve storage for 3 lists of varying length designated A, B, and I; and two 6 by 6 matrix, designated C and D:

100 DIM A (25), B (100), I (1000), C (6,6), D (6, 6)

As illustrated, lists may be of almost any length. In dimensioning matrices, the number of rows are specified first, and the number of columns second. The largest matrix that can be accommodated by microcomputers is about 50 by 50.

One of the more valuable applications for a list of subscripted variables is a result of the ability to define one variable by another variable. Let us consider list "A" with subscript "I", and use it to obtain a new list, "B". We could read data into the A-list sequentially by varying I from one to 26.

100 DM A(26)
110 LET I = 1 to 26
120 READ A (I)
130 NEXT I
140 DATA 16, 12, 7, 8 .....19, 23

Lists such as this are conveniently employed to perform certain repetitive operations.

For example, we could obtain the mean between each successive pair of values in the above list and record these in a new list called "B".

```
150 PRINT "I", "A(I)", "MEAN"
160 DIM B(25)
170 FOR I = 1 to 25
180 LET B(I) = A(I) + A (I + 1))/2
190 Next I
```

with the subscripted variable, it is also possible to store data and print it out at the end of all calculations. For example, we could print out the previous data as follows:

```
200 For I = 1 to 25
210 PRINT I, A(I), B (I)
220 NEXT I
```

Subscripted variables may also be used to sort numbers in order to produce a histogram. Assume we have 1000 integers between 1 and 25, which have been listed as data or produced during a simulation. These may be defined as I, and counted in the following way:

```
100 PRINT "NUMBER", "FREQUENCY"
110 DIM A (25)
120 FOR N = 1 TO 1000
130 READ I
140 LET A (I) = A (I) + 1
150 NEXT N
160 FOR I = 1 to 25
170 PRINT I, A (I)
180 NEXT I
190 DATA 17, 21, 3, ...., 6, 22
200 END
```

Sorting of counting routines of this type are very valuable for Monte Carlo simulations. These simulations normally must be run a great many times to give meaningful results, and thus usually require some data analysis.

4.3.2.2.19 Matrix Subroutine

The following section provides some standard matrix subroutines.

4.3.2.2.19.1 Inputting Data to a Matrix

```
100 DIM A (R,C), B(R,C), C(R,C), Y(C), X (C)
110 FOR I = TO R: REM ROWS
120 FOR J = 1 to C: REM COLUMNS
```

```
130 READ A (I, J)
140 NEXT J
150 NEXT I
160 DATA 6,1,3,8, ....
```

4.3.2.2.19.2 Printing a Matrix

```
200 FOR I = 1 to R
210 FOR J = 1 to C
220 PRINT A(I, J); " ";
230 NEXT J: PRINT " "
240 NEXT I
```

4.3.2.2.19.3 Scalar Multiplication by a Constant, K

```
300 FOR I = 1 to R
310 FOR J = 1 to C
320 B(1, J) = A(I, J)* K
330 NEXT J
340 NEXT I
```

4.3.2.2.19.4 Post-multiplication of a Matrix by a Vector, X (C)

```
400 FOR I = 1 to R
410 Y (I) = 0
420 FOR J = 1 to C
430 Y (I) = Y(I) + A (I, J)* x (J)
440 NEXT J
450 NEXT I
```

When multiplying a matrix by a Vector, R must equal C.

4.3.2.2.20 Important Command Mode Instructions for Apple II and TRS-80

4.3.2.2. 20.1 Apple II Plus

<RETURN> Causes any keyboard entry to be transferred to memory
<RESET> or < CTRL > C Stops program execution and places the computer in command mode.
HOME Clears text display, and returns the cursor to top left position.

LOAD NAME   Loads Program called NAME from diskette to memory.

SAVE NAME Saves program called NAME from memory to diskette.

RUN NAME Loads program called NAME from diskette to memory.

RUN Clears all variables and causes program to be executed.

RUN 100 Starts run at line 100.

LIST Displays complete program from the beginning.

LIST 10-500 Displays lines 10 through 500 only.

LIST 300 Displays program starting at line 300.

DEL 10, 200 Deletes line 10 through 200

310 Deletes line 300 only ... (gone forever)

CONT   Continues program execution following STOP, <BREAK> on the TRS-80, or <CTRL> C on the Apple-II plus.

TRACE Causes line numbers for each statement to be displayed as the program is executed.

NO TRACE Turns off TRACE.

<ESC> I, J, K, M Allows you to edit a program statement. Refer to manuals for full details on editing program statements.

Radio Shack TRS-80

<ENTER> Causes any keyboard entry to be transferred to memory.

<BREAK> Stops program execution and places computer in command mode.

<CLEAR> Clears text display, and returns cursor to top left position.

LOAD "NAME" Loads program called NAME from diskette to memory.

SAVE "NAME" Saves program called NAME from memory to diskette.

RUN "NAME" Leads and runs program called NAME.

RUN Clears all variables and causes program to be executed.

RUN 100 Starts RUN at line 100.

List Displays complete program from the beginning.

List 10-500 Displays lines 10 through 500 only.

List 300 Display program starting at line 300

Delete 10-200 Deletes lines 10 through 200.

300   Deletes line 300 only ... (gone forever).

Cont   Continues program execution following STOP, <BREAK> on the TRS-80, or <CTRL> C on the Apple II plus.

TRON   Causes line numbers for each statement to be displayed as program is executed.

TROFF   Turns off TRACE

EDIT   Allows you to edit a program statement. Refer to manual for full details on editing program statements.

REMARK   See relevant reference manuals for details.

4.3.3 Data Structure

4.3.3.1 Object Data Structure

In OO (object-oriented) databases, the state (current value) of a complex object may be constructed from other objects (or other values) by using certain type of constructions. One formal way of representing such objects is to view an object as a triple (i, c, v) where i is a unique object identities (The OID), c is a type constructor (that is, an indication of how the object state is constructed, and v is the object state (or current value). The data model will typically include several type of constructors. The three most basic constructors are atom, tuple, and set. Other commonly used constructors include list, bag, and array. The atom constructor is used to represent all basic atomic values such as integers, real numbers, character string, Boolians, and any other basic data types that the system supports directly.

The object state v of an object (i, c, v) is interpreted based on the constructor c. If $c =$ atom, the state (value) v is an atomic value from the domain of basic values supported by the system. If $c =$ set, the state v is a set of object identifier $(i_1, i_2, ..., i_n)$, which are the OIDs for a set of objects that are typically of the same type. If $c =$ tuple, the state v is a tuple of the form $<a_1: i_1, a_2, i_2, ..., a_n, i_n$, where each a_i is an attribute name (instance variable name) and each i_j is an OID. If $c =$ list, value v is an ordered list $[i_1, i_2, ..., i_n)$ of OIDs of objects of the same type. A list is similar to the set except that the OIDs in a list are ordered, and hence we can refer to the first, second, or jth object in a list. For c = array, the state of the object is a single-dimensional array of object identifiers. The main difference between array and list is that a list can have an arbitrary number of elements, whereas an array typically has a maximum size. The difference between set and bag is that all elements in a set must be distinct, whereas a bag can have duplicate elements.

The model of objects allows arbitrary nesting of the set, list, tuple, and other constructors. The state of an object that is not of atom type will refer to other objects by their object identifier.

The type constructors set, list, array, and bag are called collection types (or bulk types), to distinguish them from basic types and tuple types. The main characteristic of a collection type is that the state of the object will be a collection of objects that may be unordered (such as a set or bag) or ordered (such as a list or an array). The tuple type constructor is often called a structured type, since it corresponds to the struct construct in the C and C++ programming languages.

EXAMPLE 1: A complex object.

DEPT-LOCATION	DNUMBER	DLOCATION
	1	Houston
	4	Stafford
	5	Bellaire
	5	Sugarland
	5	Houston

DEPARTMENT	DNAME	DNUMBER	MGRSSN	MGR STRTDATE
	Research	5	333445555	1988-05-22
	Administration	4	987654321	1995-01-01
	Headquarters	1	888665555	1981-06-19

Fig. 4.17. Relational Database Schema.

We now represent some objects from the relational database shown in Figure 4.17, using the model where an object is defined by a triple (OID, Type constructor, State) and the available type constructors are atom, set, and tuple. We use $i_1, i_2, i_3, ...$ to stand for unique system-generated object identifiers. Consider the following objects.

$O_1 = (i_1,$ atom, "HOUSTON")

$O_2 = (i_2,$ atom, "Bellaire")

$O_3 = (i_3,$ atom, 'Sugarland')

$O_4 = (i_4,$ atom, 5)

$O_5 = (i_5,$ atom, 'Research')

$O_6 = (i_6,$ atom, '1988-05-22')

$O_7 = (i_7,$ set, $\{i_1, i_2, i_3\})$

$O_8 = (i_8,$ tuple, <DNAME: $i_5,$ DNUMBER: i_4>)

The first six objects (O_1-O_6) listed here represents the atomic values. There will be many similar objects, one for each distinct constant atomic value in the database. Object O_7 is a set-valued object that represents the set of location for department 5; the set $\{i_1, i_2, i_3\}$ refers to the atomic objects with values {'Houston', 'Bellaire', 'Sugarland'}. Here, object O_8 is a tuple-valued object that represents department 5 itself, and has the attributes DNAME, DNUMBER.

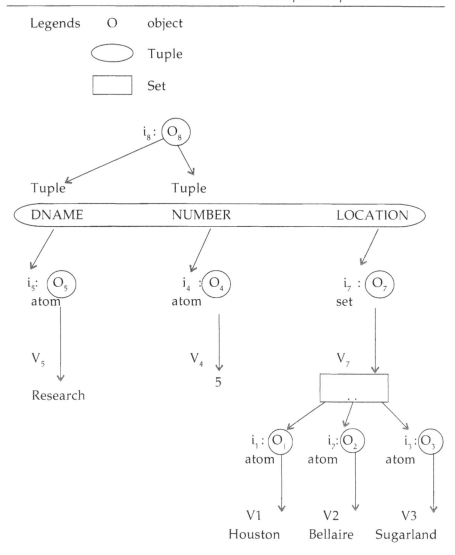

Fig. 4.18 Representation of a department complex object as a graph.

The attributes DNAME and DNUMBER have the atomic objects O_5 and O_4 as their values.

In this model, an object can be represented as a graph structure constructed by recursively applying the type constructors. The graph representing an object O_1 can be constructed by first creating a node for the object O_1 itself. The node for O_1 is labelled with the OID and the object constructor c. We also create a node in the graph for each basic atomic value. If an object O_1 has an atomic value, we draw a directed arc from the node representing O_1 to the node representing its basic value. If the object value is constructed, we draw a directed arc from the object node to a node that represents the constructed value. Figure 4.18 shows the graph for the example DEPARTMENT object O_8 given earlier.

EXAMPLE 2: Identical versus Equal objects.

An example can illustrate the difference between the two definitions for comparing object states for equality. Consider the following objects O_1, O_2, O_3, O_4, O_5 and O_6:

$$O_1 = \{i_1, \text{tuple}, <a_1: i_4, a_2: i_6>\}$$
$$O_2 = (i_2, \text{tuple}, <a_1: i_5, a_2: i_6>)$$
$$O_3 = (i_3, \text{tuple}, <a_1: i_4, a_2: i_6>)$$
$$O_4 = (i_4, \text{atom}, 10)$$
$$O_5 = (i_5, \text{atom}, 10)$$
$$O_6 = (i_6, \text{atom}, 20)$$

The objects O_1 and O_2 have equal states, since their states at the atomic level are the same but the values are reached through distinct objects O_4 and O_5. However, the states of objects O_1 and O_3 are identical, even though the objects themselves are not so because they have distinct OIDs. Similarly, although the states of O_4 and O_5 are identical, the actual objects O_4 and O_5 are equal but not identical, because they have distinct OIDs.

4.3.3.2 *The Relational Data Structure*

The relational model was first introduced by Ted Codd of IBM Research in 1970 in a classic paper (Codd 1970; as quoted in Elmasri and Navathe, 2000) and attracted immediate attention due to its simplicity and mathematical foundations. The model uses the concept of a mathematical relation—which looks somewhat like a table of values—as its basic building block, and has its theoretical basis in set theory and first order predicate logic. The model has been implemented in a large number of commercial systems over the last twenty or so years.

The SQL quarry language is the standard for commercial relational DBMS. There are two commercial relational DBMSs—ORACLE and Microsoft ACCESS.

Data structures that preceded the relational data structure include the hierarchical and network models. They were proposed in the 1960s and implemented in early DBMSs during the 70s and 80s. These have their historical importance and large existing user base for these DBMSs. These models and systems will be with us for many years and are today being called legacy systems.

4.3.3.2.1 Relational Model Concepts

The relational model represents the database as a collection of relations. Informally, each relation resembles a table of values or, to some extent, a "flat" file of records. For example, the database of files that is shown in Figure 4.19 is considered to be in the relational model.

STUDENT	Name	Student Number	Class	Major
	Smith	17	1	CS
	Brown	8	2	CS

COURSE	Course Name	Course Number	Credit Hour	Department
	Introduction to Computer Science	CS1310	4	CS
	Data Structures	CS3320	4	CS
	Discrete Mathematics	MATH2410	3	MATH
	Database	CS 3380	3	CS

SECTION	Section Identifer	Course Number	Semester	Year	Instructor
	85	MATH 2410	Fall	98	King
	92	CS 1310	Fall	98	Anderson
	102	CS 3820	Spring	99	Knuth
	112	MATH 2410	Fall	99	Chang
	119	CS 1310	Fall	99	Anderson
	135	CS 3380	Fall	99	Stone

GRADE REPORT	Student Number	Section Identifier	Grade
	17	112	B
	17	119	C
	8	85	A
	8	92	A
	8	102	B
	8	135	A

(Contd.)

PREREQUISITE	Course Number	Prerequisite Number
	CS 3380	CS 3320
	CS 3380	MATH 2410
	CS 3320	CS 1310

Fig. 4.19. An example of a database that stores students' records and their grades (*Source:* Elmasri and Navathe, 2000).

When a relation is thought of as a table of values, each row in the table represents a collection of related data values. We have introduced entity types and relationship types as concepts for modeling real-world data in Figure 4.20. In the relational model, each row in the table represents a fact that typically corresponds to real-world entity or relationship. The table name and column names are used to help in interpret the meaning of the values in each row. For example, the first table of Figure 4.19 is called STUDENT because each row represents facts about a particular student entity. The column names—Name, Student Number, Class, Major—specify how to interpret the data values in each row, based on the column each value is in. All the values in a column are of the same data type. In the formal relational model terminology, a row is called a tuple, a column header is called an attribute, and the table is called a relation. The data type describing the types of values that can appear in each column is called a domain.

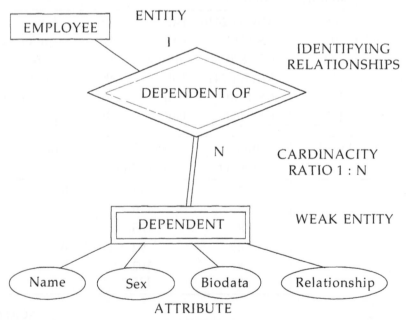

Fig. 4.20 ER (Entity relationships) schema diagram for the company database (*Source :* from Elmasri Navathe, 2000).

4.3.3.2.1.1 Domains, Attributes, Tuples, and Relations : A domain D is a set of atomic values. By atomic values we mean that each value in the domain is indivisible as far as the relational model is concerned. A common method of specifying a domain is to specify a data type from which the data values forming the domain are drawn. It is also useful to specify a name for the domain, to help interpret its values. Some examples of domain follow:

* USA phone numbers: The set of 10-digit phone numbers valid in the United States.
* Local Phone numbers: The set of 7-digit phone numbers valid within a particular area code in the United States.
* Social security numbers: The set of valid 9-digit social security numbers.
* Name: The set of names of persons.
* Grade point average: Possible values of computed trade point averages. Each must be a real (Floating point) number between 0 and 4.
* Employee ages: Possible ages of employees of a company; each must be a value between 15 and 80 years old.
* Academic department name: The set of academic department names—such as Computer Science, Economics, and Physics—in a university.
* Academic department code: The set of academic department codes such as CS, ECON, and PHYS, in an university.

The preceding statements are called logical definitions of domains. A data type or format is also specified for each domain. For example, the data type for the domain USA Phone number can be declared as a character string of the form (ddd) ddd-ddd, where each d is a numeric (decimal) digit and the first three digits form a valid telephone area code. The data type for employee–age is an integer number between 15 and 80. For academic department name, the data type is the set of all character strings that represent valid department names. A domain is thus given a name, data type, and format. Additional information for interpreting the values of a domain can also be given; for example, a numeric domain such as person weights should have the units of measurements—pounds or kilograms.

A relation schema R_1 denoted by $R (A_1, A_2, ..., A_n)$ is made up of a relation name R and a list of attributes $A_1, A_2, ..., A_n$. Each attribute A_1 is the name of a role played by some domain D in the relation schema R. D is called the domain of A_1 and is denoted by dom (A_1). A relation schema is used to describe a relation; R is called the name of this relation. The degree of a relation is the number of attributes n of its relation schema.

An example of a relation schema for a relation of degree 7, which describes university students, is the following:

Student (Name, SSN, Home Phone, Address, Office phone, Age, GPA)

For this relation schema, student is the name of the relation, which has seven attributes. We can specify the following previously defined domains for some of the attributes of the STUDENT relation: dom (Name) = Names; dom (SSN) = Social Security numbers; dom (Home Phone) = Local phone numbers; dom (Office Phone) = Office-phone-numbers; and dom (GPA) = Grade point averages.

A relation (or relation state) r of the relation schema R $(A_1, A_2, ..., A_n)$, also denoted by R (R), is a set of n-Tupels $r = \{t_1, t_2, ..., t_m\}$. Each n-tuple t is an ordered list of n values $t = <v_1, v_2, ..., v_n>$, where each value V_1 $1 \le i \le n$, is an element of dom (A_1) or is a special null value. The ith value in tuple t, which corresponds to the attribute A_1, is referred to as t $[A_1]$. The term relation **intension** for the schema R and relation **extension** for a relation state r (R) are also used.

STUDENT	Name	SSN	Home Phone	Address	Office Phone	Age	GPA
	Benjamin Bayer	305612435	3731616	2918 Blue Bonnet Lane	Nil	19	3.21
Tuples	Katherine Ashly	381621245	3754409	125 Kity Road	Nil	18	2.89
		422112320	Nil	3452 Elgin Road	7491253	25	3.53
	Dick Davidson						

Fig. 4.21. The attributes and tuples of a relation student.
(Source : Elmasri and Navathe, 2000).

Figure 4.21 shows an example of a STUDENT relation, which corresponds to the STUDENT schema specified above. Each tuple in the relation represents a particular student entity. We display the relation as a table, where each tuple is shown as a row and each attribute corresponds to a column header indicating a role or interpretation of the values in that column. Nill values represents attributes whose values are unknown or do not exist for some individual STUDENT tuples.

The above definition of a relation can be restated as follows. A relation r (R) is a mathematical relation of degree n on the domain dom (A_1), dom (A_2), ..., dom (A_n) which is a subset of the Cartesian product of the domain that defines R:

$$r\,(R) \leq (\text{dom}\,(A_1) \times \text{dom}\,(A_2) \times \ldots \times \text{dom}\,(A_n)$$

The Cartesian product specifies all possible combinations of values from the underlying domain. Hence, if we denote the number of values N cardinality of a domain D by IDI, and assume that all domains are finite, the total numbers of tuples in the Cartesian product is:

$$|\text{dom}\,(A_1)\,| * |\,\text{dom}\,(A_2)\,| * \ldots * |\,\text{dom}\,(A_n)\,|$$

Out of all these possible combinations, a relation state at a given time—the current relation state—reflects only the valid tuples that represent a particular state of the real world. In general, as the state of the real world changes, so does the relation, by being transformed into another relation state. However, the schema R is relatively static and does not change except very infrequently, for example, as a result of adding an attribute to represent new information that was not originally stored in the relation.

It is possible for several attributes to have the same domain. The attributes indicate different roles, or interpretations, for the domain. For example, in the student relation, the same domain local phone numbers plays the role of Homophone, referring to the "home phone of a student", referring to the "office phone of the student".

In the subsection of the Relational Data Structure, we have included Relational Model Concept, Domains, Attributes, Tuples, and Relations. For other details related to Relational Data Structure, consult Elmasri and Navathe (2000).

4.3.3.3 Network Data Structure

The original network model and language were presented in the CODASYL Data Base Task Group's 1971 report (as quoted in Elmasri and Navathe, 2000); hence it is sometimes called the DBTG model. Revised reports came in 1978 and 1981 incorporated relatively more recent concepts.

The original CODASYL/DBTG report used COBOL as the host language. Regardless of the host programming language, the basic database manipulation commands of the network model remain the same. Although the network model and the object-oriented data model are both navigational in nature, the data structuring capability of the network model is much more elaborate and allows for explicit insertion/deletion/modification semantic specification. However, it lacks some of the desirable features of the object models such as inheritance and encapsulation of structure and behavior (as quoted in Elmasri and Navathe, 2000).

4.3.3.3.1 Network Data Modeling Concepts

There are two basic data structures in the network model: records and sets.

4.3.3.3.1.1. Records, Record Types, and Data Items: Data are stored in records. Each record consists of a group of related data values. Records are classified into record types, where each record type describes the structure of a group of records that store the same type of information. We give each record type a name and we also give a name and format (data type) for each data item (or attribute) in the record type. Figure 4.22 shows a record type STUDENT with data items NAME, SSN, ADDRESS, MAJOR DEPT, and BIRTH DATE.

We can declare a virtual data item (or derived attribute) AGE for the record type shown in Figure 4.22 and write a procedure to calculate the value of AGE from the value of the actual data item BIRTH DATE in each record.

STUDENT				
NAME	SSN	ADDRESS	MAJOR DEPT	BIRTH DATE

Data item name	Format
NAME	CHARACTER 30
SSN	CHARACTER 09
ADDRESS	CHARACTER 40
MAJOR DEPT	CHARACTER 10
BIRTH DATE	CHARACTER 09

Fig. 4.22. A record type student.

A typical database application has numerous record types—from a few to a few hundred. To represent relationships between records, the network model provides the modeling construct called set type, which we shall discuss next.

4.3.3.3.1.2 Set Types and Their Basic Properties : A set type is a description of a 1: N relationship between two record types. Figure 4.23 shows how we represent a set type diagrammatically as an arrow. This type of diagrammatic representation is called a Bachman diagram. Such a set type definition consists of three basic elements:

* A name for the set type
* A owner record type
* A member record type

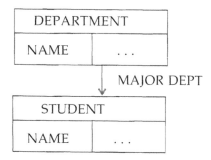

Fig. 4.23 The set type MAJOR DEPT.

The set type in Figure 4.23 is called MAJOR DEPT.; DEPARTMENT is the owner record type, and STUDENT is the member record type. This represents the 1: N relationship between academic departments and students majoring in those departments. In the database itself, there will be many set occurrences (or set instances) corresponding to a set type. Each instance relates one record from the owner record type—a DEPARTMENT record in our example—to the set of records from the member record type related to it, the set of STUDENT records for students who major in that department. Hence, each set occurrence is composed of:

* One owner record from the owner record type.
* A number of related members records (zero of more) from the member record type.

A record from the member record type **cannot exist in more than one set occurrence** of a particular set type. This maintains the constraint that a set type represents a 1: N relationship. In our example, a STUDENT record can be related to atmost one major DEPARTMENT and hence is a member of atmost one set occurrences of the MAJOR DEPT set type.

A set occurrence can be identified either by the **owner record** or by **any of the member records**. Figure 4.24 shows four set occurrences (instances) of the MAJOR DEPT set type. Notice that each set instance must have one owner record but can have any number of member records (zero or more). Hence, we usually refer to a set instance by its owner record. The four set instances in Figure 4.24, can be referred as the 'Computer Science', 'Mathematics', 'Physics', and 'Geology' sets. It is customary to use a different representation of a set instance (Figure 4.25) where the records of the set instance are shown linked together by pointers, which corresponds to a commonly used technique for implementing sets.

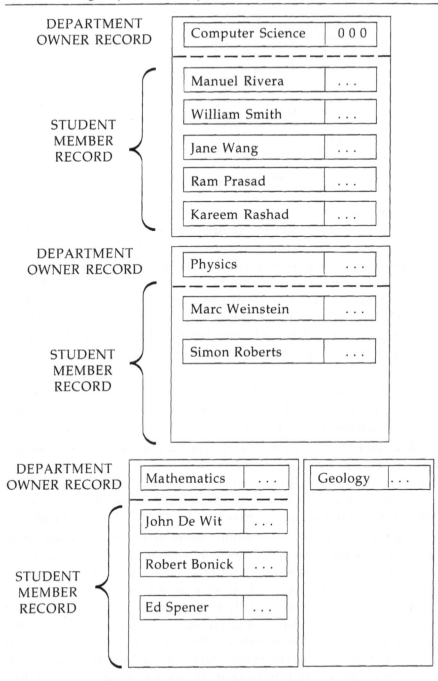

Fig. 4.24 Four set instances of the set type MAJOR DEPT.
(*Source* : Elmasri and Navathe, 2000).

In the network model, a set instance is not identical to the concept of a set in mathematics. There are two principal differences:

* The set instance has one distinguished element, the owner record, whereas in a mathematical set there is no such distinction among the elements of a set.

* In the network model, the member records of a set instance are ordered, whereas the order of elements is immaterial in a mathematical set. Hence, we can refer to the first, second, *i*th, and last member records in a set instance. Figure 4.25 shows an alternate "linked" representation of an instance of the set MAJOR DEPT. In Figure 4.25, the record of "Manuel Rivera" is the first STUDENT (member) record in the 'Computer Science' set, and that of 'Kareem Rashad' is the last member record. The set of the network model is sometimes referred as an owner-coupled set or co-set, to distinguish it from a mathematical set.

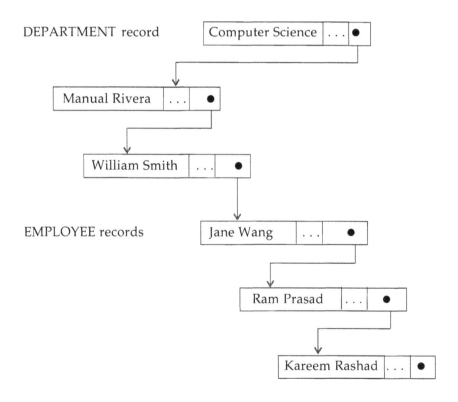

Fig. 4.25 Alternate representation of a set instance as a linked list
(*Source* : Elmasri and Navathe, 2000).

4.3.3.3.2. Special Type of Sets

System-owned (singular) sets

A system owned set is a set with no owner record type; instead, the system (DBMs software) is the owner. We can think of the system as a special "virtual" owner record type with only a single record occurrence. System-owned sets serve two main purposes in the network model:

- They provide entry points into the database via the records of the specified member record type. Processing can commence by accessing members of that record type, and then retrieving related records via other acts.

- They can be used to order the records of a given record type by using the set ordering specifications. By specifying several system-owned sets on the same record type, a user can access its records in different orders.

A system-owned set allows the processing of records of a record type by using the regular set operations. This type of set is called a singular set because there is only one set occurrence of it. The diagrammatic representation of the system-owned set ALL-DEPTS is shown in Figure 4.26, which allows DEPARTMENT records to be accessed in order of some field, say, NAME, with an appropriate set-ordering specification. Other special set types include recursive set types, with the same record serving as an owner and a member, which are mostly disallowed; multimember sets containing multiple record types as member in the same set type are allowed in some systems.

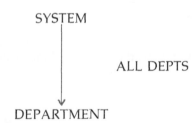

Fig. 4.26 A singular (SYSTEM-owned) set ALL-DEPTS
(*Source:* Elmasri and Navathe, 2000).

4.3.3.3.3 Stored Representations of Set Instances

A set instance is commonly represented as a ring (circular linked list) linking the owner record and all member records of the set, as shown in Figure 4.25. This is also sometimes called a circular chain. The ring representation is symmetric with respect to all records; hence, to

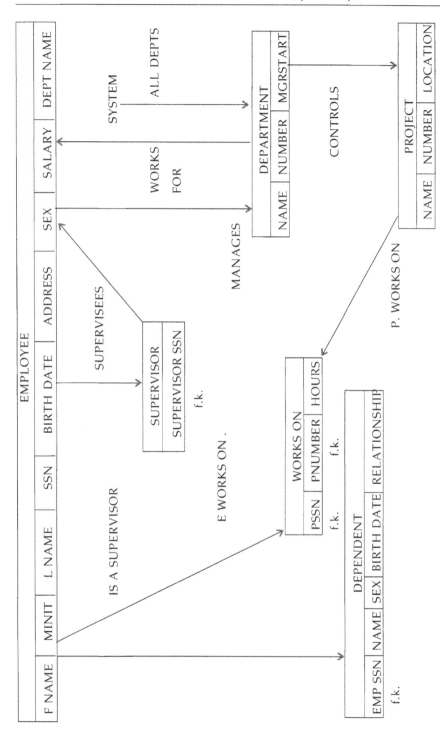

Fig. 4.27 A network schema diagram for the COMPANY database (*Source:* Elmasri and Navathe, 2000).

distinguish between the owner record and the member records, the DBMS includes a special field called the type field that has a distinct value (assigned by the DBMS) for each record type. By examining the type field, the system can tell whether the record is the owner of the set instance or is one of the member records. This type field is hidden from the user and is used only by the DBMS.

In addition to the type field, a record type is automatically assigned a **pointer field** by the DBMS for each set type in which it participates as owner or member. This pointer can be considered to be labelled with set type name to which it corresponds; hence, the system internally maintains the correspondence between these pointer fields and their set types. A pointer is usually called the next pointer in a member record and the first pointer in an owner record because these point to the next and first member records, respectively. In Figure 4.25, each student record has a NEXT pointer to the next student record within the set occurrence. The NEXT pointer of the last member record in a set occurrence points back to the owner record. If a record of the member record type does not participate in any set instance, its NEXT pointer has a special **nil** pointer. If a set occurrence has an owner but no member records, either the FIRST pointer points right back to the owner record itself or it can be **nil**.

The preceding representation of sets is one method for implementing set instances. In general, a DBMS can implement sets in various ways. However, the chosen representation must allow the DBMS to do all the following operations:

* Given an owner record, find all member records of the set occurrence.

* Given an owner record, find the first, ith, or last member record of the set occurrence. If no such record exists, return an exception code.

* Given a member record, find the next (or previous) member record of the set occurrence. If no such record exists, return an exception code.

* Given a member record, find the owner record of the set occurrence.

The circular linked list representation allows the system to do all of the preceding operations with varying degrees of efficiency. In general, a network database schema has many record types and set types, which means that a record type may participate as owner and member in numerous set types. For example, in the network schema that appears as Figure 4.27, the EMPLOYEE record type participates as an owner in four set. TYPES—MANAGES, IS_A_SUPERVISOR, E._WORKSON, and DEPENDENTS_OF—and participates as members in two set types WORKS_FOR and SUPERVISEES. In the circular linked list representation, six additional pointer fields are added to the EMPLOYEE record type. However, no confusion arises, because each pointer is labelled by the system and plays the role of FIRST or NEXT pointer for a *specific set type*.

4.3.3.3.4 *Using Sets to Represent M : N Relationships*

A set type represents a 1: N relationship between two record types. This means that a record of the member record type can appear in only one set of occurrence. This constraint is automatically enforced by the DBMS in the network model. To represent a 1:1 relationship, the extra 1:1 constraint must be imposed by the application program.

(a) - (c) Incorrect representations.

(d) Correct representation using a linking record type.

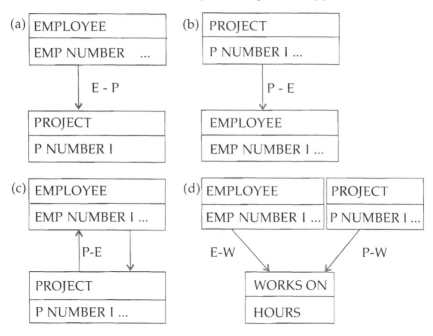

Fig. 4.28 Representing M : N relationships.

An M : N relationship between two record types cannot be represented by a single set type. For example, consider the WORKS ON relationship between EMPLOYEES and PROJECTS. Assume that an employee can be working on several projects simultaneously and that a project typically has several employees working on it. If we try to represent this by a set

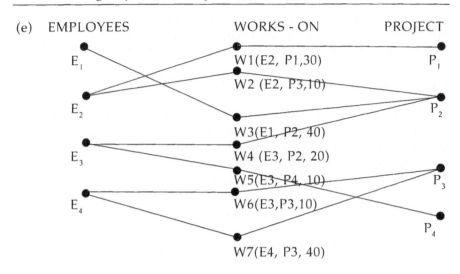

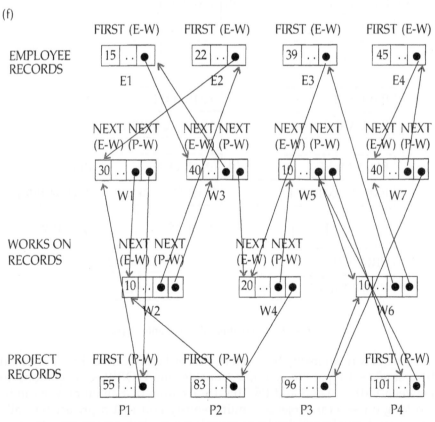

Fig. 4.28 (*Continued*) Representing M : N relationships (e) Some instances (f) using linked representation.

type, neither the set type in Figure 4.28(a) nor that in Figure 4.28(b) will represent the relationship correctly. Figure 4.28(a) enforces the incorrect constraint that a PROJECT record is related to only one EMPLOYEE record, whereas Figure 4.28(b) enforces the incorrect constraint that an EMPLOYEE record is related to only one PROJECT record. Using both set types E.P. and P.E. simultaneously as in Figure 4.28(c), leads to the problem of enforcing the constraint that P.F. and E.P. are mutually consistent inverses, plus the problem of dealing with relationship attributes.

The correct method for representing an M : N relationship in the network model is to use two set types and an additional record type, as shown in Figure 4.28(d). The additional record type WORKS ON, in the example, is called a linking (or Dummy) record type. Each record of the WORKS ON record type must be owned by one EMPLOYEE record through the E-W set and by one PROJECT record through the P-W set and serves to relate these two owner records. This is illustrated conceptually in Figure 4.28(e).

Figure 4.28(f) shows an example of individual record and set occurrences in the linked list representation corresponding to the schema in Figure 4.28(d). Each record of the WORKS ON record type has two ensuing pointers: the one marked NEXT (E-W) points to the next record in an instance of the E-W set, and the one marked NEXT (P-W) points to the ensuing record in an instance of the P-W set. Each WORKS ON record relates its two owner records. Each WORKS ON record also contains the number of hours per week that an employee works on a project. The same occurrences in Figure 4.28(f) are shown in Figure 4.28(e) by displaying the W records individually, without showing the pointers.

To find all projects that a particular employee works on, we start at the EMPLOYEE record and then trace through all WORKS ON records owned by that EMPLOYEE, using the FIRST (E-W) and NEXT (E-W) pointers. At each WORKS ON record in the set occurrence, we find its owner PROJECT record by following the NEXT (P-W) pointers until we find a record of type PROJECT record by following the NEXT (P-W) pointers until we find a record of type project. For example, for the E_2 EMPLOYEE record, we follow the FIRST (E-W) pointer in E_2 leading to W_1, the NEXT (E-W) pointer in W_1 leading to W_2 and the NEXT (E-W) pointer in W_2 leading back to E_2. Hence, W_1 and W_2 are identified as the member records in the set occurrence of E-W owned by E_2. By following the NEXT (P-W) pointer in W_1, we reach P_1 as its owner; and by following the NEXT (P-W) pointer in W_2 (and through W_3 and W_4), we reach P_2 as its owner. Notice that the existence of direct OWNER pointers for the P-W set in the WORK-ON records would have simplified the process of identifying the owner PROJECT record of each WORKS ON record.

In a similar fashion, we can find all EMPLOYEE records related to a particular project. In this case, the existence of owner pointers for the E-W set would simplify processing. All this pointer tracing is done automatically by DBMS; the programr has DML (Data manipulating language) commands for directly finding the owner or the next member.

Also notice that we could represent the M : N relationship as in Figure 4.28(a) or (b) if we were allowed to duplicate the PROJECT (or EMPLOYEE) records. In Figure 4.28(a), a PROJECT record would be duplicated as many times as there were employees working on the project. However, duplicating records creates problems in maintaining consistency among the duplicates whenever the database is updated and it is not recommended in general.

4.3.3.4 Hierarchial Data Structure

There are no original documents that describe the hierarchial model as there are for the relational and network models. The principles behind the hierarchial model are derived from Information Management System (IMS), which is the dominant hierarchial system in use today by a large number of banks, insurance companies, and hospitals as well as several government agencies. Another popular hierarchial DBMS is MRI's system 2000 (which was later sold by SAS Institute, as quoted in Elmasri and Navathe, 2000).

In this section, we shall include hierarchial database structures.

4.3.3.4.1 Hierarchial Database Structures

4.3.3.4.1.1 Parent-Child Relationships and Hierarchial Schemas : The hierarchial model employs two main data structuring concepts: records and parent-child relationships. A record is a collection of **field values** that provide information on an entity or a relationship instance. Records of the same type are grouped into records types. A record type is given a name, and its structure is defined by a collection of named **fields** or data items. Each field has a certain data type, such as integer, real or string.

A parent-child relationship type (PCR type) is a 1:N relationship between two record types. The record type on the one side is called the **parent record** type, and the one on the N-side is called the **child record type** of the PCR type. An occurrence (or instance) of the PCR type consists of one record of the parent record type and a number of records (zero or more) of the child record type.

A hierarchial database schema consists of a number of hierarchial schemas. Each hierarchial schema (or hierarchy) consists of a number of record types and PCR types.

A hierarchial schema is displayed as a hierarchial diagram, in which record type names are displayed in rectangular boxes and PCR types are displayed as lines connecting the parent record type to the child record type. Figure 4.29 shows a simple hierarchial diagram for a hierarchial schema with three record types and two PCR types. In record types are DEPARTMENT, EMPLOYEE, and PROJECT. Field names can be displayed under each record type name, as shown in Figure 4.29. In some diagrams, for brevity, we display only the record type names.

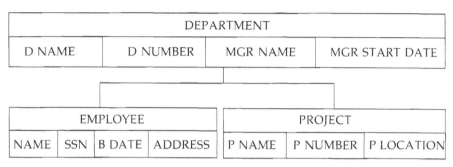

Fig. 4.29. A hierarchial schema.

We refer to a PCR type in a hierarchial schema by listing the pair (Parent record type, child record type) between parenthesis. The two PCR types in Figure 4.29 are DEPARTMENT, EMPLOYEE and DEPARTMENT, PROJECT. Notice that PCR types do not have a name in the hierarchial model. In Figure 4.29, each occurrence of the DEPARTMENT, EMPLOYEE PCR type relates one department record to the records of the many (zero or many) employees who work in that department. An occurrence of the DEPARTMENT, PROJECT PCR type relates a department records to the records of project controlled by that department. Figure 4.30 shows two PCR occurrences (or instances) for each of these two PCR types.

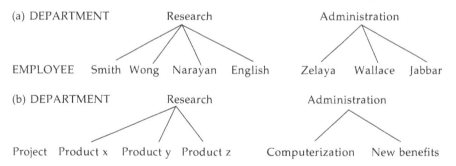

Fig. 4.30 Occurrences of Parent-Child Relationships.
(a) Two occurrences of the PCR type DEPARTMENT, EMPLOYEE.
(b) Two occurrences of the PCR type DEPARTMENT, PROJECT
(*Source* : Almasri and Navathe, 2000).

4.3.3.4.1.2 Properties of a Hierarchial Schema : A hierarchial schema of record types and PCR types must have the following properties:

1. One record type, called the root of the hierarchial schema, does not participate as a child record type in any PCR type.
2. Every record type except the root participates as a child record type in **exactly one** PCR type.
3. A record type can participate as a parent record type in any number (zero or more) of PCR types.
4. A record type that does not participate as a parent record type in any PCR type is called a leaf of the hierarchial schema.
5. If a record type participates as parent in more than one PCR type, then its child record types are ordered. The order is displayed, by convention, from left to right in a hierarchial diagram.

The definition of a hierarchial schema defines a **tree data structure**. In the terminology of tree data structure, a record type corresponds to a **node** of tree and a PCR type corresponds to an **edge (or arc)** of the tree. We use the term *node* and *record type*, and *edge* and *PCR type*, inter-changeably. The usual convention of displaying a tree is slightly different from that used in hierarchial diagrams, in the sense that each tree edge is shown separately from other edges (Figure 4.31). In hierarchial diagrams, the convention is that all edges emanating from the same parent node are joined together (as in Figure 4.29). We use this latter heirarchial diagram convention.

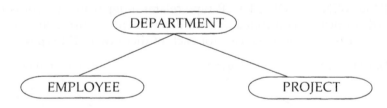

Fig. 4.31 A tree representation of the hierarchial schema in Figure 4.29.

The preceding properties of a hierarchial schema mean that every node except the root has exactly one parent node. However, a node can have several child nodes, and in this case they are ordered from left to right. In Figure 4.29, EMPLOYEE is the first child of DEPARTMENT, and PROJECT is the second child. The previously identified properties also limit the types of relationships that can be represented in a hierarchial schema. In particular, M : N relationships between record types cannot be directly represented, because parent-child relationships are 1: N relationships, and record type cannot participate as a child in two or more distinct parent-child relationships.

An M : N relationship may be handled in the hierarchial model by allowing the duplication of child record instances. For example, consider an M : N relationship between EMPLOYEE and PROJECT, where a project can have several employees working on it, and an employee can work on several projects. We can represent the relationship as a (PROJECT, EMPLOYEE) PCR type. In this case, a record describing the same employee can be duplicated by appearing once under each project that the employee works for. Alternatively, we can represent the relationship as an EMPLOYEE, PROJECT PCR type, in which case, the project records may be duplicated.

Example: Consider the following instances of the EMPLOYEE-PROJECT relationship:

Project	*Employees working on the Project*
A	E_1, E_3, E_5
B	E_2, E_4, E_6
C	E_1, E_4
D	E_2, E_3, E_4, E_5

If these instances are stored using the hierarchial schema (PROJECT, EMPLOYEE) PCR type, one for each project, the employee records for E_1, E_2, E_3 and E_5 will appear twice, each as child records, however, because each of these employees works on two projects. The employee record for E_4 will appear three times—once each of project B, C, and D and may have number of hours that E_4 works on each project in the corresponding instances.

To avoid such duplication, a technique is used whereby several hierarchial schemes can be specified in the same hierarchial database schema. A relationship like the preceding PCR type can now be defined across different hierarchial schema. This techique, called **virtual relationship**, causes a departure from the "strict" hierarchial model.

4.3.3.4.1.3 Hierarchial Occurrence Trees : Corresponding to a hierarchial schema, many hierarchial occurrences, also called occurrence trees, exist in the database. Each one is a tree structure whose root is a single record from the root record type. The occurrence tree also contains all the children record occurrences of the root record and continues all the way to records of the leaf record types.

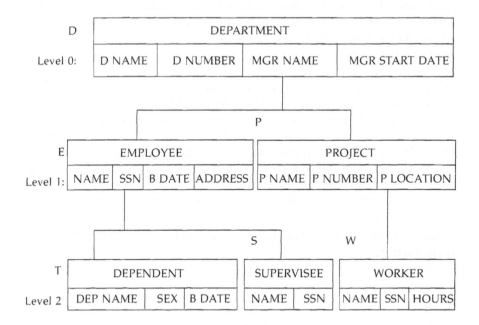

Fig. 4.32 A hierarchial schema for part of the COMPANY database
(*Source:* Elmasri and Navathe, 2000).

For example, consider the hierarchial diagram shown in Figure 4.32, which represents part of the COMPANY database. Figure 4.33 shows one hierarchial occurrence tree of this hierarchial schema. In the occurrence tree, each **node** is a **record occurrance**, and each arc represents a Parent-child relationship between two records. In both Figures 4.32 and 4.33, we use the characters D, E, P, T, S and W to represent **type indicators** for the record types DEPARTMENT, EMPLOYEE, PROJECT, DEPARTMENT, SUPERVISEE and WORKER, respectively. A node N and all its descendent nodes forms a subtree of node N. An **occurrence tree** can be defined as the subtree of a record type whose type is of the root record type.

4.3.3.4.1.4 Linearized Form of a Hierarchial Occurrence Tree : A hierarchial occurence tree can be represented in storage by using any of a variety of data structure. However, a particular simple storage structure that can be used is the **hierarchial record**, which is a linear ordering of the records in an occurrence tree in the **preorder traversal** of the tree. This order produces a sequence of record occurrences known as the **hierarchial sequence (or hierarchial record sequence)** of the occurrence tree; it can be obtained by applying a recursive procedure called the **pre-order traversal**, which visits nodes depth first and in a left-to-right fashion.

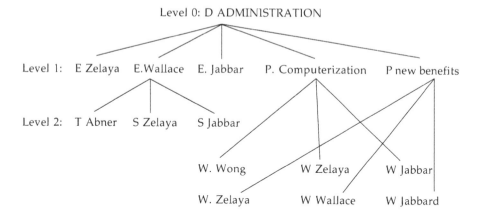

Fig. 4.33 An occurrence tree of the schema in Figure 4.32.

The occurrence tree in Figure 4.33 gives the hierarchial sequence shown in Figure 4.34. The system stores the type indicator with each record so that the record can be distinguished within the hierarchial sequence. The hierarchial sequence is also important because hierarchial data-manipulation languages such as that used in IMS (International Management System), use it as a basis for defining hierarchial database operations. The HDML (Hierarchial Data Manipulation Language) is a simplified version of DL/1 (Data Language one) based on the hierarchial sequence. A **hierarchial path** is a sequence of nodes $N_1, N_2, ..., N_i$, where N_1 is the root of a tree and N_1 is a child of N_{1-1} for $J = 2, 3 ...., i$. A hierarchial path can be defined as either a hierarchial schema or on an occurrence tree. We can now define a **hierarchial database occurrence** as a sequence of all the occurrence trees of a hierarchial schema. For example, a hierarchial database occurrence of the hierarchial schema shown in Figure 4.32 would consist of a number of occurrence trees similar to the one shown in Figure 4.33, one for each distinct department.

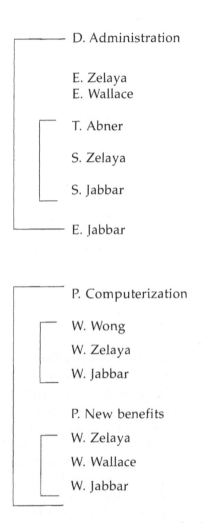

Fig. 4.34 Hierarchial sequence for the occurrence tree in Figure 4.33.

4.3.3.4.1.5 Virtual Parent-Child Relationships : The hierarchial model has problems in modeling certain types of relationships. These include the following relationships and situations:

1. M : N relationships.
2. The case where a record type participates as a child in more than one PCR type.
3. N-ary relationships with more than two participating record types.

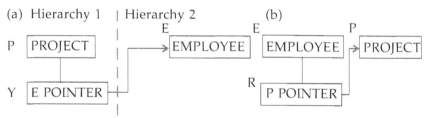

Fig. 4.35 Representing M : N relationships using virtual parent-child relationships
(a) EMPLOYEE as virtual parent.
(b) Project as virtual parent.

Notice that the relationship between EMPLOYEE and E POINTER in Figure 4.35(a) is a 1: N relationship and hence qualifies as a PCR type. Such a relationship is called a **virtual parent-child relationship** (VPCR) type. The EMPLOYEE is called the virtual parent of E POINTER; and conversely, E POINTER is called a **virtual child** of EMPLOYEE. Conceptually, PCR types and VPCR types are similar. The main difference between the two lies in the way they are implemented. A PCR type is usually implemented using the hierarchial sequence, whereas a VPCR type is usually implemented on establishing a pointer, (a physical one containing an address, or a logical one containing a key) from a virtual child record to its virtual parent record. This mainly affects the efficiency of certain queries.

Figure 4.36 shows a hierarchial database schema of the COMPANY database that uses some VPCRs and has no redundancy in its record occurrences. The hierarchial database schema is made up of two hierarchial schemas—one with root DEPARTMENT, and the other with root EMPLOYEE. Four VPCRs, all with virtual parent EMPLOYEE, are included to represent the relationships without redundancy. Notice that IMS may not allow this because an implementation constraint in IMS limits a record to being a virtual parent of atmost one VPCRs; to get around this constraint, one can create dummy children record types of EMPLOYEE in Hierarchy 2 so that each VPCR points to a distinct virtual parent record type.

In general, there are many feasible methods of designing a database using the hierarchial model. In many cases, performance considerations are the most important factor in choosing one hierarchial database schema over another. Performances depends on the implementation options available; for example, whether certain types of pointers are provided by the system and whether certain limits on number of levels are imposed by the DBA (Database Administrators).

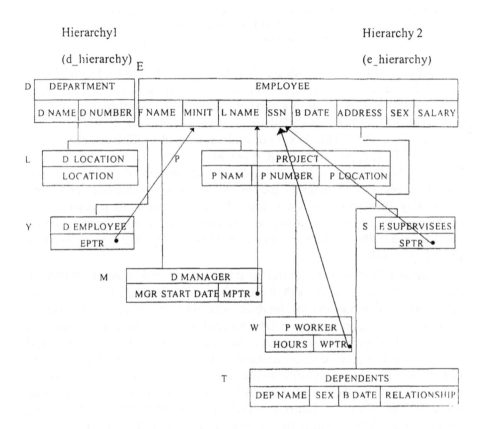

Fig. 4.36 Using VPCR types to eliminate redundancy in the COMPANY database.

4.4 DATA SYSTEMS

4.4.1 Centralized Data System

4.4.1.1 *Centralized DBMS (Database Management System) Architect*

Architecture for DBMSs follows trends similar to those of general computer systems architectures. Earlier, architectures used mainframe computers to provide the main processing for all functions of the system, including user applications programs, user interface programs, as well as all the DBMS functionality. The reason was that most users accessed such systems via computer terminals that did not have processing power and only provided display capabilities. So, all processing was performed by remote, and only display information and controls were sent from the computer to the display terminals, which were connected to the central computer via various types of communications networks.

As the prices of hardware declined, most users replaced their terminals with personal computers (PCs) and workstations. At first, database systems used these computers in the same way as they had used display terminals, so that the DBMS itself was still a centralized DBMS where all the DBMS functionality, application program execution, and user interface processing were carried out in one machine. Figure 4.37 illustrates the physical components in a centralized architecture. Gradually DBMS systems started to exploit the available processing power at the user side, which led to client-server DBMS architectures (*Source*: Elmasri and Navathe, 2000).

4.4.1.2 *Client-Server Architecture*

The client-server architecture was developed to deal with computing environments where a large number of PCs, workstations, file servers, printers, database servers, web servers and other equipments are connected together via a network. The idea is to define specialized servers with specific functionalities. For example, it is possible to connect a number of PCs, or small workstations as clients to a file server that maintains the files of the client machines. Another machine could be designated as a printer server by being connected to various printers; thereafter, all print requests by the clients are forwarded to the machine. Web servers or E-mail servers also fall into the specialized server category. In this way, the resources provided by specialized servers can be accessed by many client machines. The client machines provide the user with the appropriate interfaces to utilize these servers as well as with local processing power to run local applications. This concepts can be carried over to a software with specialized technology such as a DBMS or a CAD (Computer aided design) package—being stored on specific server machines and being made accessible to multiple clients. Figure 4.38 illustrates client-server architecture at the logical level, and Figure 4.39 is a simplified diagram that shows how the physical architecture would look. Some machines would be only client sites (for example, diskless workstations or workstations/PCs with disks that have only client software installed). Other machines would be dedicated servers. Still other machines would have both client and server functionality.

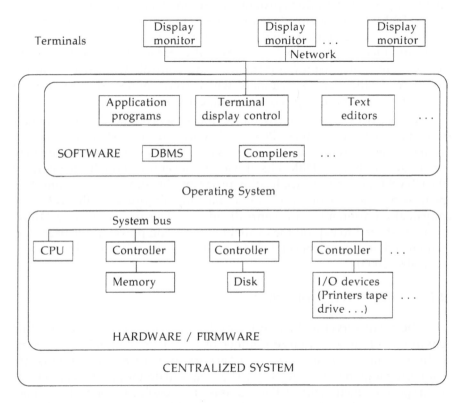

Fig. 4.37. Diagrams to illustrate different architectures.
Physical centralized architecture.

The concept of client-server architecture assumes an underlying framework that consists of many PCs and works-stations as well as a number of mainframe machines connected via local area networks and other types of computer networks. A **client** in this framework is typically a user machine that provides user interface capabilities and local processing. When a client requires access to additional functionality—

such as database access—that does not exist at that machine, it connects to a server that provides the needed functionality. A server is a machine that can provide services to the client machines such as printing, archiving, or database access.

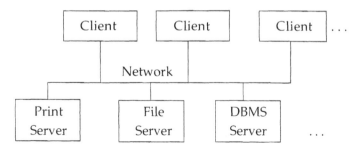

Fig. 4.38. Diagram to illustrate different architectures.
Simplified logical client-server architecture.

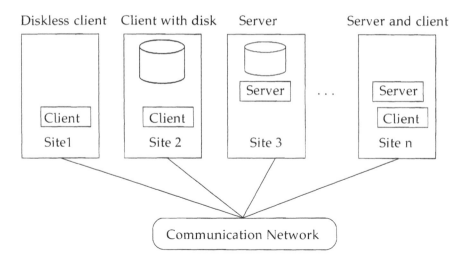

Fig. 4.39. Diagrams to illustrate different architectures. Simplified physical client-server architecture (*Source* : Elmasri and Navathe, 2000).

4.4.1.3 *Client-Server Architectures for DBMSs*

The client-server architecture is being increasingly incorporated into commercial DBMS package. In relation DBMS—many of which started as centralized systems—the system components that were first moved to the client side were the user iterface and application programs. Because SQL

(Standard query language) provided a standard language for RDBMSs, it created a logical dividing point between a client and a server. Hence, the query and transaction functionality remained at the server side. In such an architecture, the server is often called a **query server** or transaction server, because it provided these two functionalities. In RDBMSs, the server is also often called an **SQL server**, since most RDBMS servers are based on SQL language and standard.

In such a client-server architecture, the user interface programs and application programs can run at the client's side. When DBMS access is required, the program establishes a connection to the DBMS—which is on the server side—and once the connection is started, the client program can communicate with the DBMS. A standard called Open Database Connectivity (ODBC) provides an Application Programming Interface (API), which allows client-side programs to call the DBMS, as long as both client and server machines have the necessary software installed. Most DBMS vendors provide ODBC drivers for their systems. Hence, a client program can actually connect to several RDBMSs and send query and transaction requests using the ODBC API, which are processed at the server sites. Any query results are sent back to the client program, which can process or display the results as needed. Another related standard for the JAVA programming language, called JDBC, has also been defined. This allows JAVA client programs to access the DBMS through a standard interface.

The second approach to client-server was taken by some object-oriented DBMSs. Because many of those systems were developed in the era of client-server architecture, the approach taken was to divide the software modules of the DBMS between client and server in a more integrated way. For example, the server level may include the part of the DBMS software responsible for handling data storage on disk pages, local concurrency control and recovery, buffering and caching of disk pages, and other such functions. Meanwhile, the **client level** may handle the user interface, data dictionary functions, DBMS interaction with prog-ramming language compilers, global query optimization/concurrency control/ recovery, structuring of complex objects from the data in the buffer, and other such functions. In this approach, the client-server interaction is more tightly coupled and is done internally by the DBMS modules—some of which reside in the client rather than by the users. The exact division of functionality varies from system to system. In such a client-server architecture, the server has been called **data server** because it provides data in disk pages to the client, which can then be structured into objects for the clients programs by the client-side DBMS software itself.

4.4.2 Hierarchial Data System

Hierarchial Data Structure has been explained in subsection 4.3.3.4. Now we will discuss on Integrity Constraints and Data Definition in the Hierarchial model.

4.4.2.1 *Integrity Constraints in the Hierarchial Model*

A number of built-in **inherent constraints** exist in the hierarchial model whenever we specify a hierarchial schema. These include the following constraints:

1. No record occurrences except root records can exist without being related to a parent record occurence.This has the following indications:

    (a) A child record cannot be inserted unless it is linked to a parent record.

    (b) A child record may be deleted independently of its parents. However, deletion of a parent record automatically results in deletion of all its child and descendent records.

    (c) The above rules do not apply to virtual child records and virtual parent records.

2. If a child record has two or more parent records from the same record type, the child record must be duplicated once under each parent record.

3. A child record having two or more parent records of *different* record types can do so only by having at most one real parent, with all the others represented as virtual parents. IMS limits the number of virtual parents to one.

4. In IMS, a record type can be the virtual parent to only one VPCR type . That is, the number of virtual children can be only one per record type in IMS.

4.4.2.2 *Data Definition in the Hierarchial Model*

HDDL (Hierarchial Data Definition Language) is not a languge of any specific hierarchial DBMS but is used to illustrate the language concepts for a hierarchial database. The HDDL demonstrates how a hierarchial database schema can be defined. To define a hierarchial database schema, we must define the fields of each record type, the data type of each field, and any key constraints in the fields. In addition, we must specify a root record type as such; and for every non-root record type, we must specify its (real) parent in the PCR type. Any VPCR types must also be specifed.

In Figure 4.40, either each record type is declared to be of type root or a single (real) parent record type is declared for the record type. The data items of the record are then listed along with their data type.

```
SCHEMA NAME = COMPANY
HIERARCHIES = HIERARCHY 1, HIERARCHY 2
RECORD
            NAME = EMPLOYEE
            TYPE = ROOT OF HIERARCHY 2
DATA ITEMS
            F NAME        CHARACTER 15
            MINIT         CHARACTER 1
            L NAME        CHARACTER 15
            SSN           CHARACTER 9
            B DATE        CHARACTER 9
            ADDRESS       CHARACTER 30
            SEX           CHARACTER 1
            SALARY        CHARACTER 10

            KEY = SSN
            ORDER BY L NAME, F NAME
RECORD
            NAME = DEPARTMENT
            TYPE = ROOT OF HIERARCHY 1
            DATA ITEM =
                D NAME   CHARACTER 15
                D NUMBER          INTEGER
            KEY = D NAME
            KEY = D NUMBER
            ORDER BY D NAME
RECORD
            NAME = D LOCATIONS
            PARENT = DEPARTMENT
            CHILD NUMBER = 1
            DATA ITEM =
            LOCATION    CHARACTER 15
RECORD
            NAME = D MANAGER
            PARENT = DEPARTMENT
```

Fig. 4.40 (continued)

```
                CHILD NUMBER = 3
                DATA ITEM =
                    MGR START DATE        CHARACTER 9
                    MPTR POINTER  WITH VIRTUAL PARENT
                    = EMPLOYEE
RECORD
                NAME = PROJECT
                PARENT = DEPARTMENT
                CHILD NUMBER = 4
                DATA ITEM =
                        P NAME        CHARACTER 15
                        P NUMBER      INTEGER
                        P LOCATION  CHARACTER 15
                KEY = P NAME
                KEY = P NUMBER
                ORDER BY P NAME
RECORD
                NAME = P WORKER
                PARENT = PROJECT
                CHILD NUMBER = 1
                DATA ITEM =
                        HOURS         CHARACTER 4
                        WPTA, POINTER WITH VIRTUAL PARENT
                        = EMPLOYEE
RECORD
                NAME = D EMPLOYEES
                PARENT = DEPARTMENT
                CHILD NUMBER = 2
                DATA ITEM =
                        EPTR POINTER WITH VIRTUAL PARENT
                        = EMPLOYEE
RECORD
                NAME OF SUPERVISEES
                PARENT = EMPLOYEE
                CHILD NUMBER = 2
                DATA ITEM =
                        SPTR POINTER WITH VIRTUAL PARENT
                        = EMPLOYEE
RECORD
                NAME = DEPARTMENT
                PARENT = EMPLOYEE
                CHILD NUMBER = 1
```

Fig. 4.40 (continued)

DATA ITEM =	
DEP NAME	CHARACTER 15
SEX	CHARACTER 1
BIRTH DATE	CHARACTER 9
RELATIONSHIP	CHARACTER 10
ORDER BY DESC BIRTH DATE	

Fig. 4.40 HDDL declarations for the hierarchial schema in Figure 4.36
(*Source*: Elmasri and Navathe, 2000).

We must specify a virtual parent for data items that are of type pointer. Data items declared under the KEY clause are constrained to have unique values for each record. Each KEY clause specifies a separate key; in addition, if a single KEY clause lists more than one field, the combination of these field values must be unique in each record. The CHILD NUMBER clause specifies the left to right order of a child record type under its (real) parent record type. The ORDER BY clause specifies the order of individual records of the same record type in the hierarchial sequence. For non-root record types, the ORDER BY clause specifies how the records should be ordered *within each parent record* by specifying a field called a **sequence key**. For example, PROJECT records controlled by a particular DEPART-MENT have their subtrees order alphabetically within the same parent DEPARTMENT record by P NAME, according to Figure 4.40.

4.4.2.3 *Data Manipulation Language for the Hierarchial Model*

Hierarchial Data Manipulation Languages (HDML) is a record-at-a-time language for manipulating hierarchial databases. Its strcuture has been based on IMS's DL/1 language. It is introduced to illustrate the concepts of a hierarchial database manipulation language. The commands of the language must be embedded in a general-purpose programming language called the **host language**.

The HDML is based on the concept of *hierarchial sequence* defined in Section 4.3.3.4. Following each database command, the last record accessed by the command is called the current **database record**. The DBMS maintains a pointer to the current record. Subsequent database commands *proceed from the current record* and may define a new current record, depending on the type of command.

4.4.2.3.1 *The GET Command*

The HDML command for retrieving a record is the GET **command**. There are many variations of GET; the structure of the two of these variations is as follows, with optional parts enclosed in brackets [...]:

* GET FIRST <record type name>

[WHERE   <condition>]

* GET NEXT <record type name>

[WHERE   <condition>]

The simplest variation is the GET FIRST **command**, which always starts searching the database from the beginning of the *hierarchial sequence* until it finds the first record occurrence of <record type name> that statisfies <condition>. This record also becomes the current of database, current of hierarchy, and current of record type and is retrieved into the corresponding program variable. For example, to retrieve the "first" EMPLOYEE record in the hierarchial sequence whose name is John Smith, EX1.

EX 1: **GET FIRST** EMPLOYEE **WHERE**

F NAME = 'John' and LNAME = 'Smith';

The DBMS uses the condition following WHERE to search for the first record in order of the hiererchial sequence that satisfies the condition and is of the specified record type. If more than one record in the database satisfies the WHERE condition and want to retrieve all of them, one must write a looping construct in the host program and use the GET NEXT command. Assume that the GET NEXT starts its search from the *current record of the record type* specifies in GET NEXT, and it searches forward in the hierarchial sequence to find another record of the specified type satisfying the WHERE condition. For example, to retrieve records of all EMPLOYEEs whose salary is less than $20,000 and obtain a printout of their names, one can write the program segment shown in EX2:

EX 2: **GET FIRST** EMPLOYEE **WHERE**

SALARY <'20000.00';

while DB-STATUS = 0 do

Begin

writeln (P - EMPLOYEE. FNAME, P-EMPLOYEE. LNAME);

$ GET NEXT EMPLOYEE **WHERE** SALARY < '20000.00'

end;

In EX2, the while loop continues until no more EMPLOYEE records in the database satisfy the WHERE condition; hence, the search goes through to the last record in the database (hierarchial sequence). When no more records are found, DB-STATUS becomes non-zero, with a code indicating "end of database reached", and the while loop terminates.

4.4.2.3.2 The GET PATH and GET NEXT WITHIN PARENT

Retrieval Commands

So far it has considered retrieving single records by using the GET command. But when it has to locate a record deep in the hierarchy, the retrieval may be based on a series of conditions on records along the entire hierarchial path. To accommodate this, GET PATH command is introduced:

GET (FIRST | NEXT) PATH < hierarchial path > [WHERE <condition>]

Here, <hierarchial paths > is a list of record types that starts from the root along a path in the hierarchial schema, and <condition> is a Boolean expression specifying conditions on the individual record types along the path. Because several record types may be specified, the field names are prefixed by the record type names in <condition>. For example, consider the following query: "List the last name and birth dates of all employee-dependent pairs, where both have the first name John". This is shown in EX3:

EX 3 $ GET FIRST PATH EMPLOYEE, DEPENDENT

WHERE EMPLOYEE. FNAME = "John' and DEPENDENT. DEPNAME = 'John';

while DB-STATUS = 0 do

begin

writeln (P - EMPLOYEE. FNAME, P-EMPLOYEE. BDATE;

P-DEPENDENT. BIRTHDATE);

$GET NEXT PATH EMPLOYEE, DEPENDENT WHERE EMPLOYEE.FNAME 'John' and DEPENDENT, DEPNAME = 'John'

end;

Assume that a GET PATH command retrives *all records along the specified paths* into the user work area variables and the last record along the path becomes the current database record. In addition, all records along the path become the current records of their respective record types.

Another common type of query is to find all records of a given type that have *the same parent record*. In this case, it needs the GET NEXT WITHIN PARENT command, which can be used to loop through the child records of a parent record and has the following format;

GET NEXT <Child record type name > WITHIN [VIRTUAL] PARENT [<parent record type name >]

[WHERE <condition>]

This command retrieves the next record of the child record type searching forward from the current of the child record type for the next child record owned by the current parent record. If no more child records are found, DB STATUS is set to a non-zero value to indicate that "there are no more records of the specified child record type that have the same parent as the current parent record". The <parent record type name> is *optional*, and the default is the immediate (real) parent record type of <child record type name >. For example, to retrieve the names of all projects controlled by the "Research' department, we can write the program segment shown in EX4.

> EX4: $GET FIRST PATH DEPARTMENT, PROJECT
>
> WHERE D NAME = 'RESEARCH';

(This establishes the 'Research' DEPARTMENT record as current parent of type DEPARTMENT, and retrieves the first child project record under that DEPARTMENT record.)

> while DB - STATUS = 0 do
>
> begin
>
> writeln (P- PROJECT.PNAME);
>
> $GET NEXT PROJECT WITHIN PARENT
>
> end;

4.4.2.3.3 HDML Commands for Update

The HDML commands for updating a hierarchial database are shown in Table 4.3, along with the retrieval command. The INSERT command is used to insert a new record. Before inserting a record of a particular record type, we must first place the field values of the new record in the appropriate user work area program variable.

Table 4.3 Summary of HDML Commands

RETRIEVAL	
GET	Retrieve a record into the corresponding program variable and make it the current record. Variations include GET FIRST, GET NEXT, GET NEXT WITHIN PARENT, and GET PATH.
RECORD UPDATE	
INSERT	Store a new record in the database and make it the current record.
DELETE	Delete the current record (and its subtree) from the database.
REPLACE	Modify some fields of the current record.
CURRENCY RETENTION	
GET HOLD	Retrieve a record and hold it as the current record so it can subsequently be deleted or replaced.

(*Source* : Elmasri and Navathe, 2000)

The INSERT command inserts a record into the database. The newly inserted record also becomes the current record for the database, its hierarchial scheme, and its record type. If it is a root record, it creates a new hierarchial occurrence tree with the new record as root. The record is inserted in the hierarchial sequence in the order specified by any ORDER BY fields in the schema definition.

To insert a child record, one should make its parent, or one of the sibling records, the current record of the hierarchial schema before issuing the INSERT command. One should also set any virtual parent pointers before inserting the record.

To delete a record from the database, one must first make it the current record and then issue the DELETE command. The GET HOLD is used to make the record the current record, where the HOLD key word indicates to the DBMS that the program will delete or update the record just retrieved. For example, to delete all male EMPLOYEEs, one can use EX5, which also lists the deleted employee names <before> deleting their records:

EX5:   **$GET HOLD FIRST** EMPLOYEE
         WHERE SEX = 'M';
         while DB-STATUS = 0 then
         begin
         writeln (P - EMPLOYEE.LNAME,
         P - EMPLOYEE.FNAME);
         $DELETE EMPLOYEE;
         $ **GET HOLD NEXT** EMPLOYEE
         WHERE SEX = 'M';
         end;

4.4.2.3.4 IMS—A Hierarchial DBMS

IMS, one of the earliest DBMSs, ranks as the dominant system in the commercial market for support of large-scale accounting and inventory systems. IBM (International Business Machines) manuals refer to the full product as IMS/VS (Virtual storage), and typically the full product is installed under the MVS operating system. IMS DB/DC is the term used for installations that utilize the product's own subsystems to support the physical database (DB) and to provide data communication (c).

However, the other important versions exist that support only the IMS data language—Data Language one (DL/1). Such DL/1 - only configurations can be implemented under MVS, but they may also use the DOS/VSE operating system. These systems issue their calls to VSAM files and use IBM's

Customer Information Control System (CICS) for data communications. The trade off is a sacrifice of support features for the sake of simplicity and improved throughout.

A number of versions of IMS have been marketed to work with various IBM operating system, including (among the recent systems) OS/VS1, OS/VS2, MVS, MVS/XA, and ESA. The system comes with various options. IMS runs under different versions on the IBM370 and 30 XX family of computers. The data definition and manipulation language of IMS is DL/1. Application programs written in COBOL, PL/1, FORTRAN, and BAL (Basic Assembly Language) interface with DL/1.

REFERENCES

Boillot, M.H., Gleason, G.M., and Horn, L.W. (1975). Essentials of flow-charting. 114p. Dubuque: W.C. Brown Co.

Chapin, N. (1971). Flowcharts. 173p. Princeton: Aurbach Publishers.

Codasyl. (1978). Data Description language journal of development. Canadian Government Publishing Centre, 1978.

Codd, T. (1970). A relational model for large shared data bank, CACM, 13-6, June, 1970.

DBTG, (1971). Report of the CADASYL Data Base Task Group, ACM, April 1971.

Dent, J.B. and Blackie, M.J. (1979). System simulation in agriculture. 180p. London; Applied Science Publishers Ltd.

Edward, P. and Broadwell, B. (1974). Flowcharting and Basic. 214p. New York: Harcour Brace Jovanovich, Inc.

Elmasri, R. and Navathe, S.B. (2000). Fundamentals of database systems. 955p. + 30 p. (Bibliography) + 24 p. (Index). Mass: Addison Wesley Publishing Co., an imprint of Pearson Education.

Fridgen, J.J., Kitchen, N.R., Sudduth, K.A., Drummond, S.T., Wiebold, W.J. and Fraisse, C.W. (2004). Management zone analyst (MZA): software for subfield management zone delineation. Agron. J. **96** (1): 100-108.

Gold, H.J. (1977). Mathematical modeling of biological system: an introductory guidebook. 357p. New York: John Wiley & Sons.

Jones, J.W. and Luyten, J.C. (1998). Simulation of biological processes. In: Agricultural Systems Modeling and simulation. (eds. Peart, R.M. and Curry, R.B.) pp. 19-62. New York: Marcel Dekker, Inc.

Liebig, M.A., Miller, M.E., Varvel, ·G.E., Doran, J.W. and Hanson, J.D. (2004). AEPAT: Software for assessing agronomic and environmental performances of management practices in long-term agroecosystem experiments. Agron. J. **96** (1): 109-115.

Sack, J. and Meadows, J. (1973). Entering Basic. 133p. Chicago: Science Research Associates, Inc.

Spain, J.D. (1982). Basic microcomputer models in biology. 354p. London: Addison-Wesley Publishing Co., Advanced Book Program/World Science Division, Reading, Massachusetts.

Spencer, D.D. (1975). A guide to Basic Programming. 242p. Reading: Addison-Wesley Publishing Co.

Watt, K.E.F. (1968). Ecology and Resource Management. 450p. New York; McGraw-Hill Co.

5

Model Testing and Validation

5.1 SENSITIVITY ANALYSIS

The purpose of sensitivity analysis is to study the behavior of the model. Sensitivity analysis can also be structured to determine important subsystems, relationships, and inputs. Sensitivity analysis should be designed with respect to the objectives of the study. A sensitivity analysis is the process by which parameters or inputs are evaluated with regard to their effects on simulated results. For example, rainfall would be an environmental input to a crop water balance model for estimating crop irrigation requirements. Soil characteristics such as soil water-holding capacity or rootzone depth are examples of model parameters. One may be interested in the sensitivity of estimated irrigation requirements to changes in rainfall, changes in soil charactersitics or both (Jones and Luyten, 1998).

Computer graphics are highly useful in sensitivity analysis. Graphical display of simulated results for a number of parameter values or inputs provides a visual image of model behavior over a range of parameter values. A mathematical approach to sensitivity analysis may also be an useful way to organize and present model behavior. This approach compares the changes in one or more simulated outputs relative to the change in one or more parameters by approximating partial derivatives using numerical results. Absolute sensitivity, $\sigma\,(y\,|\,k)$, of some model output y to a variable k (a parameter of the model or input of the system) is defined by Jones and Luyten (1998).

$$SI\,(y\,|\,k) = \sigma\,(y\,|\,k) = \frac{\partial y}{\partial k} = \frac{y\,(k+\Delta k/2) - y\,(k-\Delta k/2)}{\Delta k} \tag{5.1}$$

Relative sensitivity $\sigma_r\,(y\,|\,k)$, is often used to provide a normalized measure to compare the sensitivity of a model to several variables. Relative sensitivity is defined by Jones and Luyten (1998).

$$\sigma_r \, (y \,|\, k) = SI_r \, (y \,|\, k) = \frac{\partial y \,/\, y}{\partial k \,/\, k} = \sigma \, (y \,|\, k) \, \frac{k}{y} \tag{5.2}$$

For example, if $SI_r \, (y \,|\, k) = 2.0$ and k is changed by 3% ($\Delta k / k = 0.03$), we can expect y to change by 6%.

$$\left[\frac{\partial y}{y} = 2 \frac{\partial k}{k} = 2 \times 3 = 6 \right]$$

Also, if $\sigma_r \, (y \,|\, k_1) = 0.5$ and $\sigma_r \, (y \,|\, k_2) = 1.5$,

then y is more sensitive to a percentage change in k_2 than the same percentage change in k_1 (3 times as sensitive).

$$\left[\frac{\partial y}{y} = 0.5 \frac{\partial k_1}{k_1} = 0.5 \times 3 = 1.5; \frac{\partial y}{y} = 1.5 \frac{\partial k_2}{k_2} = 1.5 \times 3 = 4.5 \right]$$

In simulation studies, Δk is used to represent a small change in k so that changes in y, Δy can be simulated. Then equations (5.1 and 5.2) can be used to approximate $SI \, (y \,|\, k)$ and $SI_r \, (y \,|\, k)$ using the discrete changes in k and y, or $\partial y / \partial k \cong \Delta y / \Delta k$. By varying parameters through their expected range of extremes, one can use this approach to compare the sensitivity of model results.

Sensitivity analysis usually begins with the selection of model output results that are considered to be crucial to the study. A set of "base" conditions are then established usually comprising the set of the best estimates of each parameter and input. The base results are the simulation outputs obtained when base condition values are used. A range of values are then selected representing the extreme conditions associated with each parameter to be evaluated. Simulation runs are made using each value while holding all other base conditions constant. A comparison between changes in the base condition values and change in the base results provides an indication of the relative importance of the variable. Results can be compared in graphs or Tables or by computing the sensitivity variable values.

Regression analysis are also useful techniques in sensitivity analysis. Computer experiments are conducted by varying parameters over a range of interest followed by regressing simulated results against one or more parameters to determine whether outputs are effected by changes in the parameter, and if so, to what extent (Jones and Luyten, 1998).

Kanneganti et al. (1998) calculated the sensitivity index (SI_r, Equation 5.2) for annual forage dry matter yield to a ± 25% change in selected model parameters and inputs for three different alfalfa cultivars at a fall

Table 5.1. Sensitivity index (SI, Eq. [5.2]) for annual forage dry matter yield to a ± 25% change in selected model parameters and inputs for three different cultivars at a FGS of 2, 3 and 4 during three productive years: *Py1*, *Py2*, and *Py3*. (Source: Kanneganti et al, 1998)

Variable definition unit	Variable name	Scenario	Reference value	SI					
				Negative change (−25%)			Positive change (+25%)		
				Py1	*Py2*	*Py3*	*Py1*	*Py2*	*Py3*
Potential crop death coefficient, d^{-1}	PDFMX	CV2	0.109	0	−0.05	−0.05	0	−0.05	−0.05
		CV3	0.121	0	−0.1	−0.1	0	−0.1	−0.1
		CV4	0.133	0	−0.7	−0.8	0	−1.3	−1.4
Potential rate of cold hardening, $°C\ d^{-1}$	CHRMX	CV2	0.184	.05	0.3	0.35	0	0.1	0.15
		CV3	0.162	2.45	3.85	3.95	−0.05	0.25	0.3
		CV4	0.139	4.0	4.00	4.00	0.05	1	1.1
Potential rate of dehardening	CDRMX	CV2	0.82	.2	0.2	0.3	0.05	0.05	0.05
		CV3	0.795	.2	0.05	0.1	0.05	0	0.05
		CV4	0.770	0	−0.15	0.6	0	−0.2	−0.2
Lowest temperature tolerance °C	CTMX	CV2	−22.4	−0.05	−0.05	−0.05	0	0	−0.15
		CV3	−20.4	−0.05	0	−0.05	0	0	−0.05
		CV4	−18.4	−0.2	1.35	1.4	0	0	0.0
Snowfall, mm	SNOD	CV2	dly inp	0	0	0	0	0	0
		CV3	dly inp	0.05	0	0	0.05	0	0
		CV4	dly inp	0.25	1.7	1.7	0.1	0.55	0.9
Initial plant density, plant m^{-2}	POP ini	CV2	160	0.25	0.3	0.35	0.1	0.3	0.35
		CV3	160	0.2	0.7	0.8	0.15	0.3	0.3
		CV4	160	0.2	1.8	1.8	0.2	0.8	0.3

growth stage (FGS) of 2,3, and 4 during three production years $Py1$, $Py2$, and $Py3$. (An absolute—without considering the plus or minus signs— SI value >1 indicates that the model is very sensitive to the particular change. The SI values are shown in Table 5.1.

From Table 5.1, it is clear that the sensitivity of calculated forage dry matter yield to a $\pm 25\%$ change in selected model parameters was measured in cultivars of FGS 2, 3 and 4 during three production years of a 4-yr alfalfa crop. A cultivar's rate of cold hardening and lowest temperature tolerance during autumn and winter did more to influence forage yield more than did the rate of dehardening in the spring.

Sensitivity of crop productivity to climate change, and marpho-physiological variation for characters in mung bean (*Vigna radiata*) under drought conditions, and sensitivity of wheat productivity to morphological variation for characters under global environmental change have been studied by Singh, 1995a, b and c.

Dobermann and Ping (2004) developed the yield sensitivity vegetative indices. Utilizing the most suitable vegetative index at each site and SKLM (Simple kriging with varying local means) as interpretation method, the root mean squared error of yield prediction was increased by nearly 20% over OK (Ordinary Kriging).

5.2 STABILITY ANALYSIS

A sensitivity analysis also provides a mechanism for testing the simulation in the extremes, i.e., using the extreme values of parameters will rigorously test the model in terms of mathematical logic and stability (Jones and Luyten, 1998).

5.3 VALIDATION

Validation is the process of comparing simulated results to real system data not previously used in any calibration or parameter estimation process.

5.3.1 Types of Goals of Validation

    (a) General screening: It includes a test to detect anomalies during the course of model development.

    (b) Research problem: Very detailed and vigorous statistical tests are used. The goal is to exhibit any deviation which can be found.

    (c) Control problem: The basic goal is to see if the model may be successfully used to manage the real system.

(d) Design problem: Nothing to compare it with. This includes the original model, e.g. calculator, nuclear reactor, economic policy.

The purpose of validation is to determine if the model is sufficiently accurate for its application as defined by objectives of the simulation study. Simulated state variables are compared with measured values of state variables. Usually, in crop simulation studies, only a few state variables out of many possibilities are measured, and thus a complete comparison is usually not possible. Validation involves subjective judgement. First, the areas of comparison must be selected. Then a measure of "accuracy" or "closeness of fit" must be established, such as the final crop yield. The choice of criteria and important variables for comparison should be based on model objectives. Validation efforts are essential in the application of crop models.

5.3.2 Validation Test Based on Confidence Limits

In simulating crop growth with a process-based model, comparison between the model output and the measurement is an important activity to test the model accuracy and locate any room for further improvements. The comparison is often based on the correlation between the calculated and measured values, and any regression of the measured on calculated values. For example, when Kiniry et al. (1997), compared the measured and simulated yields of maize for 10 years from 1983 to 1992 in nine locations in the USA, the measured yields were plotted against the simulated ones, the correlation coefficient was calculated and regression lines were fitted. This has been common practice while comparing between calculated and measured values (e.g. Teo et al., 1992; Chapman et al., 1993; and Retta et al., 1996). In this approach, the correlation is a criterion of the predictive accuracy of the model along with the requirement for the regression line (i.e. the intercept is not significantly different from zero and the slope is not significantly different from unity). Statistical testing of these requirements for the regression line has been established (Mayer et al., 1994).

Simulated values for a growth trait is denoted as x, and measured value is denoted as y. It is assumed that y is the sum of the true mean (μ) and the random error (ε) associated with the measurement, namely

$$y = \mu + \varepsilon \qquad (5.3)$$

In regressing y on x, a linear relationship is assumed between x and μ, namely,

$$\mu = bx + a \qquad (5.4)$$

where b is the slope and a is the y-interest of the regression line. Then the null hypothesis

$$H_0: b = 1 \text{ and } a = 0 \tag{5.5}$$

is tested against the alternative hypothesis (Mayer et al. 1994)

$$H_1: b \neq 1 \text{ and/or } a \neq 0 \tag{5.6}$$

These hypothesis can be translated into relationships between x (simulated value) and μ (true mean value).

$$H_0: \mu = x \tag{5.7}$$

$$H_1: \mu = bx + a \ (b \neq 1 \text{ and/or } a \neq 0) \tag{5.8}$$

Any contrast in the direct comparison between x and y, the null (H_0) and alternative (H_1) hypothesis are

$$H_0: \mu = x \tag{5.9}$$

$$H_1: \mu \neq x \tag{5.10}$$

The difference between the regression and the direct comparison is in the alternative hypothesis (Eqs.5.8 vs 5.10). The regression analysis assumes the linear relationship betwen x and μ under the alternative as well as the null hypothesis, but the assumption is not guaranteed and should not be taken for granted. If each measurement is based on replicated measurements, the variance of the error term (ε of Eq. (5.3) can be estimated independently from the assumption of the linear relationship (Draper and Smith, 1981, p. 33-38). The error variance is then used to test the assumption. If the linear assumption (Eq. 5.4) is rejected, the linear relationship is inadequate. A curvilinear relationship may be sought, but it is possible that no continuous function fits the relationship between x and y. However, the user's concern lies more in the comparison between x and y than in the functional relationship between the two. The direct comparison between x and y can always be made by testing the equality hypothesis (Eq. 5.9) against the non-equality hypothesis (Eq. 5.10).

Example 1 : Table 5.2 shows the simulated and observed values of seed yield ha[-1] of peanut in various seasons sowing dates and moisture regimes (Source: Singh et al., 1994).

Find the regression line for the data of Table 5.2.

$$\sum_{i=1}^{10} x_i = 10.6 \qquad \sum_{i=1}^{10} y_i = 13.5 \qquad \sum_{i=1}^{10} x_i y_i = 16.17$$

$$\sum_{i=1}^{10} x_i^2 = 13.14 \qquad \sum_{i=1}^{10} y_i^2 = 20.77 \qquad \bar{x} = 1.06, \bar{y} = 1.35$$

Table 5.2. Simulated (S) and observed (O) seed yield (ton/ha) of groundnut in various season, sowing dates and moisture regimes (Source: Singh et al. 1994).

S	O
x	y
0.4	0.7
1.1	2.1
1.2	1.5
1.5	1.8
0.8	1.0
1.6	2.0
0.2	0.5
1.2	1.2
1.2	1.2
1.4	1.5

Therefore,

$$b = \frac{n\sum_{i=1}^{n} x_i y_i - \left(\sum_{i=1}^{n} x_i\right)\left(\sum_{i=1}^{n} y_i\right)}{n\sum_{i=1}^{n} x_i^2 - \left(\sum_{i=1}^{n} x_i\right)^2} \tag{5.11}$$

$$= \frac{(10)(16.17) - (10.6)(13.5)}{(10)(13.14) - (10.6)^2} = 0.976$$

$$a = \bar{y} - b\bar{x} \tag{5.12}$$

$$= 1.35 - 0.98\,(1.06) = 0.3112$$

The regression line is then given by

$$y_p = a + bx$$

$$y_p = 0.3112 + 0.976\,(x) \text{ (where } y_p \text{ is predicted } y)$$

By substituting any two of the given values of x into this equation, say $x_1 = 0.2$ and $x_2 = 1.6$, we obtain the ordinates $y_{1p} = 0.506$ and $y_{2p} = 1.873$. The regression line in Figure 5.1 was drawn by connecting these two points with a straight line.

An unbiased estimate of σ^2 with $n - 2$ degree of freedom is given by the formula (Walpole, 1982).

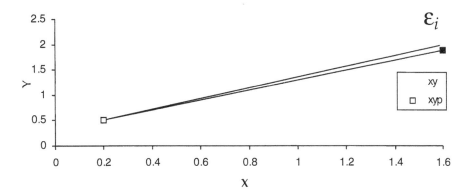

Fig. 5.1 Showing residual ε_t

$$S_e{}^2 = \frac{n-1}{n-2} \; (S_y{}^2 - b^2 \, S_x{}^2)$$

Example 2: Calculate S^2 for the data of Table 5.2.

Solution: In Example 1, we found that

$$\sum_{i=1}^{10} x_i = 10.6 \qquad \sum_{i=1}^{10} x_i^2 = 13.14 \qquad \sum_{i=1}^{10} y_i = 13.5$$

Referring to the data of Table 5.2, we now find

$$\sum_{i=1}^{10} y_i^2 = 20.77$$

Therefore,

$$S_x^2 = \frac{n \sum\limits_{i=1}^{n} x_i^2 - \left(\sum\limits_{i=1}^{n} x_i \right)^2}{n\,(n-1)}$$

$$S_y^2 = \frac{n \sum\limits_{i=1}^{n} y_i^2 - \left(\sum\limits_{i=1}^{n} y_i \right)^2}{n\,(n-1)}$$

and

$$S_x^2 = \frac{10(13.14)-(10.6)^2}{10(9)} = 0.2115$$

$$S_y^2 = \frac{10(20.77)-(13.5)^2}{90} = 0.2827$$

Recall that $b = 0.976$

$$S_e^2 = \left(\frac{n-1}{n-2}\right)\left(S_y^2 - b^2 \, S_x^2\right)$$

$$= \frac{10-1}{10-2} \,(0.2827) - (0.976)^2 \,(0.2115)$$

$$= 0.1165$$

Example 3: Find a 95% confidence interval for α in the regression line. $\mu_{y|x} = \alpha = \beta x$, based on the data in Table 5.2.

Solution: In Example 2, we found that

$$S_x^2 = 0.2115 \text{ and } S_e{}^2 = 0.1165$$

Therefore, taking square roots we obtain $S_x = 0.4598$ and $S_e = 0.3413$ From Example 1, we had

$$\sum_{i=1}^{n} x_i^2 = 13.14 \text{ and } a = 0.3112.$$

Using a table (from any textbook on statistics) showing the critical values of t distribution, we find $t_{0.025}$ for 8 degrees of freedom equals 2.306. Therefore, a 95% confidence interval for α is given by

$$a - \frac{t_{\alpha/2} \, S_e \sqrt{\sum_{i=1}^{n} x_i^2}}{S_x \sqrt{n(n-1)}} < \alpha < a + \frac{t_{\alpha/2} \, S_e \sqrt{\sum_{i=1}^{n} x_i^2}}{S_x \sqrt{n(n-1)}}$$

$$0.3112 - \frac{(2.306)(0.3413)\sqrt{13.14}}{0.4598\sqrt{10(9)}} < \alpha < 0.3112 + \frac{(2.306)(0.3413)\sqrt{13.14}}{0.4598\sqrt{10(9)}}$$

Thus, it simplifies to

$$- 0.3428 < \alpha < 0.9652$$

Example 4: Using the estimated values $a = 0.3112$ in Example 1, test the hypothesis that $\alpha = 0.7$ at the 0.05 level of significance

Solution:

1. H_0: $\alpha = 0.7$
2. H_1: $\alpha \neq 0.7$
3. Choose a 0.05 level of significance
4. Critical region $T < - 2.306$ and $T > 2.306$
5. Computation

$$t = \frac{(a - \alpha_0)S_x \sqrt{n(n-1)}}{S_e \sqrt{\sum_{i=1}^{n} x_i^2}}$$

$$t = \frac{(0.3112 - 0.7)(0.4598) \sqrt{90}}{0.3413 \sqrt{13.14}} = -1.371$$

6. Decision: Accept H_0

Example 5. Find a 95% confidence interval for β in the regression line. $\mu_{y|x} = \alpha + \beta x$, based on the data in Table 5.2.

Solution: From Example 3, we note that

$S_x = 0.4598$ and $S_e = 0.3413$ and $t_{.0.025} = 2.306$ for 8 degrees of freedom.

In Example 1, the slope was calculated to be 0.976. Therefore, a 95% confidence interval for β is given by

$$b - \frac{t_{\alpha/2} S_e}{S_x \sqrt{n-1}} < \beta < b + \frac{t_{\alpha/2} S_e}{S_x \sqrt{n-1}}$$

$$0.976 - \frac{(2.306)(0.3413)}{0.4598 \sqrt{10-1}} < \beta < 0.976 + \frac{(2.306)(0.3413)}{0.4598 \sqrt{10-1}}$$

which simplifies to

$$0.4054 < \beta < 1.5466$$

Example 6: Using the estimated values $b = 0.976$ in Example 1, prove the hypothesis that $\beta = 0$ at the 0.01 level of significance against the alternative that $\beta > 0$.

Solution:

1. H_0: $\beta = 0$
2. H_1: $\beta > 0$
3. Choose a 0.01 level of significance.
4. Critical region: $T > 2.896$
5. Computation:

$$t = \frac{S_x \sqrt{n-1}\,(b-\beta)}{S_e}$$

$$= \frac{0.4598 \sqrt{10-1}\,(0.9760-0)}{0.3413}$$

$$= 3.9446$$

6. Decision: Reject H_0 and conclude that $\beta > 0$.

Example 7: Using the data of Table 5.2, construct a 95% confidence interval of y_0 when $x=1.6$.

Solution: We have $n=10$, $x_0=1.6$, $\bar{x} = 1.06$, $\bar{y} =1.35$, $y_p = 1.876$, $S^2_x = 0.2115$, $S_e = 0.3413$ and $t_{0.025} = 2.306$ for 8 degrees of freedom. Therefore, a 95% confidence interval for y_0 when $x = 1.6$ is given by

$$y_p - t_{\alpha/2}\,S_e \sqrt{1+\frac{1}{n}+\frac{(x_0-\bar{x})^2}{(n-1)S^2_x}} < y_0 < y_p + t_{\alpha/2}\,S_e \sqrt{1+\frac{1}{n}+\frac{(x_0-\bar{x})^2}{(n-1)S^2_x}}$$

$$1.873 - (2.306)\,(0.3413) \sqrt{1+1/10+\frac{(1.6-1.06)^2}{9\,(0.2115)}}$$

$$< y_0 < 1.873 + (2.306)\,(0.3413) \sqrt{1+1/10+\frac{(1.6-1.06)^2}{9\,(0.2115)}}$$

which simplifies to

$$0.9919 < y_0 < 2.7541$$

5.3.2.1 Test for linearity of regression

Example 8: Use the data in Table 5.2 to test the hypothesis that the regression is linear at the 0.05 level of significance.

Solution:

1. H_0: the regression is linear.

2. H_1: the regression is nonlinear.
3. Choose a 0.05 level of significance.
4. Critical region $F > 19.3$ with 6 and 2 degrees of freedom.
5. Computation from Table 5.2, we have

x_1	0.4	$n_1 = 1,$	$y_1 = 0.7$
x_2	1.1	$n_2 = 1,$	$y_2 = 2.1$
x_3	1.2	$n_3 = 3,$	$y_3 = 3.9$
x_4	1.5	$n_4 = 1,$	$y_4 = 1.8$
x_5	0.8	$n_5 = 1,$	$y_5 = 1.0$
x_6	1.6	$n_6 = 1,$	$y_6 = 2.0$
x_7	0.2	$n_7 = 1,$	$y_7 = 0.5$
x_8	1.4	$n_8 = 1,$	$y_8 = 1.5$
			13.5

Therefore,

$$\chi_1^2 = \sum \frac{y_i^2}{n_i} - \frac{\left(\sum y_{ij}\right)^2}{n} - b^2 (n-1)\left(S_x^2\right)$$

y_{ij} = The jth value of the random variable Y_i (Table 5.2)
y_i = The sum of the values of Y_i in our sample.

$$= \left[\frac{0.7^2}{1} + \frac{(2.1)^2}{1} + \frac{3.9^2}{3} + \frac{1.8^2}{1} + \frac{1^2}{1} + \frac{2^2}{1} + \frac{0.5^2}{1} + \frac{1.5^2}{1}\right]$$

$$- \frac{(13.5)^2}{10} - 0.976^2\,(9)\,(0.2115)$$

$$= 0.6717$$

$$\chi_2^2 = \sum y_{ij}^2 - \sum \frac{y_i^2}{n_i}$$

$$= 20.77 - 20.71$$
$$= 0.06$$

Hence
$$f = \frac{\chi_1^2 / (k - 2)}{\chi_2^2 / (n - k)}$$

$$= \frac{0.6717/6}{0.06/2}$$

$$f = 3.7316$$

6. Decision: Accept H_0 and conclude that the data may be fitted by a linear equation.

5.3.3 Test Based on Least Square Procedure

A more relevant criterion for the direct comparison than regression is the deviation (d) of the model output (x) and the measurement (y), namely,

$$d = (x - y) \tag{5.13}$$

when the comparison is made for n measurements, d can be computed for each measurement, namely

$$d = x_i - y_i, \text{ for } i = 1, 2, ..., n \tag{5.14}$$

where x_i and y_i are the simulated and measured values, respectively, for the ith data. The n deviations can be summarized with statistics of the overall deviation. The most commonly used statistics among these based on the deviation is the RMSD (e.g. Jamieson et al.,), namely

$$RMSD = \sqrt{\frac{1}{n} \sum_{i=1}^{n} (x_i - y_i)^2} \tag{5.15}$$

Another commonly used criterion is the MD, namely

$$MD = \frac{1}{n} \sum_{i=1}^{n} (x_i - y_i) \tag{5.16}$$

In literature, RMSD in often referred to as root mean squared error (RMSE) (Retta et al., 1996) and MD is often called bias (Retta et al., 1997; Jamieson et al., 1998). Of these two statistics, RMSD represents the mean distance between simulation and measurement; MD is the difference between the means of simulation and measurement. Root mean squared deviation MD, thus, represents the aspects of the overall deviation but the relationship between the two has not been well defined.

In literature, these deviation-based statistics are often used in conjunction with correlation and regression coefficients (Addiscoti and Whitmor, 1987; Retta et al., 1996; Kiniry et al., 1999; Jamieson et al., 1998). Although these different statistics may represent somewhat different aspects of the model-measurement discrepancy, it is not clear how the different statistics relate to each other, and if these statistics cover all the aspects of the discrepancy sufficiently. It is also noteworthy that the deviation-based statistics (e.g. RMSD) and the correlation-based statistics (e.g. the correlation coefficients) are not really consistent with each other in their assumption (Kobayashi and Salam, 2000).

Example 1: Using Table 5.2, calculate the RSMD and MD.

x	y	$x - y$	$(x - y)^2$	$x - \bar{x}$	$y - \bar{y}$	$(x - \bar{x})^2$	$(y - \bar{y})^2$
0.4	0.7	−0.3	0.09	−0.66	−0.65	0.4356	0.4225
1.1	2.1	−1	1	0.04	0.75	0.0016	0.5625
1.2	1.5	−0.3	0.09	0.14	0.15	0.0196	0.0225
1.5	1.8	−0.3	0.09	0.44	0.45	0.1936	0.2025
0.8	1.0	−0.2	0.04	−0.26	−0.35	0.0676	0.1225
1.6	2.0	−0.4	0.16	0.54	0.65	0.2916	0.4225
0.2	0.5	−0.3	0.09	−0.86	−0.85	0.7396	0.7225
1.2	1.2	0	0	0.14	−0.15	0.0196	0.0225
1.2	1.2	0	0	0.14	−0.15	0.0196	0.0225
1.4	1.5	−0.1	0.01	0.34	0.15	0.1156	0.0225

$\bar{x} = 1.06 \quad \bar{y} = 1.35 \qquad -2.9 \qquad 1.57 \qquad\qquad\qquad\qquad 1.904 \qquad 2.545$

$$0.1904 \qquad 0.2545$$
$$0.4363 \qquad 0.5044$$
$$SD_s \qquad\quad SD_m$$

$$RMSD = \sqrt{\frac{1}{n} \sum_{i=1}^{n}(x_i - y_i)^2}$$

$$= \sqrt{.157} = 0.4$$

$$MD = \frac{1}{n} \sum_{i=1}^{1}(x_i - y_i)$$

$$= \frac{1}{10}(-2.9) = -0.29$$

Example 2: Using Table 5.2, calculate MSD (mean squared deviation).

$$MSD = \frac{1}{n} \sum_{i=1}^{n}(x - y)^2 \tag{5.17}$$

$$= \frac{1}{10}(1.57) = 0.157$$

MSD can be participated into two components, namely

$$MSD = (\bar{x} - \bar{y})^2 + \frac{1}{n} \sum_{i=1}^{n}[(x_i - \bar{x}) - (y_i - \bar{y})]^2 \tag{5.18}$$

$$= 0.0841 + 0.0729 = 0.154$$

Example 3: Calculate the bias of the simulation

$$SB = (\bar{x} - \bar{y})^2 \tag{5.19}$$

$$= 0.0841$$

The squared bias is the square of the MD.

Example 4: Calculate the mean squared variation (MSV).

$$MSV = \frac{1}{n} \sum_{i=1}^{n} [(x_i - \bar{x}) - (y_i - \bar{y})]^2 \tag{5.20}$$

$$= 0.0729$$

A biggest MSV indicates that the model failed to simulate the variability of the measurements around the mean. Note that these two components SB and MSV are orthogonal and can be addressed separately.

Example 5: Calculate the standard deviation of the simulation as SD_s, standard deviation of the measurement as SD_m and correlation coefficient as r based on Table 5.2.

$$SD_s = \sqrt{\frac{1}{n} \sum_{i=1}^{n} (x_i - \bar{x}_i)^2} \tag{5.21}$$

$$= 0.4363$$

$$SD_m = \sqrt{\frac{1}{n} \sum_{i=1}^{n} (y_i - \bar{y}_i)^2} \tag{5.22}$$

$$= 0.5044$$

$$r = \left[1/n \sum_{i=1}^{n} (x - \bar{x})(y - \bar{y}) \right] / (SD_m\, SD_s) \tag{5.23}$$

$$= 1/10 \ (1.86)/(0.4363) \ (0.5044)$$

$$= 0.186/0.22 = 0.845$$

After some arrangement, MSV can be rewritten as

$$MSV = (SD_s - SD_m)^2 + 2\, SD_s\, SD_m\, (1-r) \tag{5.24}$$

$$= 0.00463761 + (2) \ (0.22) \ (0.155)$$

$$= 0.07284$$

The first term of the right side is the difference in the magnitude of fluctuation between the simulation and measurement, denoted by SDSD

$$\text{SDSD} = (\text{SD}_s - \text{SD}_m)^2 \tag{5.25}$$
$$= [0.4363 - 0.5044]^2$$
$$= 0.00463761$$

A larger SDSD indicates that the model has failed to simulate the magnitude of fluctuation among the n measurements. The second term of the right side of Eq. (5.24) is essentially the lack of positive correlation weighted by the standard deviation, and is denoted as LCS, namely

$$\text{LCS} = 2\text{SD}_s\,\text{SD}_m\,(1-r) \tag{5.26}$$
$$= 2\,(0.4363)\,(0.5044)\,(1 - 0.845)$$
$$= 0.06822$$

The bigger LCS means that the model has failed to simulate the pattern of the fluctuation across the n measurements.

With all the above terms combined, the MSV and MSD can be written as

$$\text{MSV} = \text{SDSD} + \text{LCS} \tag{5.27}$$
$$\text{MSD} = \text{SB} + \text{SDSD} + \text{LCS} \tag{5.28}$$
$$\text{MSV} = 0.00463761 + 0.06822 = 0.07285$$
$$\text{MSD} = 0.0841 + 0.00463761 + 0.06822 = 0.157$$

The above components of MSD can be calculated from the coefficient of regression. As in Eq. (5.11), b is the slope of the regression line and in Eq. (5.12) a is the y–intercept, then the component of MSD are given as

$$\text{SB} = \left[\left(1-\frac{1}{b}\right)\bar{x} + \left(\frac{a}{b}\right)\right]^2 \tag{5.29}$$

$$\text{SDSD} = \left[1-\frac{r^2}{b}\right]\text{SD}_m^2 \tag{5.30}$$

$$\text{LCS} = 2\,(r/b)\,(1-r)\,\text{SD}_m^2 \tag{5.31}$$

Example 6: Calculate SB, SDSD and LCS based on regression coefficient of Table 5.2.

Solution:

$$y_p = a + b\,(x)$$
$$= 0.31 + 0.976\,(x)$$

$$SB = \left[\left(1 - \frac{1}{0.976}\right)1.06 + \left(\frac{0.31}{0.976}\right)\right]^2$$

$$= 0.0845$$

$$SDSD = \left[1 - \frac{r}{b}\right]^2 SD_m^2$$

$$= \left[1 - \frac{0.845}{0.976}\right]^2 = (0.5044)^2$$

$$= 0.004582$$

$$LCS = 2\ (r/b)\ (1 - r)\ (SD_m)^2$$

$$= 2\ (0.845\ /0.976)\ (1 - 0.845)\ (0.5044)$$

$$= 0.06828$$

$$MSD = SB + SDSD + LCS$$

$$MSD = 0.084 + .0045 + 0.068$$

$$= 0.157$$

In the special case when $b = 1$ and $a = 0$

$$SB = 0 \tag{5.32}$$

$$SDSD = (1 - r)^2\ (SD_m)^2 \tag{5.33}$$

$$LCS = 2r\ (1 - r)\ (SD_m)^2 \tag{5.34}$$

$$MSD = (1 - r)^2\ (SD_m)^2 \tag{5.35}$$

Thus, any comparison based on correlation becomes equivalent to the comparison based on MSD in this special case, as long as the comparison is made within the same data set (i.e. fixed SD_m).

5.3.3.1 *Comparison Between Test, Based on Confidence Limit and Least Square Procedure*

Kobayashi and Salam (2000) compared these two tests in a very clear fashion. When the output (x) of a mechanistic model is compared with measurement (y), it is common practice to calculate the correlation coefficient between x and y, and to regress on y. There are, however, problems in this approach. The assumption of the regression that y is related to x is not guaranteed and, so, is unnecessary for the x–y comparison. The correlation and regression coefficients are not explicitly related to other commonly used statistics [e.g., root mean squared deviation (RMSD)]. Kobayashi and Salam (2000) present an approach based on the mean squared deviation (MSD = RMSD^2) and show that it is better suited to x–y comparison than regression. Mean squared deviation is the sum of three components: squared bias (SB), squared difference between standard deviation (SDSD), and lack of correlation weighted by the standard deviation (LCS). For example, the MSD-based analysis was applied to simulation vs. measurement comparisons in literature, and the results were compared with those from regression analysis. The analysis of MSD clearly identified the simulation vs. measurement contrast with larger deviation than others. The correlation regression approach tended to focus on the contrasts with lower correlation and regression lines far from the equality line. It was also shown that results of MSD-based analysis were easier to interpret than those of the regression analysis. This is because the three MSD components are simply additive and all constituents of the MSD components are explicit. This approach will be useful to quantify the deviation of calculated values obtained with the model from measurements.

5.3.4 Tests Based on Probability Distribution

The test presented in this section is useful in situations where two samples drawn from each of the two possibly different populations are identical or not. While other tests such as the median test, the Mann-Whitney test, or the parametric test may also be appropriate, they are sensitive to differences between the two means or medians; although they may not detect differences of other types, such as the differences in variances. One of the two two-sided tests presented in this section is that both tests are consistent against all types of differences that may exist between the two distribution functions.

Here, we shall consider the Smirnov test (Smirnov, 1939). This test is based on the probability distribution function. It is a two-sample version of the Kolmogorov test and is sometimes called the Kolmogorov-Smirnov two sample test, while, the Kolmogorov test is sometimes called the Kolmogorov-Smirnov one sample test. The Smirnov test is presented in both the one-sided and two-sided versions. There is another two-sided test, the Cramer-von Mises test for two samples. This test is slightly more different to compute than the Smirnov test. Actually, there is probably little difference in power between the two tests. For the Cramer-von Mises test, Conover (1980) may be referred.

5.3.4.1 The Smirnov Test

5.3.4.1.1 Data

The data consists of two independent random samples, each of size n, $x_1, x_2 \ldots x_n$ and the other $y_1, y_2, \ldots, y_n$. Let $f(x)$ and $g(x)$ represent their respective, unknown, distribution functions.

5.3.4.1.2 Assumption

1. The samples are random samples.
2. The two samples are mutually independent.
3. The measurement scale is atleast ordinal.
4. For this test to be exact, the random variables are assumed to be continuous.

If the random variables are discrete the test is still valid but becomes conservative (Noether, 1967).

5.3.4.1.3 Hypotheses

A. (Two-sided test)
H_0: $F(x) = G(x)$ for all x from minus ∞ to plus ∞.
H_1: $F(x) \neq G(x)$ for at least one value of x.
B. (One-sided test)
H_0: $F(x) \leq G(x)$ for all x from minus ∞ to plus ∞
H_1: $F(x) > G(x)$ for at least one value of x.
C. (One-sided test)
H_0: $F(x) \geq G(x)$ for all x from minus ∞ to plus ∞.
H_1: $F(x) < G(x)$ for at least one value of x.

5.3.4.1.4 Test Statistics

Let $S_1(x)$ be the empirical distribution function based on the random sample $x_1, x_2, \ldots, x_n$, and let $S_2(x)$ be the empirical distribution function based on the other random sample $y_1, y_2 \ldots, y_n$. The test statistics is defined differently for three different sets of hypothesis.

A. Two-sided test: Define the test statistics T_1 as the greatest vertical distance between the two empirical distribution functions.

$$T_1 = \text{Sup} \left[S_1(x) - S_2(x) \right] \tag{5.36}$$

B. One-sided test: Denote the test statistics by T_1^+ and let it equal the greatest vertical distance attained by $S_1(x)$ above $S_2(x)$.

$$T_1^+ = \text{Sup } [S_1(x) - S_2(x)] \qquad (5.37)$$

C. One-sided test: For the One-sided hypothesis in C above, let the test statistics, denoted by T_1^-, be the greatest vertical distance attained by $S_2(x)$ above $S_1(x)$

$$T_1^- = \text{Sup } [(S_2(x) - S_1(x_1)] \qquad (5.38)$$

5.3.4.1.5 Decison Rule

Reject H_o at the level of significance α if the appropriate test statistics T_1, T_1^+, or T_1^-, as the case may be, exceeds the $1-\alpha$ quantile as given by the standard table (Table A 20 of Conover, 1980) of the quantile of the Smirnov test statistics for ten samples of equal size n.

Example 1: Simulated values of size 10, $x_1, x_2 \ldots, x_{10}$ and observed values of size 10, $y_1, y_2, \ldots, y_{10}$ obtained from Singh et al., 1994, given in Table 5.2. The null hypothesis is that the two populations have identical probability distribution functions. If the respective distribution functions are denoted by $F(x)$ and $G(x)$, then the null hypothesis may be written as

H_o: $F(x) = G(x)$ for all x from minus ∞ to plus ∞.

The alternative hypothesis may be written as

H_1: $F(x) \neq G(x)$ for as least one value of x.

The two samples are ordered from the smallest to the largest for convenience, and their values, along with other pertinent information about their empirical distribution function, are given:

Table 5.3. Data related to functions $S_1(x)$ and $S_2(x)$ (cumulative probability of sample -1, observed and 2 predicted values).

$x1_1$	$x2_1$	$S_1(x)-S_2(x)$	$x1_1$	$x2_1$	$S_1(x)-S_2(x)$
0.2		0.1 − 0 = 0.1	1.2	1.2	
04		0.2 − 0 = 0.2	1.2	1.2	
	0.5	0.2 − 0.1 = 0.1	1.2		0.7 − 0.5 = 0.2
	0.7	0.2 − 0.2 = 0	1.4		0.8 − 0.5 = 0.3
0.8		0.3 − 0.2 = 0.1	1.5	1.5	0.9 − 0.6 = 0.3
	1.0	0.3 − 0.3 = 0		1.5	
1.1		0.4 − 0.3 = 0.1	1.6		1.0 − 0.7 = 0.3
				1.8	1.0 − 0.8 = 0.2
				2.0	1.0 − 0.9 = 0.1
				2.1	1.0 − 1.0 = 0

$$T_1 = \text{Sup}[S_1(x) - S_2(x)] = 0.3$$
$$\phantom{T_1 = \text{Sup}}_x$$

Empirical distribution function is always a step function, where each step is of height $1/n$ and occurs only at the sample values.

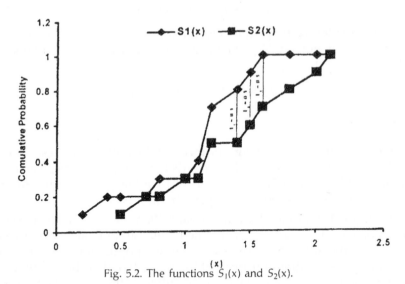

Fig. 5.2. The functions $S_1(x)$ and $S_2(x)$.

H_0: $F(x) = G(x)$ for all x from minus ∞ to plus ∞

The test statistic for the two-sided test is given by equation

$$T_1 = \sup_x |S_1(x) - S_2(x)| = 0.3$$

which is read as "T_1 equals the supremum, over all x, of the absolute value of the difference $S_1(x) - S_2(x)$.

$$T_1 = 0.3$$

0.3 is the largest absolute vertical difference between $S_1(x)$ and $S_2(x)$, (Figure 5.2 and Table 5.3).

The table value at 0.95 quantile of T_1 for the two-sided test for $n = 10$ is given in Table A20 of Conover (1980) as $W_{0.95} = 6/10$. For the above test, T_1 equals 3/10. Therefore, H_0 is accepted at the 0.5 level. Hence, simulated measurements are not different from each other and the model is validated.

5.3.4.2 Spearman's Rho Test (as quoted in Conover, 1980)

5.3.4.2.1 Data

The data may consist of a bivariate random sample of size n, (x_1, y_1), (x_2, y_2), ... (x_n, y_n). Let $R(x_i)$ be the rank of x_i as compared with the other x values for $i = 1, 2, ..., n$. That is, $R(x_i) = 1$ if x_i is the smallest of $x_1, x_2,$ ....., x_n, $R(x_i) = 2$ if x_i is the second smallest and so on, with rank n being assigned to the largest of the x_i. Similarly, let $R(y_i)$ equal 1, 2, ..., or n, depending on the relative magnitude of y_i as compared with $y_1, y_2,$ ....., y_n, for each *i*. In case of ties, assign to each tied value the average of ranks that would have been assigned if there had been no ties.

5.3.4.2.2 *Measure of correlation*

The measure of correlation is given:

$$\rho = 1 - \frac{6 \sum_{i=1}^{n} \left[R(x_i) - R(y_i) \right]^2}{N(N^2 - 1)} \tag{5.39}$$

$$= 1 - \frac{6T}{n(n^2 - 1)} \tag{5.40}$$

where *T* represents the entire sum in the numerator. If a moderate number of ties is present in the data, equation (5.39) is recommended for computational simplicity, since the difference between the given equation and Pearson's equation will be slight.

Example 1: Refer to the data on Table 5.2. The emphasis is on examination of the degree of similarity between simulated and observed data.

The data were measures of yield (ton ha^{-1}) of groundnut seed (Singh et al., 1994).

Yield set	*i*	1	2	3	4	5	6	7	8	9	10
Simulated	x_i	0.4	1.1	1.2	1.5	0.8	1.6	0.2	1.2	1.2	1.4
Observed	y_i	0.7	2.1	1.5	1.8	1.0	2.0	0.5	1.2	1.2	1.5

The simulated yields were ranked among themselves, the observed yields were ranked among themselves, with the following results.

Yield set	*i*	1	2	3	4	5	6	7	8	9	10
R	x_i	2	4	6	9	3	10	1	6	6	8
R	y_i	2	10	6.5	8	3	9	1	4.5	4.5	6.5
$[R(x_i) - R(y_i)]$		0	-6	-0.5	1	0	1	0	-0.5	1.5	1.5
$[R(x_i) - R(y_i)]^2$		0	36	.25	1	0	1	0	2.25	2.25	2.25

Sum $[R(x_i) - R(y_i)]^2 = 45$

The ties were given the average ranks.

First, the statistic *T* is computed

$$T = \sum_{i=1}^{n} [R(x_i) - R(y_i)]^2 = 45$$

Then ρ is obtained

$$\rho = 1 - \frac{6T}{n(n^2 - 1)} = 1 - \frac{6(45)}{10(99)} = 0.7273$$

Hypothesis Test

The Spearman rank correlation coefficient is often used as a test statistic to test for independence between two random variables. The hypotheses take the following form:

A. *Two-tailed test*

H_0: The x_i and y_i are mutually independent.

H_1: Either (a) there is a tendency for the larger values of x to be paired with the larger values of y_1, or (b) there is a tendency for the smaller values of x to be paired with the larger values of y.

B. *(One-tailed test for positive correlation)*

H_0: The x_i and y_i are mutually independent.

H_1: There is a tendency for the larger values of x and y to be paired together.

C. *(One-tailed test for negative correlation)*

H_0: The x_i and y_i are mutually independent.

H_1: There is a tendency for the smaller values of x to be paired with the larger values of y_1 and vice versa.

The alternative hypotheses given here state the existence of correlation between x and y, so that a null hypothesis of "no correlation x and y" would be more accurate than the statement of independence between x and y just given.

Spearman's rho may be used as a test statistic for the preceding hypotheses. The standard table (A10 of Conover, 1980) gives the quantiles of rho under the assumption of independence, the null hypothesis. Then H_0 in B is rejected if rho is too large (at a level $\propto$ if rho exceeds the $1-\alpha$ quantiles), H_0 in C is rejected if rho is too small, and the two-tailed test involves rejecting H_0 if rho exceeds the $1-\alpha/2$ quantile or if rho is less than the $\alpha/2$ quantile.

Instead of Spearman's rho, it is usually more convenient to directly use the statistic T, where T is defined explicitly as

$$T = \sum_{i=1}^{n} \left[R(x_i) - R(y_i) \right]^2$$

Note that if the number of ties is moderate, the use of T eliminates some of the arithmetics involved in computing rho. The test in this form is called the Hotelling-Pabst test, which uses the standard of Table A11 of Conover, 1980. Quantiles of T are given in the standard table. Note that T is large when rho is small, and vice versa. Therefore, the H_0 in B is rejected at the level $\propto$ if T is less than its α quantile. Also, H_0 in C is rejected if it exceeds its $1-\alpha$ quantile.

Since rho had already been computed, it would have been easier to use rho as a test statistic. The rho in example equals 0.7273, which exceeds the 0.975 quantile given by the standard (Table as A10 of Conover, 1980) as 0.6374. So, H_0 is rejected (simulated values and observed values are not independent and model is validated).

5.3.4.3 Kendall's Tau Test (as quoted in Conover, 1980)

The Kendall's Tau test resembles Spearman's rho in the sense that it is based on the order (ranks) of the observations rather than the numbers themselves, and the distribution of the measure does not depend on the distribution of x and y if x and y are independent and continuous. The Kendall's Tau (τ) measure is considered to be more difficult to compute than Spearman's rho. The chief advantage of Kendall's Tau is that its distribution approaches the normal distribution quite rapidly so that the normal approximation is better for Kendall's Tau than it is for Spearman's rho, when the null hypothesis of independence between x and y is true. Another advantage of Kendall's Tau is its direct and simple interpretation in terms of probabilities of observing concordant and discordant pairs.

5.3.4.3.1 Data

The data may consist of a bivariate random sample of size n, (x_i, y_i) for $i = 1, 2, ..., n$. Two observations, for example, (1.3, 2.2) and (1.6, 2.7), are called concordant if both members of one observation are larger than the respective numbers of the other observation. Let Nc denote the number of concordant pairs of observations, out of the $\binom{n}{2}$ total possible pairs. A pair of observations, such as (1.3, 2.2) and (1.6, 1.1) are called discordant if the two numbers in one observation differ in opposite directions (one negative and one positive) from the respective members in the other observation. Let Nd be the total number of discordant pairs of observations. Pairs with ties between respective members are neither concordant nor discordant. Because the n observations may be paired

$(\begin{smallmatrix} n \\ 2 \end{smallmatrix}) = n\ (n{-}1)/2$ different ways, the number of concordant pairs N_c plus the number of discordant pairs N_d plus number of pairs with ties should add up to $n\ (n{-}1)/2$.

The data may also consist of non-numeric observations occurring in n pairs if the observations and such that N_c and N_d just described may be computed.

5.3.4.3.2 Measure of Correlation

The measure of correlation proposed by Kendall (1938) is

$$\tau = \frac{Nc - Nd}{n(n-1)/2} \qquad (5.41)$$

If all pairs are concordant, Kendall's Tau equals 1.0. If all pairs are discordant, the value is −1.0.

The computation of Tau is simplified if the observations $(x_i,\ y_i)$ are arranged in a column according to increasing values of x. Then each y may be compared only with those below it, and the number of concordant and discordant comparisons is easily determined. Also, each pair of observations is considered only once. The procedure is illustrated in the following example.

Example 1: Again we will use the data in Table 5.2 for the purpose of illustration. Arrangement of the data $(x_i,\ y_i)$ according to increasing values of x gives the following:

$x_i\ y_i$	Concordant pairs below $(x_i,\ y_i)$	Discordant pairs below $(x_i,\ y_i)$
(0.2, 0.5)	7	0
(0.4, 0.7)	6	0
(0.8, 1.0)	5	0
(1.1, 2.1)	4	0
(1.2, 1.5)	3	0
(1.2, 1.2)	3	0
(1.2, 1.2)	3	0
(1.4, 1.5)	2	0
(1.5, 1.8)	1	0
(1.6, 2.0)	0	0
	Nc = 34	Nd = 0

Kendall's Tau is given by

$$\tau = \frac{Nc - Nd}{n(n-1)/2} = \frac{34 - 0}{10(9)/2} = \frac{34}{45} = 0.7555$$

There is a positive rank correlation between simulated and observed yield as measured by Kendall's Tau.

Hypothesis Test:

Kendall's Tau may also be used as a test statistics to test the null hypothesis of independence between x and y, with possible one-tailed or two-tailed alternatives, as described with Spearman's rho, described in Section 5.3.4.2. Some arithmetic may be saved by using $Nc-Nd$ as a test statistic without dividing by $n(n-1)/2$ to obtain Tau. Therefore, we use T as the Kendall's test statistic, where T is defined as

$$T = Nc - Nd \tag{5.42}$$

Quantiles of T are given in Table A12 of Conover (1980) showing the Kendall's Test statistics (refer to Conover, 1980). If T exceeds the $1 - \alpha$ quantile, one can reject H_0 in favour of the one-sided alternative of positive correlation, at level α. Values of T less than the α quantile lead to acceptance of the alternative of negative correlation.

Example 2: In Example 1, Kendall's Tau was computed by first finding the value of

$$T = Nc - Nd = 34 - 0 = 34$$

In Table A12 of Conover (1980), the quantiles for a two-tailed test of size $\alpha = 0.05$ are found for $n = 10$, to be

$$W_{.975} = 21 \quad \text{(Upper quantile)}$$

and

$$W_p = W_{1-p} \quad \text{(Lower quantile)}$$

$$W_{.025} = -W_{1-.025}$$

$$W_{.025} = -21$$

For $T = 34$, the null hypothesis of independence is rejected in favour of alternative hypothesis which states that these is a positive correlation between simulated and observed data.

5.3.5 Risk-to-User Criteria

For crop production, the two most important uncontrolled factors are climate and prices of the crop products. Farmers around the world have to cope with apparently unpredictable climatic hazards. Drought and floods, heat-waves and cold snaps that are merely inconvenience to most people are of much greater concern to farmers. Their livelihoods are affected, and commonly their very lives are at stake. In order to survive, they must farm efficiently under conditions of climatic risk (Clements, 1990).

Once the model is validated, one can perform the experiments on computers having different varieties and different agronomic practices as treatments. Past data on climate of 100 or more years of a location may be used as input variables. One can perform the risk analysis and choose the variety or agronomic practice which can be practiced by the risk-aversive farmers. Farmers can make their own decisions based on the risk-to-user criteria.

To make the idea more clear, the work of Muchow et al. (1990) is quoted here. The yield data are plotted as cumulative distribution functions in Figure 5.3. The models were run for the cultivars and agronomic practices using 100 years of daily temperature, radiation and rainfall data for Katherine (Latitude 14.5⁰ S) in northern Australia.

Table 5.4. Mean and quartile cumulative probability for simulated maize and sorghum grain yields at Katherine from 1889 to 1988. Crops were sown no-till after 15 December when 30 mm of rain occurred in a 5-day period (Source: Muchow et al., 1990)

Variable	*Mean*		*Cumulative Probability*			
		0	0.25	0.50	0.75	1.00
Maize yield	5012	0	3271	5569	7014	8514
Sorghum yield	4499	0	3874	5042	5449	6254

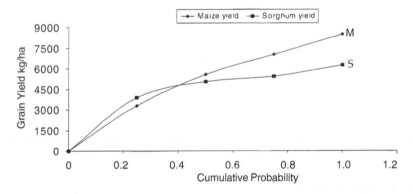

Fig. 5.3. CDFs for grain yield of maize (m, line) and sorghum (s, line) at Katherine when sown no-till after 15 December when 30 mm rain occurred in 5 days (Adapted from Muchow et al., 1990).

The yield data were plotted as cumulative distribution function (CDFs) in Figure 5.3. Grain yield of sorghum is likely to be greater than maize in 35% of years (Figure 5.3). The CDFs cross at a yield level of 4000 kg/ha. However, this does not mean that sorghum yield is always higher than that of maize upto this yield level. In fact, the plot of maize yield against sorghum yield for individual years shows that even in very poor seasons, there can be little difference in yield between maize and sorghum (Muchow et al., 1990). However, in probabilistic terms for the 100 years, it is likely that sorghum will outyield maize in poor season, whereas in the better seasons, sorghum yield is relatively stable, white maize is more variable with a higher yield potential (Figure 5.3).

The gross margin was estimated to be greater for sorghum production in 30% of the years, but it is less than $170/ha in those years (Figure 5.4). The choice of maize or sorghum by farmers depends on the individual level of risk aversion. In the long term, maize would be more profitable than sorghum, but it would also be more risky with a 25% likelihood of a negative gross margin (Figure 5.4).

Table 5.5 CDFs for gross margin of maize and sorghum at Katherine when sown no-till after 15 December when 30 mm of rain occurred 5 days (adapted from Muchow et al. 1990).

Variable	Cumulative			Probability			
	0.01	0.1	0.2	0.4	0.6	0.8	1.0
Maize	−350	−100	0	210	400	520	650
Sorghum	−250	0	75	200	210	220	240

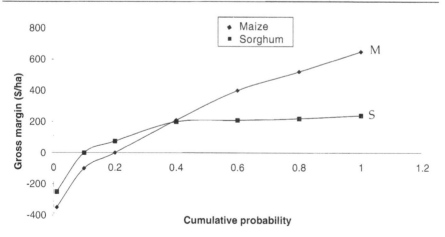

Fig. 5.4. CDFs for gross margin of maize (m, line) and sorghum (S line) at Katherine when sown no-till after 15 December, when 30 mm rain occurred in 5 days (adapted from Muchow et al., 1990).

Ram and Singh (1995) studied the low temperature and high vapour pressure deficit risks to cool season crops in the subtropics. Low temperature and high vapour pressure deficit risks had been made using the cumulative percentage frequency distribution method for judicious crop planning in the cool season, on the basis of 24-year (1971-72 to 1993-94) meteorological data for Hisar, India. During the period, 1 to 28 January, the risk of $<2^0C$ temperature at the screen level is 45 to 65%. This may reduce crop growth. The risk of vapour pressure deficit (>35mb), at the ear-emergence/anthesis stage, emerging from 19 March to 2 April is 50-92%. As such, for the Hisar area, cultivars that reach double ridge stage/floral bud initiation after 28 January and maturity (harvesting before 19 March) may be selected.

REFERENCES

Addiscoti, T.M. and Whitmore, A.P. (1987). Computer simulation of changes in soil mineral nitrogen and crop nitrogen during autumn, winter and spring. J. Agric. Sci. (Cambridge) **109**: 141-157.

Chapman, S.C., Hammer, G.L. and Meinke, H. (1993). A sunflower simulation model: I. Model development. Agron. J. **85**: 725-735.

Clements, R.J. (1990). Preface. In: Climate Risk in Crop Production: Models and Management for the Semiarid Tropics and Subtropics, (eds) Muchow, R.C. and Bellamy, J.A.) 548p. Wallingford: C.A.B. International.

Conover, W.J. (1980). Practical Non-parametric Statistics, 493p. New York: John Wiley & Sons.

Dobermann, A. and Ping, J.L. (2004). Geostatistical integration of yield monitor data and remote sensing improves field maps. Agron. J. **96** (1): 285-297.

Draper, N. and Smith, H. (1981). Applied Regression Analysis. 2nd ed. New York: John Wiley & Sons.

Jamieson, P.D., Porter, J.R., Goudriaan, J., Ritchie, J.T., Keulen, H.V. and Stol, W. (1998). A comparison of the models AFRCWHEAT 2, CERES-Wheat, Sirius, SUCROS 2 and SWHEAT with measurements from wheat grown under drought. Field Crops Res. **55**: 23-44.

Jones, J.W. and Luyten, J.C. (1998). Simulation of biological processes. In: Agricultural Systems Modeling and Simulation, (eds). Peart, R.M. and Curry, R.B.), pp.19-62. New York: Marcel Dekker, Inc.

Kanneganti, V.R., Rotz, C.A. and Walgenbach, R.P. (1998). Modeling freezing injury in alfalfa to calculate forage yield: I. Model development and sensitivity analysis. Agron J. **90**: 687-697.

Kiniry, J.R., Williams, J.R., Vanderlip, R.L., Atwood, J.D., Reicosky, D.C., Mulliken, J., Cox, W.J., Mascagni, H.J., Hollinger, S.E. and Wiebold, W.J. (1997). Evaluation of two models for nine U.S. locations. Agron. J. **89**: 421-426.

Kobayashi, K. and Salam, M.U. (2000). Comparing simulated and measured values using mean squared deviation and its components. Agron. J. **92**: 345-352.

Mayer, D.G., Stuart, M.A. and Swain, A.J. (1994). Regression of real world data on model output: An appropriate overall test of validity. Agric. Syst. **45**: 93-104.

Muchow, R.C., Hammer, G.L. and Carberry, P.S. (1990). Optimising crop and cultivar selection in response to climatic risk. In: Climatic Risk in Crop Production: Models and Management for the Semiarid Tropics and Subtropics, (eds). Muchow, R.C. and Bellamy, J.A. pp. 235-262. Wallingford: C.A.B. International.

Noether, G.E. (1967). Elements of nonparametric statistics, (2.4, 3.3, 5.1, 5.7, 5.8, 4.9), New York: John Wiley and Sons.

Ram, D. and Singh, P. (1995). Low temperature and high vapour pressure deficit risks in cool season crops in the subtropics. Annals of Arid Zone **34** (4): 267-271.

Retta, A., Vanderlip, R.L., Higgins, R.A. and Moshier, L.J. (1996). Application of SORKAM to simulate growth using forage sorghum. Agron. J. **88**: 596-601.

Singh, P., Boote, K.J., Rao, A.Y., Iruthayaraj, M.R., Sheikh, A.M., Hundal, S.S., Narang, R.S. and Singh, P. (1994). Evaluation of the groundnut model PNUTGRO for crop response to water availability, sowing dates, and seasons. Field Crops Res. **39**: 147-162.

Singh, P. (1995a). Sensitivity of crop productivity to climate change. In: Impact of modern agriculture on environment. (eds) Arora, Behl, Tauro and Joshi. **3**: 1-7. New Delhi: Soc. Sust. Agric. and N.R.M. and Max Mueller Bhavan.

Singh, P. (1995b). Sensibility of crop productivity of morphophysiological variation for characters in mungbean (Vigna radiata L.) under drought conditions. Proceedings, 2nd European Conference in Grain Legumes, 9-13 July, 1995, 484p. Copenhagen (Denmark).

Singh, P. (1995c). Modeling on water-limited productivity of mungbean. Management of stress environment for sustainable crop production. Indian Society of Agro-Physics, Agrophysics Monograph **2**: 138-140.

Smirnov, N.V. (1939). Estimate of deviation between empirical distribution functions in two independent samples (Russian) Bulletin Moscow University, **2** (2): 3-16 (6.1, 6.3).

Teo, Y.H., Beyrouty, C.A. and Gbur, E.E. (1992). Evaluating a model for predicting nutrient uptake by rice during vegetative growth. Agron. J. **84**: 1064-1070.

Walpole, R.E. (1982). Introduction to statistics. New York: Macmillan Publishing Co.

6

Biological Application of Models

As quoted in Dent and Blackie (1979), system concepts consider the multidisciplinary approach. A system consists of many elements. These elements make up the system. To understand the complex system, a set of studies of the various elements separately is essential. The whole of the system is more complex than the sum of the individual parts. One should define the precise location of the system's boundary. The environment outside the boundary of the system is uncertain and difficult to predict. Inside the boundary, the modeler can perceive a hierarchial structure of subsystems and sub-subsystems of the major system under study. Thus, the subsystems are considered to interact only through their act of joining in a higher system.

The system theory can be expressed through the techniques of modeling. It is not necessary that the model should have a symbolic form in order to describe the system. The computer-based simulation model has important advantages in this regard. The methods for the construction of a model has been developed. Appropriate experimental procedures have been explored in using these models. In this chapter, the feasibility of application of models will be dealt with. There are three broad functions of these models.

1. The model provides an objective basis for assessing and assimilating available information about the system.

2. It directs research into important areas of the system. The current knowledge of such areas is uncertain and sketchy. The model provides a platform where the results of the research can be assessed and applied.

3. The model assists in the management control or management development of the system. In agriculture, this involves the application in extension and farmer decision making. It also helps in policy direction and monitoring.

Systems Involvement in the Research Process and Management Control

6.1. ECOLOGICAL APPLICATIONS

As quoted in Poole (1974), ecology deals with the natural history. Ecology is the youngest science, but the ecologist was the oldest one ape. He first took note of the natural world surrounding him. Early ecology consisted of listing of the species of plants and animals. Afterwards, the ecology become quantitative from a descriptive one. The quantitative ecology ranges from counting the number of individuals in an area to the sophisticated and complex mathematics. Complex mathematics help in determining the causes of some ecological observed phenomena.

6.1.1 PREY-PREDATOR SYSTEM

Gold (1977) quantified the prey-predation system giving the steady-state symbol arrow graph.

$$\dot{L} \Rightarrow V \;\; \Rightarrow \;\; P \;\; \Rightarrow \;\; \pi$$
$$\leftarrow - - - \quad\quad \leftarrow - - -$$

Fig. 6.1 Steady-state symbol-arrow graph for simple prey-predator system (source: Gold 1977). Double line arrow denotes increment and dotted line arrow denotes a decrease in numbers.

Description of a steady state is very useful because it is simple. The signal-flow and symbol-arrow graph may be simplified. For example, the symbol-arrow graph for the prey-predator ecosystem simplifies to Fig. 6.1. The figure says the following:

1. A higher light level tends to increase the steady-state level of V, the vegetative matter.
2. Higher level of V tend to increase the steady state level of P, the prey.
3. Higher levels of P tend to depress the steady state level of V (negative feedback and increase the steady state level of π, the predator).
4. Higher level of π tends to depress the steady-state level of P (negative feedback).

The compartmental model does not apply when we study numbers in the population, since 'a prey does not leave the prey component and become "a predator" in the predator component'.

The prey-predator system deals with the general interaction models. The interaction between different species might be represented by two equations.

$$\frac{dN_A}{dt} = K_A N_A + K_{AB} N_A N_B \tag{6.1}$$

$$\frac{dN_B}{dt} = K_B N_B + K_{BA} N_A N_B \tag{6.2}$$

The subscript K_{AB} indicates the effect on A of B. The parameters may be either positive or negative. They represent a variety of interaction effects, depending upon the signs.

(a) Synergism $K_{AB} > 0$
            $K_{BA} > 0$
(b) Competition $K_{AB} < 0$
            $K_{BA} < 0$
(c) Prey-Predation   (A = Prey, B = Predator):
            $K_{AB} < 0$
            $K_{BA} > 0$

The cases (b) and (c) provide a built-in limit on population size, but case (a) does not. Competition terms such as $K_{AA} N^2_A$ and $K_{BB} N^2_B$ could be added to provide such limits. The general signal-flow graph showing the relation between N_A, N_B and their time derivatives is shown in Fig. 6.2.

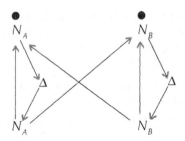

Fig. 6.2 Signal flow graph for two species interaction. The triangles indicate time delays associated with integration. $\overset{\bullet}{N_A}$ indicates the time derivative.

Equations 6.1 and 6.2 could be extended to include direct pairwise interactions involving any number of species. Numbering the species, 1, 2, ..., r.

$$\frac{dN_i}{dt} = K_i N_i + \sum_{j=1} K_{ij} N_i N_j \tag{6.3}$$

6.1.1.1 Lotka-Voltara Model for Two-Species Prey-Predator System

The following equations were rationalized as a model of direct interaction between two populations (Gold, 1977).

$$\frac{dN_A}{dt} = K_A N_A + K_{AB} N_A N_B$$

$$\frac{dN_B}{dt} = K_B N_B + K_{BA} N_A N_B$$

It was evinced that a variety of types of interaction could be ascribed depending on the signs of the coefficients. To give a specific example, we take up a prey-predation system with

P = size of prey population

π = size of predator population

α_p, α_π = Parameters expressing the intrinsic rate of increase, of each population in the absence of the other (in particular, α_p may be considered to include the influence of the environment in the form of food supply and other essentials for the growth of the prey.)

β_p, β_π = Parameters describing the influence of the interaction on each population.

The parameters are considered to be positive, so that the signs of the influences may be explicitly pointed out.

The equations were

$$\frac{dp}{dt} = \alpha_p P - \beta_p P\pi \qquad (6.4)$$

$$\frac{d\pi}{dt} = -\alpha_\pi \pi + \beta_\pi P\pi \qquad (6.5)$$

The prey and predator subsystems might each be considered as a compartment if the amount of each is quantified in terms of biomass, but not if the amounts are measured in numbers of individuals.

The first step in examining the system is to see if there is a steady state. At the steady state, both P and π become constant. So, we set equations (6.4) equal to zero, and see if we can find values for P and π that satisfy the resulting equations. That is, if P_0 and π_0 are the steady state values for P and π, then they must satisfy

$$\alpha_p P - \beta_p P\pi = 0 \qquad\qquad (6.6)$$

$$-\alpha_\pi \pi + \beta_\pi P\pi = 0 \qquad\qquad (6.7)$$

Obviously, the point $(\pi_0, P_0) = (0, 0)$ does the job. This, however, is a totally uninteresting point, since now, in the state of space, there is no system to study. We, therefore, do not consider the point $(0, 0)$ and see if there is another steady-state point that we would call (π_1, P_1).

First, however, we have to test the possibility that one of them is zero at the steady state, while the other is not. If $\pi_1 = 0$ and $P_1 \neq 0$, then equation (6.6) becomes

$$\alpha_p P_1 = 0$$

Since we are assuming $\alpha_p \neq 0$, this would not do. If $P_1 = 0$ but $\pi \neq 0$ we get

$$-\alpha_p \pi_1 = 0$$

and this would not do either. Therefore, we find that neither P_1 nor π_1 can be zero at the steady state.

Now, proceeding to divide (6.6) by P and (6.7) by π and rearranging the result

$$\pi_1 = \frac{\alpha_p}{\beta_p} \qquad\qquad (6.8)$$

$$P_1 = \frac{\alpha_\pi}{\beta_\pi} \qquad\qquad (6.9)$$

The equations (6.4 and 6.5) are unrealistic in the sense that they predict the absence of the predator, then the prey increases without bound. We cannot seriously take any prediction that a physical quantity tends to become infinite as time goes on. This defect in the model may be remedied by the inclusion of terms for intraspecies competition.

6.1.1.2 Prey-Predator Ecosystem

Gold (1977) depicted the prey-predator ecosystem, but for the benefit of students and readers, it is repeated here.

The Volterra model has been found to have an obvious defect, namely, that for zero predator population, the model predicts that prey will multiply without bound. Here, this defect would be remedied by introducing terms to account for intraspecies competition.

6.1.1.2.1 *Representation of a Simple Prey-Predator System*

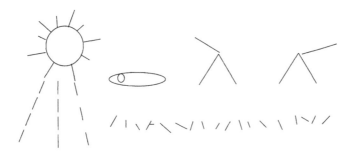

Fig. 6.3 Representation of a simple prey-predator system (*Source:* Gold, 1977).

6.1.1.2.2 *Component Diagram for Simple Prey-Predator System.*

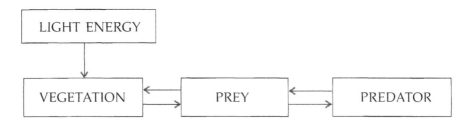

Fig. 6.4 Component diagram for simple prey-predator system (*Source*: Gold, 1977).

6.1.1.2.3 *Symbol Arrow Graph for Simple Prey-Predator System*

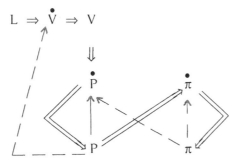

Fig. 6.5 Symbol arrow graph for simple prey-predator system (*Source*: Gold, 1977).
Single solid arrow indicates indefinite relationship.

6.1.1.2.4 *Variables for the Prey-Predator Ecosystems*

Table 6.1. Variables for the prey-predator ecosystem.

Variable	Symbol	Dimension
Light intensity	L (t)	Energy/Time
Plant material	V (t)	Mass
Prey	P (t)	Number of prey
Predator	π (t)	Number of predators

6.1.1.2.5 *Parameters for the Prey-Predator Ecosystem*

Table 6.2 shows the parameters for the prey-predator ecosystem.

Table 6.2. Parameters for the prey-predator ecosystem.

Parameter	Symbol	Dimension
Prey intrinsic rate of increase	α_r	Time^{-1}
Predator intrinsic rate of increase	α_π	Time^{-1}
Predator coefficient for prey	β_p	(Number of predator)$^{-1}$ time^{-1}
Predator coefficient for predator	β_r	(Number of prey)$^{-1}$ time^{-1}
Competition coefficient, of prey	γ_p	(Number of prey)$^{-1}$ time^{-1}
Competition coefficient, of predator	γ_π	(Number of prey)$^{-1}$ (Number of predator)$^{-1}$ time^{-1}

6.1.1.2.6 *The Model Equations of Prey-Predator Ecosystem*

$$\frac{dp}{dt} = \alpha_p P - \beta_p P\pi \qquad (6.10)$$

$$\frac{d\pi}{dt} = -\alpha_\pi \pi + \beta_\pi P\pi \qquad (6.11)$$

A direct competition term in the prey equation might be proportional to P^2, to give Equation (6.12). A direct competition term in the predator equation might be proportional to $P\pi^2$, to give equation (6.13).

$$\frac{dp}{dt} = \alpha_p P - \beta_p P\pi - \gamma_p P^2 \qquad (6.12)$$

$$\frac{d\pi}{dt} = -\alpha_\pi \pi + \beta_\pi P\pi - \gamma_\pi P\pi^2 \qquad (6.13)$$

6.1.1.2.7 Steady States of Prey-Predator Ecosystem

We have the uninteresting steady state $(\Pi_0\ P_0) = (0,0)$. We next check the possibility of a steady state $(\Pi_1\ P_1)$, with $P_1 = 0$, $\Pi_1 \neq 0$. From equation (6.13), we see that this requires $\alpha_\pi\ \Pi$ to be zero, which is implicitly assumed not to be the case. Thus, the type of steady state being sought does not exist, which simply means that a population of predators cannot be sustained without a supply of prey.

We can find a steady state (Π_1, P_1) for which $P_1 \neq 0$ while $\Pi_1 = 0$. In this case, equation (6.12) can be solved for P_1;

$$\alpha_p P - \beta_p\ P\Pi - \gamma_p P^2 = 0$$

$$\alpha_p P - \gamma_p P^2 = 0, \ \alpha_p P = \gamma_p P^2$$

$$\frac{\alpha_p}{\gamma_p} = \frac{P^2}{P}, \quad \frac{\alpha_p}{\gamma_p} = P_1$$

$$\therefore\ P_1 = \frac{\alpha_p}{\gamma_p} \tag{6.14}$$

With $\pi = 0$, this is simply the direct competition model.

Next we look for a balance point $(\Pi_2 P_2)$ with both populations as non-zero. Setting the two derivatives equal to zero and assuming that neither population size is zero, equations (6.12 and 6.13) become

$$\alpha_p P - \beta_p\ P\Pi - \gamma_p\ P^2 = 0$$

$$\alpha_p - \beta_p\Pi - \gamma_p\ P = 0 \tag{6.15}$$

$$-\alpha_\pi\Pi + \beta_\pi P\Pi - \gamma_\pi P\Pi^2 = 0$$

$$\alpha_\pi - \beta_\pi P + \gamma_\pi P\Pi = 0 \tag{6.16}$$

Solving equation (6.15) for P gives the condition under which dP/dt is zero for any fixed value of Π. Similarly, equation (6.16) gives the condition under which $d\Pi/dt = 0$ for any fixed P. We get

$$\alpha_p - \beta_p\ \Pi - \gamma_p P = 0$$

$$\gamma_p P = -\alpha_p + \beta_p\Pi$$

$$\gamma_p P = \alpha_p - \beta_p\ \Pi$$

$$P = \frac{\alpha_p - \beta_p\Pi}{\gamma_p}$$

$$P = \frac{\alpha_p}{\gamma_p} - \frac{\beta_p}{\gamma_p} \Pi \tag{6.17}$$

$$\alpha_\pi - \beta_\pi P + \gamma_\pi P\Pi = 0$$

$$\gamma_\pi P\Pi = -\alpha_\pi + \beta_\pi P$$

$$\Pi = -\frac{\alpha_\pi}{\gamma_\pi P} + \frac{\beta_\pi P}{\gamma_\pi P}$$

$$\Pi = \frac{\beta_\pi}{\gamma_\pi} - \frac{\alpha_\pi}{\gamma_\pi} \cdot \frac{1}{P} \tag{6.18}$$

A plot of equation (6.17) is simply a straight line with a negative slope. One end is at

$$P = \frac{\alpha_p}{\gamma_p} \text{ for } \Pi = 0,$$

$$
\left[
\begin{array}{c}
P = \dfrac{\alpha_p}{\gamma_p} - \dfrac{\beta_p}{\gamma_p} \Pi \\[2mm]
\text{For } \Pi = 0 \\[2mm]
P = \dfrac{\alpha_p}{\gamma_p}
\end{array}
\right]
$$

and the other is at $\Pi = \dfrac{\alpha_p}{\gamma_p}$ for $P = 0$,

$$
\left[
\begin{array}{c}
P = \dfrac{\alpha_p}{\gamma_p} - \dfrac{\beta_p}{\gamma_p} \Pi \\[2mm]
\text{For } P = 0 \\[2mm]
\dfrac{\alpha_p}{\gamma_p} - \dfrac{\beta_p}{\gamma_p} \Pi = 0 \\[2mm]
-\Pi \dfrac{\beta_p}{\gamma_p} = -\dfrac{\alpha_p}{\gamma_p} \\[2mm]
\Pi = \dfrac{\alpha_p}{\gamma_p} \cdot \dfrac{\gamma_p}{\beta_p} \\[2mm]
= \dfrac{\alpha_p}{\beta_p} \\[2mm]
\text{and the other is at } \Pi = \dfrac{\alpha_p}{\beta_p} \text{ for } P = 0.
\end{array}
\right]
$$

From equation (6.18), we find that as the "fixed" value of P gets large, predators become limited by their own competitions, and Π approaches β_Π/γ_Π asymptomatically. As P gets smaller, the prey population is able to support fewer and fewer predators, until $P = \dfrac{\alpha_\Pi}{\beta_\Pi}$ at which point, no predator population can be supported.

$$\begin{bmatrix} \alpha_\Pi - \beta\Pi + \gamma_\Pi P\Pi = 0 \\ 0 - \beta_\Pi P + \gamma_\Pi P\Pi = 0 \\ -\beta_\Pi + \gamma_\Pi\,\Pi = 0 \\ \Pi = \dfrac{\beta_\Pi}{\gamma_\Pi} \end{bmatrix} \quad \text{As P gets larger}$$

$$\begin{bmatrix} \alpha_\Pi - \beta_\Pi\,P + \gamma_\Pi P\Pi = 0 \\ \alpha_\Pi - \beta_\Pi P + \gamma_\Pi P\Pi = 0 \\ \alpha_\Pi - \beta_\Pi\,P = 0 \\ -\beta_\Pi\,P = -\alpha_\Pi \\ P = \dfrac{\alpha_\Pi}{\beta_\Pi} \end{bmatrix} \quad \text{As P gets smaller}$$

The plots of two equations (6.17 and 6.18) are shown in Fig. 6.6.

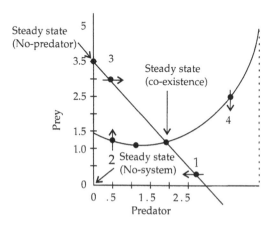

Fig. 6.6 Locating the steady states and the directions in the state space for the simple prey-predator system described by equations 6.12 and 6.13.

The point of intersection of these two curves is the coexistence steady state (Π_2, P_2). In drawing Fig. 6.6, it is assumed that $\alpha_\pi/\beta_\pi < \alpha_p/\gamma_p$ and that $\alpha_\pi/\beta_\pi < \beta_\pi/\gamma_\pi$. Unless these conditions are met, the two curves do not intersect for positive values of p and Π ; there is no coexistence steady state. The first of these inequalities simply says that the prey population must be inherently successful enough to support the predators. The second inequality says that competition between predators must not be so intense that they kill themselves off.

Accepting these assumptions, the simultaneous equations (6.18) can be solved for Π_2 and P_2. The procedure is straightforward (but disorderly), except that it involves a quadratic expression, only one of whose roots is acceptable.

An expression for the homeostatic index can be obtained directly from equations (6.18)

$$\text{H.I.} = -\left(\frac{\partial p}{\partial \Pi}\right)\left(\frac{\partial \Pi}{\partial p}\right)$$

$$= -\left(-\frac{\beta_p}{\gamma_p}\right)\left(+\frac{\alpha_\Pi}{\gamma_\Pi} \cdot \frac{1}{p^2}\right)$$

$$= +\left(\frac{\beta_p}{\gamma_p} \frac{\alpha_\Pi}{\gamma_\Pi}\right) \cdot \frac{1}{p^2} \tag{6.19}$$

We have a negative feedback loop, with a positive homeostatic index. The index is not constant. This is a result of the fact that by equation (6.18), the dependence of Π on P is non-linear; the slope of the curve (Fig. 6.6) is, therefore, not constant. In such a case, the H.I. is given by the negative of the product of the slopes at the steady state. Examination of either Fig. 6.6 or Expression 6.19 shows that the insulating capacity of the feedback loop, as measured by the homeostatic index, is less for steady states that involve large amount of prey.

Steps in calculation of Homeostatic Index (H.I.) (for students of Biology):

$$\text{H.I.} = -\left(\frac{\partial p}{\partial \Pi}\right)\left(\frac{\partial \Pi}{\partial p}\right)$$

Recall the equation (6.17)

$$P = \frac{\alpha_p}{\gamma_p} - \frac{\beta_p}{\gamma_p} \Pi$$

$$\frac{\partial p}{\partial \Pi} = \frac{\partial\left(\frac{\alpha_p}{\gamma_p}\right)}{\partial \Pi} - \left(\frac{\partial \frac{\beta_p}{\gamma_p} \Pi}{\partial \Pi} + \frac{(\partial \Pi) \cdot \frac{\beta_p}{\gamma_p}}{\partial \Pi}\right)$$

$$\frac{\partial p}{\partial \pi} = 0 - \left(0 + 1 \left(\frac{\beta_p}{\gamma_p} \right) \right)$$

$$\frac{\partial p}{\partial \pi} = - \frac{\beta_p}{\gamma_p}$$

Now recall the equation (6.18)

$$\Pi = \frac{\beta_\pi}{\gamma_\pi} - \frac{\alpha_\pi}{\gamma_\pi} \cdot \frac{1}{p}$$

$$\frac{\partial \pi}{\partial p} = \frac{\partial \left(\frac{\beta_\pi}{\gamma_\pi} \right)}{\partial p} - \left(\frac{\partial \left(\frac{\alpha_\pi}{\gamma_\pi} \right) \cdot p^{-1}}{\partial p} + \frac{\left(\partial p^{-1} \right) \cdot \frac{\alpha_\pi}{\gamma_\pi}}{\partial p} \right)$$

$$= 0 - \left(0 + \left(-1 p^{-2} \right) \left(\frac{\alpha_\pi}{\gamma_\pi} \right) \right)$$

$$= 0 - \left(\frac{\alpha_\pi}{\gamma_\pi} \left(-\frac{1}{p^2} \right) \right) = - \left(-\frac{\alpha_\pi}{\gamma_\pi} \cdot \left(\frac{1}{p^2} \right) \right) = + \left(\frac{\alpha_\pi}{\gamma_\pi} \cdot \left(\frac{1}{p^2} \right) \right)$$

$$H.J. = \left(-\frac{\beta_p}{\gamma_p} \right) \left(+\frac{\alpha_\pi}{\gamma_\pi} \cdot \frac{1}{p^2} \right) = \frac{\beta_p}{\gamma_p} + \frac{\alpha_\pi}{\gamma_\pi} \cdot \frac{1}{p^2}$$

6.1.1.2.8 *Non-Steady-State Behavior*

In Fig. 6.6, in order to arrive at the directions shown, we may reason on the basis of the curves of $\dot{P}$ (time derivative) =0 and $\dot{\Pi}$ = 0. At points 1 and 3, we have $\dot{P}$ =0, so the direction must be parallel to the Π axis. Since, it cannot be that $\dot{\Pi}$ =0 (if it were, this would be a steady state), then Π must be either increasing or decreasing. Now, in the P dimension, point 1 is below the line for $\dot{\Pi}$ =0. That is, P is less than the amount needed to maintain the predators at the indicated level. Therefore, the predator population must be decreasing; $\dot{\Pi}$ > 0. A similar argument shows that is $\dot{\Pi}$ > 0 at point 3.

At point 2 and 4, we have $\dot{\Pi}$ = 0; the direction must be parallel to the P-axis. At point 2, Π is less than that needed to limit the prey to the indicated level, and they must increase, so $\dot{P} > 0$. Similarly, at point 4, the predators are in excess of the level required to just balance the prey population, so $\dot{P} < 0$.

The state-space directions suggest the possibility of sustained oscillation. In order to see if the symbol-arrow graph confirms this, we need to know the sign of the $P \longrightarrow \dot{P}$ arrow at the steady-state. Increasing P increases the intrinsic rate of increase, leading to a positive feedback, but also increases the competition and predation effect, giving rise to a negative feedback. To be more precise, we differentiate equation (6.12)

$$\frac{\partial \dot{p}}{\partial p} = \left(\frac{(\partial \alpha_p) \cdot P}{\partial p} + \frac{(\partial p) \cdot \alpha_p}{\partial p} \right)$$

$$- \left(\frac{(\partial \beta_p) \cdot P\Pi}{\partial p} + \frac{(\partial p) \cdot \beta_p \Pi}{\partial p} + \frac{(\partial \Pi) \cdot \beta_p P}{\partial p} \right) - \left(\frac{(\partial \gamma_p) \cdot p^2}{\partial p} + \frac{(\partial p^2) \cdot \gamma p}{\partial p} \right)$$

$$= [0 + 1 \, (\alpha_p)] - [0 + 1\beta_p \Pi + 0] - [0 + 2p^1 \cdot \gamma_p]$$

$$= \alpha_p - \beta_p \Pi - 2\gamma_p \, P \qquad (6.20)$$

and try to evaluate the sign of this expression at the steady state. Substituting equation 6.17, which must hold at the steady state into equation 6.14 gives

$$\left[\begin{array}{l} \text{Recall Equation 6.17,} \\[2mm] \qquad P = \dfrac{\alpha_p}{\gamma_p} - \dfrac{\beta_p}{\gamma_p} \, \Pi \\[2mm] \text{Recall Equation 6.14,} \\[2mm] \qquad \dfrac{\partial \dot{P}}{\partial P} = \alpha_p - \beta_p \Pi - 2\gamma_p P \\[2mm] \qquad = \alpha_p - \beta_p \Pi - 2\gamma_p \left(\dfrac{\alpha_p}{\gamma_p} - \dfrac{\beta_p \Pi}{\gamma_p} \, \Pi \right) \\[2mm] \qquad = \alpha_p - \beta_p \Pi - \left(\dfrac{2\gamma_p \alpha_p}{\gamma_p} - \dfrac{2\gamma_p \beta_p}{\gamma_p} \, \Pi \right) \\[2mm] \qquad = \alpha_p - \beta_p \, \Pi - \left[2 \, \alpha_p - 2\beta_p \, \Pi \right] \\[2mm] \qquad = \alpha_p - \beta_p \, \Pi - 2\alpha_p + 2\beta_p \, \Pi \\[2mm] \qquad = -\alpha_p - \beta_p \, \Pi \\[2mm] \qquad \left(\dfrac{\partial \dot{P}}{\partial P} \right)_{ss} = -\alpha_p - \beta_p \, \Pi \end{array} \right] \qquad (6.21)$$

From Fig. 6.6, we see that steady-state value of Π must always be less than α_p/β_p, so the right hand side of [Eqs. 6.15 and 6.16] must be negative. Since all feedback loops in the system are negative, we now expect that the system spirals inward toward the steady-state and that the individual variables exhibit damped oscillations.

It is especially important to notice that even a very slight alteration to the structure of the Lotka-Voltera equations in the form of any small but non-zero competition term has switched the qualitative behavior of the model from one of sustained oscillation to one of approach to steady state. When the qualitative behavior of a model is sensitive to slight modifications in structure, small error in the specification of the functional forms can lead to large errors in predictions of system behavior. Such models are said to lack structural stability. Since one is rarely certain as to the absolute correctness of the structure of the model, models having such structural stability (such as the Lotka-Voltera model) may be heuristically useful, but must be regarded with suspicion when taken to represent actual real world systems.

The complexity of the model has increased along with the capacity of the model to treat more complex prey-predator interactions. Nevertheless, a great many idealizations remain that limit the usefulness of the model. Some of the more glaring ones are:

1. Neglect of interactions involving other species.
2. Assumptions that all individuals of a given species are stochastically identical. In particular, this leads to the neglect of the age structure of the population and the neglect of the delay between the time when an individual is added to the population and the time it becomes part of the reproducing population.
3. Assumption of stochastic independence leads to neglect of sexual pairing, and the formulation of cohort, territoriality and so on.
4. Environmental conditions, as represented by the parameters of the equations, are assumed to be constant.
5. The whole package of assumption is implied by the use of differential equations.

It is upto the individual research worker to decide which idealizations are permissible for any given system.

6.1.1.2.9 *Simulation of Prey-Predator System*

When the system is characterized by several variables, the description of the motion of the systems point in state space requires that we describe the change in each of the variable, using a set of differential equations. This can be illustrated with the prey-predator system. Including the vegetation upon which the prey feed, a set of differential equations might be rationalized of the following form

(V = Vegetation, P = Prey, Π = Predators),

$$\frac{dv}{dt} = \alpha_v - \beta_v\ VP \tag{6.22}$$

$$\frac{dp}{dt} = \alpha_p\ VP - \beta_v\ P\Pi \tag{6.23}$$

$$\frac{d\Pi}{dt} = \alpha_\pi\ \Pi + \beta_\pi\ P\Pi \tag{6.24}$$

A possible sequence of calculations would be as shown in Fig. 6.7.

6.1.1.2.9.1 Exercise

If the derivative is changing rapidly, we need to adjust more often. That is, the size of the interval Δt must be smaller. If the derivative is changing only slowly, we can take Δt to be larger and readjust less often. If the derivative does not change at all,

$$\frac{dy}{dt} = const.$$

then we do not have to readjust at all. There is a limit to how small Δt can be made. The computer is a finite machine. If Δt is taken to be too small, the computer would not be able to tell it from zero. As a result, the computed change during the interval Δt is all rounding error.

Use this procedure to obtain a graph of y versus t from the equation,

$$\frac{dy}{dt} = -ky;\ k > 0$$

Use a convenient value for y_0 and a small value for k, say $k = .01$. Compute as many points as you think you need to get the feel of the procedure. Repeat the procedure with a different choice of Δt. After a certain number of time periods, compare the result with the exact solution $y(t) = y_0\ e^{-kt}$. Observe that if Δt is too large, very peculiar result emerges. Why? (Gold, 1977).

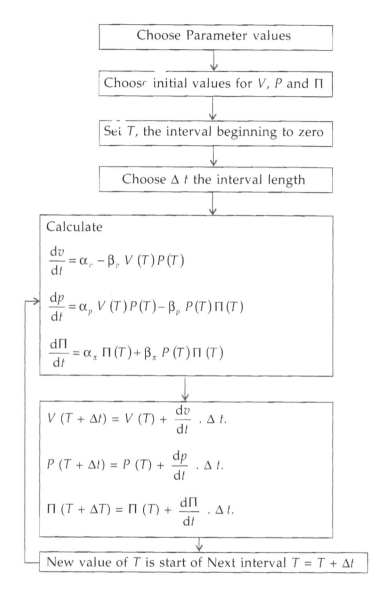

Fig. 6.7 Possible sequence of calculating the equations 6.22, 6.23 and 6.24.

6.1.1.2.10 Effects of Time Delay

France and Thormley (1984) in Chapter 9 of "Plant Diseases and Pest" described the philosophy of the time delay. The general influence of a time delay in a system is to increase instability. Familiar examples include the correcting of a skid in a car, which so easily makes the skid worse.

Our attempts to stabilize the economic system can sometimes be counter productive. Under a shower, one can easily be alternately frozen and scalded in one's efforts to adjust the water temperature. Delayed responses in biology may arise in many ways.

6.1.1.2.11 *Time Required for Developmental Event*

For the growth of a biological population, y, one sometimes assumes the exponential growth equation

$$\frac{dy}{dt} \ \mu y(t) \qquad\qquad (6.25)$$

where μ is the proportional rate of growth, and the rate of growth in y is related to the value of y at time t. However, if it takes time τ for eggs to become adults (for example), it is more appropriate to write.

$$\frac{dy}{dt} \ \mu y(t-\tau) \qquad\qquad (6.26)$$

where $y\,(t-\tau)$ is the adult population in time $t-\tau$.

6.1.1.2.12 *Delayed Effects of Environment*

The growth rate parameter μ of equation (6.25) may depend upon environment variables $E\,(t)$, which might include substrate supply. There may be delays in their effects upon the organism, in which case it is more accurate to write.

$$\frac{dy}{dt}(t)=\mu\left[E\left(t-\tau\right)\right]y\left(t\right) \qquad\qquad (6.27)$$

where μ is calculated from the values of the environmental variable E time τ ago. For instance, a reduced food supply may have little effect on the rate of population growth for some considerable time.

6.1.1.2.12.1 *An Example of a System with Delays*

Let Y denote an insect population which is nourished with a constant food supply f; the adults have a maintenance requirement $m\ Y$, where m is a constant, the surplus food $f-mY$ is converted into eggs with efficiency η, so that the rate of egg production is $\eta\,(f-m\ Y)$; the eggs hatch into adult after time τ, and finally, the adults suffer from a constant death rate h. Putting these assumption together, the differential equation for $Y\,(t)$ at time t is

$$\frac{dY}{dt}(t) = \eta \left[f - mY(t - \tau) \right] - hY(t) \tag{6.28}$$

The steady-state solution occurs at y, where

$$Y(t) = Y(t - \tau) = Y_s$$

and substituting into equation 6.28 with $\dfrac{dY}{dt} = 0$, therefore

$$Y_s = \frac{\eta f}{\eta m + h} \tag{6.29}$$

$$
\left[
\begin{array}{l}
\text{Recall equation 6.28} \\[4pt]
\dfrac{dY}{dt} = \eta \left[f - mY(t - \tau) \right] - hY(t) \\[6pt]
0 = \eta \left[f - mY_{s1} \right] - hY_s \\[4pt]
0 = \eta f - \eta m Y_s - hY_s \\[4pt]
0 = \eta f - \left[\eta m Y_s - hY_s \right] \\[4pt]
0 = \eta f - Y_s - (\eta m - h) \\[4pt]
\quad + Y_s (\eta m + h) = \eta f \\[4pt]
Y_s = \dfrac{\eta f}{\eta m + h}
\end{array}
\right]
$$

In order to examine the stability of the system, small deviation y from the steady state are considered, y is defined by

$$Y = Y_s + y \tag{6.30}$$

and substituting equation 6.30 into equation 6.28, and using equation 6.29.

$$\frac{dy}{dt}(t) = -\eta m(t - \tau) - hy(t) \tag{6.31}$$

Recall equation 6.28

$$\frac{dy}{dt}(t) = \eta \left[f - my(t - \tau) \right] - hy(t)$$

$$= \eta \left[f - m(y_s + y)(t - \tau) \right] - h(y_s + y)(t)$$

$$= \eta f - \eta m y_s + \eta m y(t - \tau) - hy_s + hy(t)$$

$$= y_s(\eta m + h) - \eta m y_s - \eta m y(t - \tau) - hy_s - hy(t)$$

$$= y_\eta m + y_s h - \eta m y_s - \eta m y (t - \tau) - h y_s - h y (t)$$
$$= - \eta m y (t - \tau) - h y(t)$$

This is written more simply as

$$\frac{dy}{dt} (t) = - ay (t - \tau) - hy (t)$$

with $h = 0$, it can be shown that this equation is stable if $(a \geq 0)$

$$a < \Pi / 2\tau \tag{6.32}$$

6.1.1.2.13 Models with Age Structure: The Laslie Matrix Approach

France and Thornley (1984) explained the model with age distribution structure with the Laslie matrix approach. In population dynamics, the simplest modeling approach is in terms of total population number, N. However, total population models are unrealistically simple, and have been found to be generally too inaccurate. An age-specific distribution function $n (t, a)$ can be defined in such a manner that

$$n (t, a) \, da \tag{6.33}$$

is the number of organisms at time t with ages lying between a and $a + da$. The total population is obtained from

$$N (t) = \int_0^0 n (t, a) \, da \tag{6.34}$$

Birth rate and death rate functions $B (t, a)$ and $D (t, a)$ can be defined as the rates of these processes at time t for organisms of age a so that

$$\text{birth rate} = \int_0^0 B (t, a) n (t, a) \, da \tag{6.35}$$

$$\text{death rate} = \int_0^0 D (t, a) n (t, a) \, da \tag{6.36}$$

Birth occurs into age $a = 0$ organisms, so that

$$n (t, 0) = \text{birth rate} \tag{6.37}$$

From a group of organisms of age a at time t, organisms are lost by death and ageing. After time interval Δt, in the absence of death, organisms $n (t, a) \, da$ move into $n (t + \Delta t, a + \Delta a) \, da$, and this can be shown to lead to

$$\frac{\partial n}{\partial t} + \frac{\partial n}{\partial a} = 0 \tag{6.38}$$

and in the presence of death

$$\frac{\partial n}{\partial t} + \frac{\partial n}{\partial a} = -D\left(t, a\right) n\left(t, a\right) \tag{6.39}$$

Even greater realism may be achieved by using an age-size specific distribution function $n\,(t, a, w)$ where

$$n\,(t, a, w)\,\mathrm{d}a\,\mathrm{d}w \tag{6.40}$$

is the number of organisms with age lying between a and $a + \mathrm{d}a$ and sizes between $w + \mathrm{d}w$. Total population number N is now obtained by a double integral.

$$N = \int_0^\alpha \int_0^\alpha n\left(t, a, w\right) \mathrm{d}a\,\mathrm{d}w \tag{6.41}$$

and equations analogous to equations 6.35–6.38 can be constructed. The approach can rapidly lead to very complicated equations (Streifer, 1974), and in solving the partial differential equations numerically, it is necessary to take finite intervals for t, a, and w if it is included. The equations are then approximated by matrix equations which are identical to the matrix approach of Lewis (1942) and Leslie (1945).

In Fig. 6.8 (France and Thornley, 1984) a continuous age-specific distribution is divided into four cohorts. Using the notation defined in Fig. 6.8, a transition matrix can be written down:

$$\begin{pmatrix} n_1 \\ n_2 \\ n_3 \\ n_4 \end{pmatrix}^{(t\,=\,t\,+\,1)} = \begin{pmatrix} B_1 & B_2 & B_3 & B_4 \\ S_{12} & 0 & 0 & 0 \\ 0 & S_{23} & 0 & 0 \\ 0 & 0 & S_{34} & 0 \end{pmatrix} \begin{pmatrix} n_1 \\ n_2 \\ n_3 \\ n_4 \end{pmatrix}^{t\,=\,t} \tag{6.42}$$

The Bs are the birth rates of the four cohorts, and the Ss are survival rates – S_{12} is the fraction of the first cohort which, after one time unit, survive and constitute the second cohort. Thus, equation (6.42) denotes the equations

$$\begin{aligned} n_1\,(i + 1) &= B_1 n_1\,(i) + B_2 n_2\,(i) + B_3 n_3\,(i) + B_4 n_4\,(i) \\ n_2\,(i + 1) &= S_{12} n_1\,(i) \\ n_3\,(i + 1) &= S_{23} n_2\,(i) \\ n_4\,(i + 1) &= S_{34} n_3\,(i) \end{aligned} \tag{6.43}$$

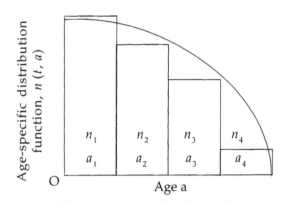

Fig. 6.8 The age-specific population distribution function shown by the continuous line is approximated by the four quartile n_1, n_2, n_3 and n_4.

The matrix approach has been widely used in practical studies, and is discussed extensively by Williamson (1972).

6.1.1.2.14 An Example of a Biological Control Model

A much simplified scheme is shown in Figure 6.9 (France and Thornley, 1984) to indicate a possible form by a biological control model. This is partially based on a study of the population dynamics of the aphid *Sitobion avenae* in winter wheat, and control of the aphid population by predation by hoverfly larvae (Rabbinge, Ankerisonit and Pak, 1979).

The aphid larvae have four instart L1 to L4. The developmental times of L1, L2 and L3 are similar; rather longer is required in L4. There are two forms of instar L4, wingless and winged, and the relative proportion of each depends on factors such as temperature, crowding and plant status. The developmental rates, which determine the time spent in each instar, are also highly dependent on temperature.

Adult aphids are consumed by the larvae of the hoverfly, and it may be assumed that the mortality of hoverfly larvae is greatly reduced by an adequate supply of aphid prey. The rate of predation, for example, may be modeled using one of the relations described in equations (used in exercise 64 and 65 page 453). Many of the rates in Figure 6.9, mortality, reproduction, development and emigration, depend on weather as well as plant.

It can be seen that models of these problems can quickly become highly complex, and they must be very carefully formulated if they are to contribute to understanding and problem solution in the area.

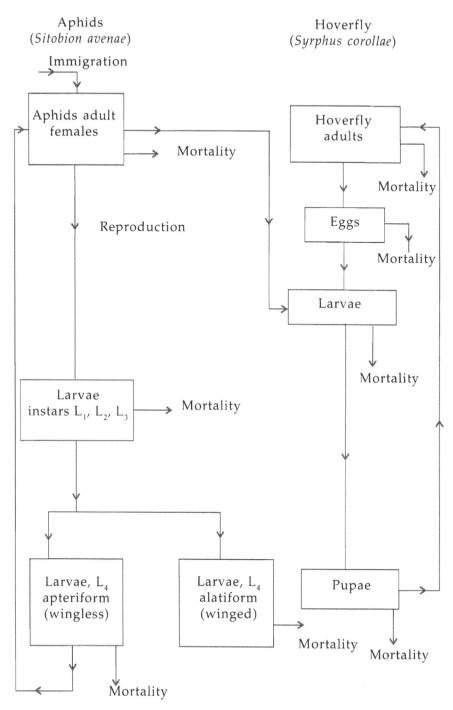

Fig. 6.9. Model for biological control of wheat aphid by predation by hoverfly larvae.

6.1.1.2.15 *Sensitivity Analysis*

As discussed by Gold (1977), in many feedback systems, it is possible to associate the feedback with some parameters of the system. In such cases, the homeostatic index can be obtained by comparing the response to environment change when the parameter is zero (open loop). The response that forms the basis of these comparisons is the displacement of the steady state position. In some cases, no comparison is possible because no steady state exists when the feed loop is opened, i.e. when the parameter is set to zero. However, it is also possible to look at the effect of small changes in the parameters.

It is always important to assess the sensitivity of the behavior of the model to variations in the values of the parameters, since the parameter values can never be known with absolute precision. In many cases, the parameter must be supposed to reflect the influences of the environment. For most models, the parameters should be thought of as random quantities whose expectations appear in the model equations but whose exact values fluctuate in some random manner. It is clearly important to understand how the behavior of the system changes as the parameter values are varied. Some discussions of sensitivity analysis (and further references) may be found in Kowal (1971), Miller (1975) and Miller, Weidhaas, and Hall (1973).

6.1.2 PLANT COMPETITION STUDIES

6.1.2.1 Intraspecific Competition in Plants

Poole (1974) described the intraspecific competition in plants. The numerical response of a population of individuals to limited resources, perhaps most often food, has been studied. The measurement of intraspecific competition in plants almost always depends on the relative distance of one plant from another rather than the overall density of the population.

Marshall and Jain (1969) have provided experimental verification of the sensitivity of plant fertility and growth to change in density. *Avena barbara* seems to be rather insensitive to increasing density in the percentage of seeds emerging as seedlings, but in *A. fatua*, there is a marked reduction in percentage of seedling emerge.

Fertility is reduced by increasing density in both species of grass, slightly less so in *A. fatua* than in *A. barbara*. They also found significant effects of density on dry matter weight per plant, number of tillers produced, plant height, and number of spikelets per plant.

6.1.2.1.1 *Pielou's Measures of Intraspecific Competition in Plants*

Gold (1977) depicted the Pielou's measures of intraspecific competition in plants in detail. Many of the studies on intraspecific competiton in plants have been concerned with field crops, proposing some specific spatial arrangement of the individual plants. Examples of such studies are those of Berry (1967), Mead (1967), Willey and Heath (1969), Pant (1979), France and Thornley (1983). France and Thornley (1984) gave a detailed account of crop yield and planting density responses and mechanistic basis for yield and density responses. In this subsection, a more general method of detecting intraspecific competition applicable to natural plant population will be discussed. In practice, the method requires that the individuals of the population occur in almost pure stand.

Pielou (1962) suggested that competition between two neighboring plants may manifest itself in two ways: (1) the distance between any plant and its nearest neighbor may be positively correlated with the sum of their sizes; and (2) there may be a lower limit to the distance between any plant and its nearest neighbors. The first sign indicates that the closer two individuals are the more they inhibit each other's growth. In the latter case, each successful plant may have a territory within which no new colonizer can establish itself. Detection of the first indication of competition is accomplished by calculating the correlation coefficient between the distance between a plant and its nearest neighbor, and the sum of the sizes of the two individuals. Detection of the second indication of competition is more complicated.

If the distance between a randomly chosen individual and its nearest neighbor is r, the frequency distribution of these distances squared is

$$f(w) = \lambda e^{-\lambda w}$$

if the population has a random dispersion pattern.

The letter w represents r^2, the squared distance between the individuals and its nearest neighbor, and λ is the mean density of the population in the number of individuals per circular area of unit radius. In an aggregated population, the number of small and large values of w will be greater than expected and the number of intermediate values too few.

Competition will manifest itself in the plants closest together, i.e. lower values of w. Therefore, some values of w is chosen, c, and only distances equal to c or smaller are considered. This eliminates the high values of w, resulting in a new, truncated frequency distribution for w between 0 and c as

$$f(w)\,|\,0 \le w \le c = \frac{\lambda e^{-\lambda w}}{1 - e^{-\lambda c}}$$

It is now necessary to divide the range of possible values of w into i equal class intervals so that the expected proportion in each class is $1/i$. The boundary values of the classes are

$$w_r \begin{cases} = 0 & r = 0 \\ = \dfrac{-1}{\lambda} \log\left(1 - \dfrac{r}{i}\left(1 - e^{-\lambda c}\right)\right) & 0 < r < i \\ = c & r = i \end{cases}$$

where r represents each class interval. Having divided the lower squared distances into each class interval, the observed frequency of observations in each class may be plotted. If competition is manifesting itself in manner 2, the number of observations falling into the class interval $r = 0$ to $r = 1$ should be less than expected because at the lower limit of distances between plants, the presence of one individual should inhibit the establishment of another. The expected frequency distribution in the transformed y scale used in creating the class interval is

$$f(y) = \frac{1 - e^{-\lambda w}}{1 - e^{-\lambda c}}$$

Dividing the truncated distribution into 10 classes y ranging from 0 to 1.0 ($i = 10$), the significance of a deviation of the frequency of the first class f_1 from the expected frequency can be tested using the standardized normal values

$$Z = \frac{N - 10\,f_1}{3\,N^{\frac{1}{2}}}$$

where f_1 is the number of observations in the first class, and N is the total number of observations, i.e. squared distances, between 0 and c. A one-tailed test is used because the hypothesis is that the observed frequency f_1 is less than the expected frequency.

Pielou (1962, as quoted in Poole (1974) studied a population of Pinus ponderosa in a 4055-square-metre area. All trees larger than 2 meter in height were considered part of the population. A total of 148 distances between a plant chosen at random and its nearest neighbor were measured. A truncated frequency distribution was created by disregarding all values of w greater than 2.00 ($c=2.00$). This left a sample size of N = 58. The range of w from 0 to 2.00 was divided into 10 equal parts, i.e. $y = 0$ to 1.0. To test the first frequency class $f_1 = 5$, the Z statistic is

$$Z = \frac{58 - (10)(5)}{(3)(58)^{\frac{1}{2}}} = 0.358$$

Comparing the Z statistic to a table of the standarized normal distribution using a one-tailed test, it is found that the probability of a deviation form zero, the expected value of Z, as large as 0.358 is about 36 per cent. It is concluded that the observed number of observations in the first class is not significantly smaller than expected. These is no reason to conclude that the competition is being exhibited by an established individual excluding invading individuals within some minimum distance. This is not to say that intraspecific competition is not taking place, because the indication number 1 is still open. Pielou (1960) found a significant positive correlation between the distance from an individual to its nearest neighbor and the sum of their circumference in these same trees. This finding indicates that close neighbors in a ponderosa pine forest inhibit each other's growth.

6.1.2.2 Competition Among Several Plant Species

Kropff and van Laar (1993) made an excellent contribution on the modeling of crop-weed interactions.

Poole (1974) reviewed the literature in competition among several plant species. Analysis of competition among plant species has usually revolved around the yield of plant grown in monoculture as compared to its yield when grown in combination with another species. So, competition is defined in terms of changes in biomass rather than changes in density. The method of analysis of multispecies competition presented here from McGilchrist (1965) and Williams (1962).

The basic experiment is to grow each species by itself in r_1 replicates and measure the yield. Williams measured the weight of plant tops in grams per one-half pot (each pot was a replicate). The data were transformed to log (weight + 1) to stabilize variances in order to test the significance of the competition effects by an analysis of variance. In addition, each possible pair of species is grown in equal numbers in a pot. These are r_2 replicates of each species pair. The yield of each species in the two-species experiment is measured. If no competition is occurring, the yield of a plant in the two species pot should be equal to the half-pot yield in the monoculture.

The experimental values of log {(weight in grams) + 1} for the two replicates (Williams, 1962) are given in Table 6.3. The transformed yield of the ith species grown in competition with the kth species in the kth replicate can be represented by the linear model

$$Y_{ijk} = \alpha_{ij} + \rho_k + T_{ijk} + e_{ijk}$$

where

λ_{ij} = the mean for the ith species grown in competition with the jth

ρ_k = The difference between replicates ($\Sigma\rho_k = 0$)

T_{ijk} = Pot effects, assumed to be an independent normal variable with zero mean.

e_{ijk} = an error term.

There are two possible measures of competition between two species of plants. The first is the increase in yield of species i when it is grown with species j over species i yield in monoculture. The other measure is the depression species i causes in the yield of species j as compared to the yield of species as j in monoculture.

McGilchrist (1965) defines the competition advantage of species i over species j as an average of these two components. If this competitive advantage is denoted as γ_{ij}

$$\gamma_{ij} = \frac{1}{2}(\alpha_{ij} - \alpha_{ii}) + \frac{1}{2}(\alpha_{jj} - \alpha_{ji})$$

He goes on to define the competitive depression of species i and species j as one-half the decrease in the total yield of species i and j when grown in competition, as compared to their total yield in monocultures. The competitive depression of species i and species j can be represented as

$$\delta_{ij} = \frac{1}{2}(\alpha_{ii} + \alpha_{jj}) - \frac{1}{2}(\alpha_{ij} - \alpha_{ji})$$

The competitive advantage of species i can be averaged, and a general competitive ability over all p species is

$$k_i = \frac{1}{P}\sum_i \gamma_{ij}$$

The quantity k_i is called the competition effect of species i. In the same way, the average depression effect is

$$\lambda_i = \frac{1}{P-2}\sum_i \delta_{ij} - \frac{1}{P}\sum_i\sum_j \delta_{ij}$$

There are also interactions between the competition effects and depression effects of each species. The interaction between the competition effects of two species is

$$\theta_{ij} = \gamma_{ij} - k_i + k_j$$

The interaction between two depression effects is

$$\tau_{ij} = \delta_{ij} - \mu - \lambda_i - \lambda_j$$

Table 6.3. Experimental values of log {(weight in grams) + 1} for the two replicates. Source : Williams, 1962.

Associates	1		2		3		4		5		6		7		V_i	
Species																
1	1.48	1.51	2.03	1.76	1.68	1.76	1.68	1.66	1.79	1.65	1.77	1.82	1.76	1.61	10.71	10.26
2.	1.60	1.69	1.78	1.80	1.98	1.82	1.88	1.89	1.97	1.96	2.01	2.01	1.81	1.73	11.25	11.10
3.	0.94	0.58	0.97	1.16	0.96	1.05	0.82	0.91	1.23	1.14	1.29	1.23	0.97	0.69	6.02	5.71
4.	0.79	0.74	1.25	1.18	1.11	1.06	1.07	1.06	1.35	1.16	1.23	1.21	1.09	1.05	6.82	6.40
5.	0.37	0.21	0.39	0.77	0.74	0.97	0.59	0.59	1.00	1.02	1.12	1.03	0.38	0.29	3.59	3.86
6.	0.14	0.23	0.42	0.47	0.37	0.58	0.38	0.35	0.77	0.93	0.93	0.57	0.20	0.18	2.28	2.74
7.	1.00	1.03	1.41	1.53	1.62	1.56	1.39	1.41	1.55	1.59	1.65	1.59	1.43	1.40	8.62	8.71
A_i	4.64	4.48	6.47	6.87	7.50	7.75	6.74	6.81	8.66	8.43	9.07	8.89	6.21	5.55	49.29	48.78

$\Sigma x_{ii} =$ 8.65   8.41

where

$$\mu = \frac{1}{P(P-1)} \sum_i \sum_j \delta_{ij}$$

McGilchrist (1965) analysed data collected by Williams (1962) on competition among seven species of plants, $P = 7$. The species mean, the competition effects, and the depression effects are listed in Table 6.4. The best competitor is species 1 and the two worst competitors are species 5 and 6. On the other hand, the species with the largest depressing effect is species 5. If species 5 were grown in mixed cultures with another species, it would tend to lower the overall yield more than any other species. Evidently, species 2 is the best species to grow in mixed cultures.

Table 6.4. Estimates of species, competition and depression effects in Williams' competition experiments among seven species of plants.

Species	Species mean (S_j)	Competition effect (K_c)	Depresion effect (S_j)
1	1.495	0.2850	0.0347
2	1.790	0.0361	−0.0978
3	1.005	−0.0189	−0.0548
4	1.065	0.0650	−0.0143
5	1.010	−0.2400	0.0697
6	.750	−0.2278	0.0177
7	1.415	0.1007	0.0447
Standard error of entry	0.052	.030	0.029

Source: From Mc Gilchrist, C.A. Analysis of competition experiments, Biometries, Vol. 21, 1965, as quoted in Poole, 1974

Although the competition and depression effects have been calculated, are there significant differences among them? Because the basic model of competition is a linear analysis of variance model, analysis of variance may be used to check for the significance of differences among the terms in Table 6.4.

6.1.3 ENVIRONMENTAL MANAGEMENT

Environmental Management is one of the important components of model application in ecology. Ecology deals with the ecosystem. An ecosystem consists of a network of functional relationship between a community of plants and animals and the physical environment, and within the community itself. The ecosystem also includes the cycles of nutrients and direction flow of energy through the community and physical environment. The primary purpose of this section is to describe the environmental management–the flow of energy and matter–through the various links of the ecosystem.

Poole (1974) described the flow of energy through an ecosystem in its simplest form by a diagram (Figure 6.10). The cycling of nutrients through an ecosystem is conceptually similar to the flow of energy through the same ecosystem, with one fundamental difference. Energy enters the ecosystem as light and is rapidly lost as heat. In contrast, nutrients are not lost or gained except to and from adjacent ecosystems. These nutrients are constantly cycled through the ecosystem, except for losses to water run off, harvesting by man, and migration of animals or plant seed and pollen. In the jargon of the system analyst, energy flow is an open system and nutrient cycling is a closed system.

As in energy flow studies, an ecosystem can be broken down into a series of components or compartments. For example, a forest can be divided into the compartments in Figure 6.11 (Poole,1974).

6.1.3.1 Example: Energy Flow and Nutrient Cycling in a Grassland Ecosystem

This example has been described by Poole (1974). Van Hook (1971) studied a grassland ecosystem in Tennessee. The grassland was dominated by two species of grass, *Festuca arundinacea* and *Andropogon virginicus*, and a few species of herbs. The herbivores of the field were represented by *Mclanoplus sanguinipas* (a grasshopper), *Conocephalus*

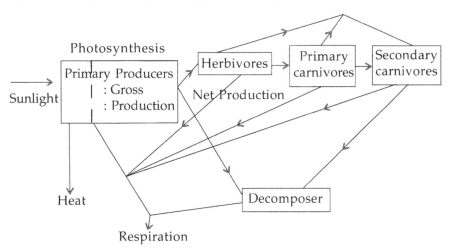

Fig. 6.10 A diagrammatic representation of the flow of energy through an ecosystem divided into a series of components (Poole, 1974).

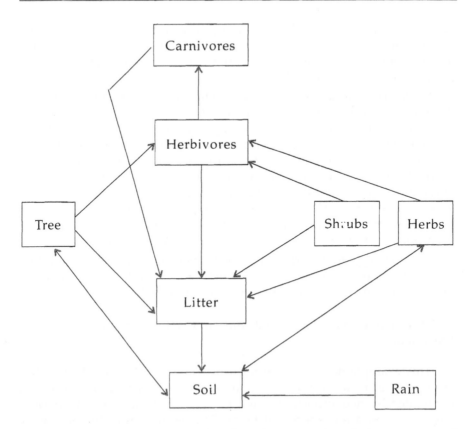

Fig. 6.11 Diagrammatic illustrations of the flow of a nutrient through the components of a simple ecosystem. The flow from one component to another may be either one-way or two-way. Each arrow represents a rate (Poole, 1974).

fasciates (a grasshopper), and a Catchall class Hemiptera-Homoptera. *Petronemobius fasciates*, a cricket, represented an omnivore component, and a wolf spider, *Lycosa punctulate*, was the primary carnivore.

Plant biomass, both live and dead, was estimated by sampling the vegetation at 6-week intervals of 1 year on 0.25-meter-square quadrants at six randomly chosen locations. The insects and spiders were sampled by quadrant cages at weekly intervals from April to December.

6.1.3.1.1 *Energy Flow*

This grassland system was divided into a series of 12 components. Figure 6.12 illustrates these twelve components and all the possible routes of energy transfer. Each arrow represents a rate. The problem is to determine the energy in each component and the rates of transfer from one component to another.

The total biomass of the grass material in the grassland over a period of a year were determined. For each arthopod species, an energy budget was constructed. Ingestion was estimated by the radioactive tracer method by allowing the herbivores to feed on radioactively tagged plants. The energy flux from a plant component to an herbivore component is complicated because of food preferences of the herbivores. Van Hook calculated the energy flux per day from component i to component j, EF_{ij}, as

$$C_i \lambda_{ji} = EF_{ij} = \frac{(X_i)(W_{ji})(P_i)(X_j)(CE_i)}{(\Sigma X_i)(W_{ji})} \qquad (6.44)$$

where

X_i = the proportion of the total available food composed of component i

W_{ji} = the feeding preference of the jth component for the ith component

P_i = the quantity of food ingested by a consumer or predator expressed as per cent of dry body weight per day.

X_j = the biomass dry weight of the jth compartment.

CE_i = the caloric equivalent of the food component consumed.

The change in energy in any of the compartments in Figure 6.13 is equal to the total gains from the other components minus the losses, or

$$\Delta C_i = \Sigma \lambda_{ij} - \Sigma \lambda_{ji} C_i \qquad (6.45)$$

The transfer rates between producers and consumers and the predator were estimated with equation (6.44) and expressed in calories per square metre per day. The energy transferred to egestion, compartment 10, was estimated by multiplying the proportions of dry body weight egested per day by the biomass of the compartment in grams per metre square times the caloric equivalents in calories per gram. The rates of energy loss to respiration of the consumers and the predator were estimated from the equation

$$O_2 = a \, W^B$$

Because temperature affects the rate of metabolism, the rate of respiration was corrected for temperature and expressed as

$$O_2 = aW^B e^{0.0693(T - 20)}$$

where T is in degree centrigrade with 20°C as a reference point.

The mortality compartment includes energy losses to predatory mortality other than *Lycosa exuviae*, and reproductive products. The size of the component was estimated as

$$C_{12} = \sum_{i=5}^{9} \lambda_{i,\,12}\, C_i$$

where λ_i, $12\, C_i$ equals the net flux to compartment 12 from compartment i. This flux was estimated to be

$$\lambda_{i,12}\, C_i = \sum_{i=1}^{3} \lambda_{ij} C_i - \left(\lambda_{i,9}\, C_9 + \sum_{j=10}^{11} \lambda_{ij} C_i \right)$$

where $\lambda_{ij} C_i$ equals the total input from components 1 through 3, $\lambda_{i,9} C_9$ equals the total loss to the predators, and $\lambda_{ij} C_i$ equals losses due to excretion and respiration. In other words, each $\lambda_{i,12} C_i$ represents the energy of the arthopod population not accounted for by predators, respir-ations, or excretion.

The model, equation 6.45, was evaluated on a weekly basis for 35 weeks. If conditions remain relatively constant during a week, the weekly transfer rates will be seven times the daily rates. The total 35-week growing season transfer rates are the sum of the 35 weekly rates. A necessary amount of fudging and extrapolation of data comes in at this point. The annual energy fluxes from compartment i to j in kilocalories per square metre are listed in Table 6.5.

Table 6.5. Annual energy fluxes in kilocalories per square metre through the arthopod compartments of a grassland ecosystem

Compo-nent i	Component j							
	C_5	C_6	C_7	C_8	C_9	C_{10}	C_{11}	C_{12}
C_1	7.83	1.56	1.76	0.67	–	–	–	–
C_2	82.60	14.76	7.81	5.22	–	–	–	–
C_3	–	–	–	–	–	–	–	–
C_4	–	–	12.64	–	–	–	–	–
C_5	–	–	–	–	3.04	38.82	24.71	23.14
C_6	–	–	–	–	1.12	10.27	5.12	1.22
C_7	–	–	–	–	1.96	12.99	6.24	1.00
C_8	–	–	–	–	0.64	3.19	1.38	0.07
C_9	–	–	–	–	–	0.64	3.84	2.26
Total	90.43	16.30	22.21	5.89	6.76	65.91	41.29	27.69

– Means transfer not existing or not pertinent.
Source: From Van Hook, 1971.

Utilizing the harvest method, Van Hook estimated the net primary productibility of the grassland to be 12741 kilocalories per square metre during 1968. The net secondary productivity of the herbivores was calculated for 1 week as

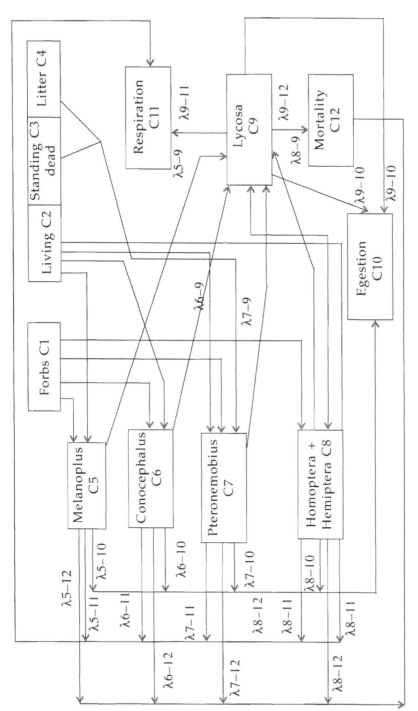

Fig. 6.12 A model of the grassland arthropod community showing the major compartments (C_i) and the pathways of energy flow (λ_{ij}). The arrows indicate the direction of flow (From Van Hook, 1971).

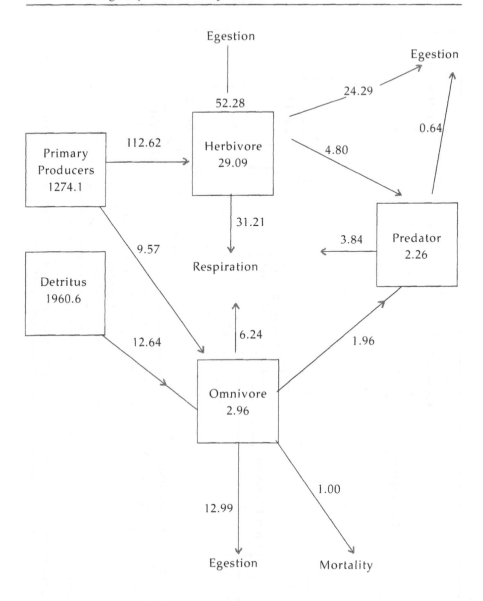

Fig. 6.13 Annual energy budget of energy fluxes and net production in kilo-calories per square metre by the arthopod and vegetation components of an eastern Tennessee grassland ecosystem. The values inside components represent the net production except for the detritus component. The detritus value equals standing crop of standing dead and litter. The values on the arrows represent annual energy fluxes. Arrows indicate the direction of flow (From Van Hook, 1971).

$$NP_{c,t} = \sum_{t=c}^{8} (\lambda_{1a} C_a + \lambda_{i,12} C_{12}$$

where the transfer rates are for 1 week, not for 1 day, as before. The total annual net secondary production is

$$NP_{-} = \sum_{t=1}^{15} NP_{-t}$$

The annual energy budget for the grassland is shown in Figure 6.13, the figures in the boxes are the annual net production and the numbers on the arrows are the annual transfer rates. A number of transfer efficiencies are listed in Table 6.6.

Table 6.6 shows the trophic level comparisons of energy flow and net production for the arthopod and vegetational components of a grassland ecosystem, the subscripts n and $n-1$ denote the respective trophic levels.

Table 6.6. Trophic level comparisons of energy flow and net production

Trophic level	$\dfrac{An}{In}$ (1)	$\dfrac{Rn}{In}$ (2)	$\dfrac{Pn}{In}$ (3)	$\dfrac{NPn}{In}$ (4)	$\dfrac{In}{In-1}$ (5)	$\dfrac{An}{An-1}$ (6)	$\dfrac{Rn}{Rn-1}$ (7)	$\dfrac{In}{NPn-1}$ (8)	$\dfrac{NPn}{Rn}$ (9)
Herbivore	0.54	0.28	0.04	0.26	–	–	–	0.09	0.93
Omnivore	0.41	0.28	0.09	0.13	–	–	–	0.01	0.47
Predator	0.90	0.57	–	0.33	0.05	0.09	0.10	0.21	0.59

Source: From Van Hook. – mean ratios not applicable

(1) Assimilation efficiency: ratio of assimilated energy to ingested energy.
(2) Respiration efficiency: ratio of energy expended in metabolism to ingested energy.
(3) Ecological efficiency: ratio of energy passed on to the next higher trophic level of the ingested energy.
(4) Ecological growth efficiency: ratio of the total energy available to be passed on to the next higher trophic level to the ingested energy.
(5) Transfer efficiency: ratio of energy ingested in trophic level n to that ingested in trophical level $(n-1)$.
(6) Progressive efficiency: ratio of energy assimilated in trophic level n to that assimilated in trophic level $n-1$.
(7) Respiration ratio: ratio of energy used in metabolism in trophic level n to that used in trophic level $n-1$.
(8) Consumption efficiency: ratio of energy ingested by consumers to the net production by the preceding trophic level.
(9) Secondary production/respiration: ratio of energy accumulated to the energy expended in metabolism.

6.1.3.1.2 Nutrient Cycling

Van Hook (1971) studied the cycling of potassium, calcium, and sodium in the grassland ecosystem. The same fundamental compartment model was used, but the respiration compartment was removed. Rates of nutrient transfer for pteronemobius, melanoplus, conocephalus, and lycosa were determined in the laboratory at a range of temperatures with radioactive tracers the rate of transfer per day from compartment i to compartment j, NE_{ij} are

$$NE_{ij} = \frac{(X_i)(W_{ij})(P_j)(X_j)(Q_i)}{\left(\sum X_i\right)(W_{ij})}$$

where Q_i equals the concentration of the nutrient in the ith compartment. The rate of nutrient loss through egestion in milligrams of nutrient per square metres per day NE_{jk} was estimated as

$$NE_{jk} = (K'Q_j + P_j \, Q_j \, I_j)x_j$$

Where K' = the biological elimination rate of the element.

Q_j = The concentration of the element in the consumer.

I_j = The ingestion rate for the consumer in milligrams per milligram dry weight of animal per day.

The pool of each nutrient in each compartment was estimated monthly. The monthly rates were extrapolated from the daily rates, and added to obtain an annual nutrient transfer. The annual rates for potassium are listed in Table 6.7. The annual nutrient budget is illustrated in Figure 6.14. The values in the boxes are the maximum nutrient concentrations of the standing crop. Efficiencies are presented in Table 6.8. Assimilation rates of sodium and potassium are both high compared with the assimilation rates of calcium. This observation suggest that sodium and potassium are limiting nutrients, particular to the herbivores, but that calcium is in abundant supply.

Singh and Upadhyaya (2001) reviewed the literature on utilization of biological interactions and matter cycling in agriculture.

Jorgensen (1984) edited an excellent book on modeling the fate and effect of toxic substances in the environment.

An excellent work on environment management is going on the divisions of System Simulations and Environmental Sciences, Nuclear Research Laboratory, IARI, New Delhi. Work on this aspect is also going on at the National Centre for Integrated Pest Management, ICAR, situated at IARI, New Delhi. Readers and students may consult the work done by Aggarwal et al. (2001), Sehgal et al. (2001a), Sehgal et al. (2001b), Sehgal (2003), Aggarwal et al. (2000), Sankaran et al. (2000), Aggarwal et al. (1999), Aggarwal (1998), Aggarwal et al. (2001b), Aggarwal et al. (2002), Mall et al. (2002), and Pathak et al. (2001).

Table 6.7. Annual fluxes of potassium in milligrams per square metre through the arthopod compartments of a grassland ecosystem

Compart-ment i	Compartment J							
	C_5	C_6	C_7	C_8	C_9	C_{10}	C_{11}	C_{12}
C_1	4.55	1.12	1.36	0.85	–	–	–	–
C_2	47.76	13.41	7.11	6.42	–	–	–	–
C_3	–	–	–	–	–	–	–	–
C_4	–	–	10.79	–	–	–	–	–
C_5	–	–	–	–	0.45	2.69	–	49.17
C_6	–	–	–	–	0.08	0.44	–	14.01
C_7	–	–	–	–	0.47	5.36	–	13.43
C_8	–	–	–	–	0.42	?	–	6.85
C_9	–	–	–	–	–	0.55	–	0.87
Total	52.31	14.53	19.26	7.27	1.42	9.04	–	84.33

— Means transfers not existing or not pertinent.
Source: From Van Hook, R.L., 1971.

Table 6.8. Trophic level of comparisons of nutrient fluxes through arthopod and vegetation components of a grassland ecosystem. The subscripts n and n-1 denote the respective trophic levels.

Trophic Level	An/In	Pn/In	In/In-1	An/An-1	In/NPn-1
	(1)	(2)	(3)	(4)	(5)
Sodium					
Herbivore	0.91	0.03	–	–	0.08
Omnivore	0.75	0.14	–	–	0.01
Predator	0.65	–	0.04	0.03	0.05
Calcium					
Herbivore	0.16	0.01	–	–	0.11
Omnivore	0.46	0.02	–	–	0.01
Predator	0.35	–	0.01	0.02	0.04
Potassium					
Herbivore	0.95	0.01	–	–	0.02
Omnivore	0.72	0.02	–	–	0.01
Predator	0.61	–	0.02	0.01	0.02

— Means ratios not applicable.
Source: From Van Hook, 1971.

    (1) Assimilation efficiency
    (2) Ecological efficiency
    (3) Transfer efficiency
    (4) Progressive efficiency
    (5) Consumption efficiency

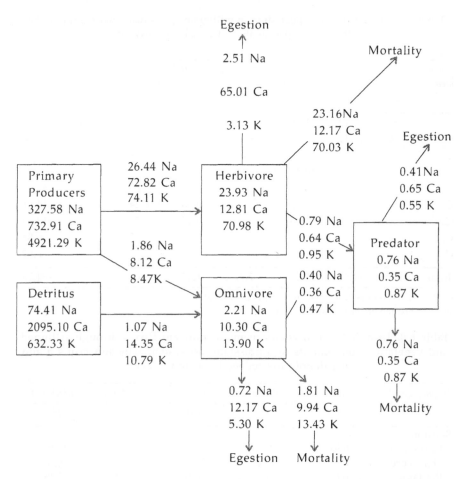

Fig. 6.14 Annual nutrient budget in milligrams per square metre for sodium, calcium, and potassium showing the maximum standing crops of nutrients inside compartments and nutrients fluxes on the arrows through the arthopod and vegetation components of an eastern Tennessee grassland ecosystem (From Van Hook, 1971).

6.2. AGRICULTURAL APPLICATION

6.2.1 CROP YIELD

6.2.1.1 Potential Crop Yield

The potential crop production is defined by Van Keulen (1986). Potential crop yield is the total dry matter production of a green crop surface that, during its entire growth period, is optimally supplied with water and all essential nutrient elements, and grows without interference from weeds, pests and diseases. A step may be made to the estimation of potential production of economically useful plant parts. For the estimation of production of economically useful plant parts, we take into account the phenological development of a particular crop species or cultivar. We associate the phenological development with the partitioning of dry matter over various organs of the plant. Here, we calculate both total dry matter production and economic yield for some crops. We further consider only the effects of radiation and temperature on crop production.

The principle of the procedure is to perform the repetitive calculations. We begin calculating at some point in time. At the starting point, the state of crop can be described in quantitative terms. The quantification of the state of the crop is made by using some experimental data or by estimating from other known relations. A suitable point in time is emergence of the crop. The emergence of the crop may be defined as the moment of transition from growth of the seedling from the reserves in the seeds to the growth originating from carbohydrates formed in the process of photosynthesis. In paddy crop, the moment of transplanting is a better starting point.

The state of the crop at the beginning of the computation is characterized by measurable quantities. We should start with the weight of the aerial plant parts, the weight of the roots and the green leaf area which is photosynthetically active. The assimilation and respiration are calculated from this state of the crop and the environmental conditions in the following period. The processes of photosynthesis and respiration, control the rates of change of the various quantities. These rates are realized over relevant time interval. The rate is added to the quantity present in the beginning of the period. This addition process results in the magnitude of the quantities at the end of the period. Mathematically, this may be symbolized as:

$$Q_{t+\Delta t} = Q_t + R_q * \Delta t \qquad (6.46)$$

where,

$Q_{t+\Delta t}$ is quantity at time $t + \Delta t$

Q_t is quantity at time t

R_q is the rate of change of quantity Q during time interval Δt.

Δt is time interval between the beginning of the period and the end of the period.

This calculation is repeated for the next time interval. This repeated calculation is done until the end of the growth period of the crop. In this way, the cumulative dry matter production is obtained. The weight of the various organs can be calculated by partitioning the dry matter produced during each time interval. The development stage of the vegetation controls the partitioning coefficients. The phenological development must also be quantified. We add the average air temperature in the course of the growing period and dividing the accumulated temperature sum at any moment by the sum required for the completion of a certain phenological phase. The ratio obtained is the required quantity defined as the developmental stage.

It is assumed that the rates of change calculated at the beginning of a time interval do not change during this interval. Here we have chosen a period of ten days. The principles of this calculation is based on the state variable approach in systems analysis and modeling. A description of this approach is given in de Wit and Goudriaan (1978) and Penning de Vries and van Laar, (1982).

A computer program in BASIC language has been written by the author as below:

Module 1: Estimation of actual global radiation

```
10      Rem calculations on actual global radiation using Apple-III
15      Dim X1 (120), X2 (120), D (120), SSIN (120), CCOS (120), X (120),
        Y (120), DAYL (120) DAYLH (120), E (120), F (120), O (120), A
        (120), B(120), H (120), RS (120)
17      Print "day length Pot.rad act.red"
20      For I = 1 to 105
30      Read X1(I), X2 (I)
35      Rem X1 (I) is Julian day and X2 (I) is actual sunshine hours
50      L = 29.17
60      D (I) = -23.45* (Cos((360 * (X1 (I) + 10)/ 365) * 3.14159/180))
70      SSIN (I) = Sin (D (I) * 3.14159/180)* Sin (L* 3.14159/180))
80      CCOS (I) = COS (D (I) * 3.14159/180)* COS (L*3.14159/180)
90      X(I) = SSin (I)/CCOS (I)
100     Y(I) = (ATN (X (I)/SQR (-X(I)*X(I)+1)))* 180/3.14159
110     DAYL (I) = 43200* (3.14159 + 2 & Y (I) (3.14159/180))/3.14159
120     DAYLH (I) = DAYL (I)/3600
140     Rem calculations on angots values in J/Sqm. d
150     C = TAN (L* 3.14159/180)
160     E (I) = TAN (D (I)*3.14159/180))
170     F(I) = C * E (I)
180     O (I) = (-ATN (F (I)/SQR (-F (I)* F (I)+1))) + 1.5704) * 180/3.14159
```

190   Z = 24/3.14159 * 4.871 E+06
200   A (I) = 1+.033 * COS (360 *X1 (I)/365 * 3.14159/180)
210   B (I) =(COS (L*3.14159/180) * COS (D (I) *3.14159/180) * SIN (O
        (I) *3.14159/180)+(2 * 3.14159 * O (I)/360 * SIN (L *3.14159/180)
        * SIN (D (I) * 3.14159/180
220   H (I) = 2 * A (I) * B (I)
221   RS (I) = H (I) * (.25+(.45 * (X2 (I)/DAYLH (I)))))
222   RS (I) = RS (I) * 1
225   Print X1 (I), DAYLH (I), H(I), RS (I)
226   Next I
400   DATA 347, 7.6, 348, 8, 349, 7.8, 350, 8.4, 351, 8.2, 352, 6.8, 353, 0,
        354, 6.1, 355, 6.5, 356, 7.9
500   DATA 357, 7.3, 358, 5.5, 359, 6.8, 360, 6.7, 361, 7.8, 362, 7.5, 363,
        1.3, 364, 5, 365, 7, 1, 7.2, 2, 7, 3, 0, 4, 3.1, 5, 7.5, 6, 5, 7, 6.2, 8, 4.1,
        9, 5.7, 10, 7.3, 11, 3.1, 12, 2.9, 13, 2, 14, 0, 15, 2.4, 16, 0, 17, 4.9, 18,
        4.6, 19, 6.7, 20, 4.6, 21, 4.5, 22, 9.2, 23, 9.1, 24, 8.7, 25, 8.1, 26, 4.5,
        27, 5.2, 28, 8.2
510   DATA 29, 6.6, 30, 9, 31, 9.2, 32, 7.3, 33, 9, 34, 4.5, 35, 8.9, 36, 7, 37,
        7.6, 38, 5.3, 39, 6.2, 40, 2.2, 41, 9, 42, 9.3, 43, 7.5, 44, 8.9, 45, 8, 46,
        5.2, 47, .3, 48, 8.4, 49, 9.1, 50, 9.8, 51, 9.8, 52, 6.8, 53, 4.4, 54, 9.3,
        55, 8.6, 56, 9.9, 57, 9.4, 58, 8.9, 59, 9.8, 60, 10.4, 61, 9.6, 62, 9.4, 63,
        8.2, 64, 9.5, 65, 10.3
520   DATA 66, 10.1, 67, 10, 68, 9.3, 69, 10.3, 70, 9.8, 71, 6.6, 72, 8.3, 73,
        7.6, 74, 4.1, 75, 2.9, 76, 8.6, 77, 9.7, 78, 9, 79, 9, 80, 8.4, 81, 10, 82,
        10.5, 83, 10.6, 84, 10.1, 85, 10.3, 86, 10.5, 87, 10.5, 88, 9.1, 89, 9.9,
        90, 8.5, 91, 10.6, 92, 10.3, 94, 6.1, 95, 2, 96, 8.9
600   REM avg calculation
620   For I = 1 to 10
630   SUM = SUM * H (I)
640   AVG = SUM/10
670   NEXT I
680   PRINT "AVG = "AVG
700   Rem Avg rs calculation
710   For I = 1 to 10
720   SUM 1 = SUM1 + RS (I)
730   AVG 1 = SUM 1/10
740   Next I
750   PRINT "AVG 1 = "AVG
800   Rem X1 (I) = Julian day

810   Rem X2 (I) = Actual sunshine hour.

820   Rem L = latitude

830   Rem D = Declination angle of sun.

840   Rem RI = RA (a$_A$ + B$_A$ nN^{-1})

850   Rem RI = radiation actually received (J m^{-2} d^{-1})

860   Rem RA = Angot's value, or the theoretical amount of radiation that would reach the earth's surface in the absence of an atmosphere (J m^{-2}) d^{-1}. The values of RA are the function of day of the year and latitude.

870   Rem n N^{-1} is the ratio of actual duration of bright sunshine (n) to the maximum possible length on a calender day (N), both in an hour.

Module 2: Estimation of C$_3$ crop potential productivity for closed canopy.

10   Rem   C$_3$ productivity during clear days

20   DIM   D (120), E (120), F (120), O (120), A (120), B (120), HO (120), V1 (120), V2 (120), V3 (120), ARCSIN (120), ARC (120), J (120)

30   DIM SSIN (120), CCOS (120), X (120), Y (120), Y1 (120), Y2 (120), DAYLE (120), RAD (120)

40   Dim SLLAE (120), Z (120), Z1 (120), Z2 (120), Z3 (120), P (120), A1 (120), PO (120), DAYL (120)

50   L = 29.17

60   Rem effe is in kg CO_2/J, amax is in kg CO_2/sqm.s

70   EFFE = 12.9 * 10^{-9} : AMAX = 1.1 * 10^{-6} : LAI = 5

80   For I = 1 to 115

90   Read   J (I)

95   DATA   347, 348, 349, 350, 351, 352, 353, 354, 355, 356

100   DATA   357, 358, 359, 360, 361, 362, 363, 364, 365, 1, 2, 3, 4, 5, 6, 7, 8, 9, 10, 11, 12, 13, 14, 15, 16, 17, 18, 19, 20, 21, 22, 23, 24, 25, 26, 27, 28, 29, 30, 31, 32, 33, 34, 35, 36, 37, 38, 39 40, 41, 42, 43, 44, 45, 46, 47, 48, 49, 50, 51, 52, 53, 54, 55,56, 57, 58, 59, 60, 61, 62, 63, 64, 65, 66, 67, 68, 69, 70, 71, 72, 73

105   DATA   74, 75, 76, 77, 78, 79, 80, 81, 82, 83, 84, 85, 86, 87, 88, 89, 90, 91, 92, 93, 94, 95, 96

110   D (I) = -23.45 * COS ((360 * (J (1) +10) / 365 * 3.14159/180)

120   C = -TAN (L * 3.14159/180)

130   E (I) = TAN (D (I) * 3.14159/180)

140   F (I) = C * E (I)

```
150     O (I) = ((-ATN (F (I)/SQR (-F (I)
                * F (I) + 1))) + 1.5704 * 180/3.14159
160     W = 24/3.14159 * 4.871 E +06
170     A (I) = 1 + .003 * COS (360 * J (I)/365 * 3.14159/180)
180     B (I) = (COS (L * 3.14159/180) * COS (D (I) * 3.14159/180) * SIN
        (O(I) *3.14159/180) + (2 * 3.14159 * O(I) /360) * SIN (L*3.14159/
        180) * SIN (D (I) *3.14159/180))
190     HO (I) = W * A (I) +B (I) * 0.2
200     V1 (I) = HO (I) * 0.73 : V2 (I) = HO (I) *0.7 : V3 (I) = HO (I) * 0.71
210     Rem SSIN Calculation
220     SSIN (I) = SIN (D (I) * 3.14159/180) * SIN (L * 3.14159/180)
230     Rem CCOS Calculations
240     CCOS (I) = COS (D (I) * 3.14159/180) * COS (L * 3.14159/180)
250     Rem X Calculation
260     X (I) = (-SIN (8*3.14159/180) + SSIN (I) / CCOS (I)
270     Rem arcsin Calculation
280     ARCSIN (I) = ATN ( X (I) /SQR (-X (I) * X (I) +1)) * 180/3.14159
290     Rem Dayle Calculation
300     DAYLE (I) = 43200 * (3.14159 * 2 * ARCSIN (I) * (3.14159/180))/
        3.14159
310     Rem Dayl Calculation
320     Z (I) = SSIN (I)/CCOS (I)
330     ARC (I) = (ATN (Z (I)/SQR (-Z (I) *Z (I) +1))) * 180/3.14159
340     DAYL (I) = 43200 * (3.14159 +2 * ARC (I) * (3.14159/180))/3.14159
350     Rem rad calculation
360     Rad (I) = 0.5 * V1 (I)/DAYLE (I)
370     A1 (I) = RAD (I) * EFFE/(AMAX * LAI)
380     P (I) = A1 (I)/(A1 (I) +1)
390     PO (I) = (LAI * AMAX * DAYLE (I) * P (I) * 1000) * 30/44
400     PRINT J (I), PO (I), DAYL (I), DAYLE (I)
410     NEXT I
420     REM calculation Avg. PO
430     For I = 1 to 10
440     SUM = SUM + PO (I)
450     AVG = SUM/10
460     Next I
470     PRINT "Avg. AVG
```

Module 3: Estimation of potential water loss

```
10      DIM X1 (120), EA (120), X4 (120), X5 (120), X6 (120), X7 (120), X8
        (120)
20      DIM RNL (120), RN (120), ES (120)
30      DIM DELTA (120), HU (120), ETO (120), MPET (120), ETPT (120),
        MPET 1 (120)
40      REM calculations on pot. water loss
50      For I= 1 to 115
60      READ X1 (I), EA (I), X4 (I), X5 (I), X6 (I), X7 (I), X8 (I)
70      NEXT I
80      FOR I = 1 to 115
90      RNL (I) = 0.0049 * (X5 (I) + 273⁴) * (0.56 - 0.079 * (EA (I)⁰·⁵)) * (0.1
        + 0.9 * X4 (I)/X8 (I))
100     PRINT "rnl = "RNL (I)
110     NEXT (I)
120     FOR I = 1 to 115
130     RN (I) = X7 (I) * (1-0.25) - RNL (I)
140     PRINT "RN=" RN (I)
150     NEXT I
160     FOR I = 1 to 115
170     ES (I) = 6.11 * EXP (17.4 * X5 (I)/ (X5 (I) + 239))
180     PRINT "es = "ES (I)
190     NEXT I
200     FOR I = 1 to 115
210     DELTA (I) = 17.4 * ES (I) * (1- X5 (I)/(X5 (I) + 239 ))/ (X5 (I) +
        239)
220     PRINT "delta = " DELTA (I)
230     NEXT I
240     FOR I = 1 to 115
250     HU (I) = 6.4 * 10⁵ * (1+(0.54 * X6 (I)))
260     PRINT "hu (I) = "HU (I)
270     NEXT I
280     FOR I = 1 to 115
290     ETO (I) = ((1/(DELTA (I) + 0.66)) * (DELTA (I) * RN (I)) * (HU (I)
        * (ES (I)-EA (I)))))/2.45 E+06
300     PRINT "eto=" ETO (I)
310     NEXT I
```

```
320    FOR I = 1 to 115
330    ETPT (I) = (1.26 * (DELTA (I)/(DELTA (I) + 0.66)) * RN (I))/2.45
       E+06
335    REM PRIESTLEY TAYLOR ET
340    PRINT "etpt = " ETPT (I)
350    NEXT I
360    FOR I = 1 to 115
365    REM modified Pan ET
370    MPET (I) = ((DELTA (I) * 0.1 * RN (I)) (10⁶) + (((0.66 * ((5360/ (X5
       (I)+273)) - 4.02))) * (X6 (I) * ((ES (I) * 0.1)- (EA (I)* 0.1))/ ((DELTA
       (I) * 0.1) + (0.66 * (1+ (0.407 * X6 (I))))
380    MPETI (I) = MPET (I)/2.45
390    PRINT X1 (I), "mpet = "MPET (I)
400    NEXT I
401    REM AVG mpet calculations
402    FOR I = 111 to 115
403    S = S + MPET1 (I)
404    AVG = S/5
405    NEXT I
406    PRINT "avg =" AVG
410    DATA 347, 8.44, 7.6, 14.35, 0.48, 1.653 e+07, 10.151
420    DATA 348, 8.31, 8, 13.15, 0.45, 1.68 e+07, 10.147
430    DATA 349, 8.31, 7.8, 12.35, 0.95, 1.644 e+07, 10.143
440    DATA 350, 7.25, 8.4, 12.5, 0.76, 1.641e + 07, 10.139
450    DATA 351, 8.05, 8.2, 14.35, 0.53, 1.638e + 07, 10.136
451    DATA 352, 8.84, 6.8, 13.5, 0.31, 1.636e+ 07, 10.134
452    DATA 353, 8.97, 0, 12.2, 0.62, 1.6349e + 07, 10.133
453    DATA 354, 7.25, 6.1, 12.2, 0.42, 1.6341e + 07, 10.132
454    DATA 355, 6.6, 6.5, 11.2, 0.53, 1.633 e + 07, 10.131
455    DATA 356, 6.6, 7.9, 12, 0.31, 1.634e + 07, 10.132
470    DATA 357, 6.86, 7.3, 12.2, 0.19, 1.635, e + 07, 10.133
480    DATA 358, 6.46, 5.50, 11.70, 0.19, 1.637 e+ 07, 10.134
490    DATA 359, 6.2, 6.8, 10.95, 0.30, 1.639 e + 07, 10.136
500    DATA 360, 7.39, 6.7, 12.45, 0.25, 1.642 e+ 07, 10.139
510    DATA 361, 7.39, 7.8, 12.25, 0.44, 1.646 e+ 07, 10.143
520    DATA 362, 9.1, 7.5, 15.1, 0.19, 1.65 e+ 07, 10.147
530    DATA 363, 9.76, 1.3, 13.5, 0.95, 1.655 e+ 07, 10.151
540    DATA 364, 7.78, 5, 12.15, 0.5, 1.661 e+ 07, 10.157
```

550	DATA 365, 7.78, 7, 13.4, 0.64, 1.667 e+ 07, 10.163
560	DATA 1,8.97, 7.2, 13.8, 0.42, 1.674 e+ 07, 10.169
570	DATA 2, 11.48, 7, 16.7, 0.58, 1.681 e+ 07, 10.176
580	DATA 3, 10.82, 0, 15.45, 0.42, 1.689 e+ 07, 10.184
590	DATA 4, 7.52, 3.1, 13.15, 0.19, 1.698 e+ 07, 10.193
600	DATA 5, 8.44, 7.5, 13.45, 0.47, 1.707 e+ 07, 10.202
610	DATA 6, 8.58, 5, 14.15, 0.19, 1.717 e+ 07, 10.211
620	DATA 7, 8.84, 6.2, 14.55, 0.53, 1.727, e+ 07, 10.221
630	DATA 8, 11.88, 4.1, 16.6, 0.30, 1.738 e+ 07, 10.232
640	DATA 9, 8.58, 5.7, 13.95, 0.72, 1.750 e+ 07, 10.243
650	DATA 10, 8.44, 7.3, 13.7, 0.75, 1.762 e+ 07, 10.255
660	DATA 11, 11.08, 3.1, 15.15, 0.95, 1.775 e+ 07, 10.268
670	DATA 12, 13.86, 2.9, 15.7, 3.05, 1.788 e+ 07, 10.28
680	DATA 13, 15.7, 0.2, 15.55, 1.98, 1.802 e+ 07, 10.28
690	DATA 14, 15.04, 0, 13.55, 0.58, 1.816 e+ 07, 10.3
700	DATA 15, 12.4, 2.4, 13.95, 0.51, 1.831 e+ 07, 10.32
710	DATA 16, 11.48, 0, 11.05, 1.23, 1.846 e+ 07, 10.33
720	DATA 17, 8.05, 4.9, 10.15, 0.58, 1.862 e+ 07, 10.35
730	DATA 18, 8.58, 4.6, 10.6, 0.44, 1.878 e+ 07, 10.36
740	DATA 19, 9.24, 6.7, 11.4, 0.30, 1,895 e+ 07, 10.38
750	DATA 20, 10.82, 4.6, 13.7, 1.17, 1.912 e+ 07, 10.4
760	DATA 21, 7.78, 4.5, 10, 1.4, 1.93 e+ 07, 10.42
770	DATA 22, 6.73, 9.2, 9.55, 1.31, 1.948 e+ 07, 10.43
780	DATA 23, 7.52, 9.1, 11.25, 0.53, 1.966 e+ 07, 10.45
790	DATA 24, 7.52, 8.7, 11.8, 0.39, 1.985 e+ 07, 10.47
800	DATA 25, 7.65, 8.1, 14.35, 0.39, 2.005 e+ 07, 10.49
810	DATA 26, 8.97, 4.5, 13.25, 0.81, 2.024 e+ 07, 10.51
820	DATA 27, 12.93, 5.2, 16.75, 0.50, 2.044 e+ 07, 10.52
830	DATA 28, 12.54, 8.2, 19.5, 0.44, 2.065 e+ 07, 10.55
840	DATA 29, 10.56, 6.6, 15.45, 1.09, 2.085 e+ 07, 10.57
850	DATA 30, 8.58, 9, 11.75, 1.34, 2.106 e+ 07, 10.59
860	DATA 31, 7.39, 9.2, 11.55, 0.98, 2.128 e+ 07, 10.61
870	DATA 32, 7.39, 7.3, 11.75, 0.56, 2.149 e+ 07, 10.64
880	DATA 33, 7.78, 9, 10.75, 0.61, 2.171 e+ 07, 10.66
890	DATA 34, 8.97, 4.5, 12.4, 0.53, 2.193, e+ 07, 10.68
900	DATA 35, 9.9, 8.9, 15.3, 0.81, 2.216 e+ 07, 10.7
910	DATA 36, 11.88, 7, 16.75, 3.47, 2.238, e+ 07, 10.73

```
920    DATA 37, 11.08, 7.6, 16.75, 1.4, 2.261 e+ 07, 10.75
930    DATA 38, 12.67, 5.3, 16.15, 1.06, 2.284 e+ 07, 10.78
940    DATA 39, 12.14, 6.2, 17.9, 2.6, 2.308 e+ 07, 10.8
950    DATA 40, 9.1, 2.7, 12.3, 1.48, 2.331 e+ 07, 10.82
960    DATA 41, 8.05, 9, 14.75, 0.89, 2.355 e+ 07, 10.85
970    DATA 42, 9.63, 9.3, 14.75, 0.47, 2.378 e+ 07, 10.88
980    DATA 43, 8.58, 7.5, 13.6, 0.86, 2.402 e+ 07, 10.9
990    DATA 44, 8.44, 8.9, 13.4, 0.64, 2.426 e+ 07, 10.93
1000   DATA 45, 9.76, 8, 16.1, 1.45, 2.45 e+ 07, 10.95
1010   DATA 46, 12.54, 5.2, 15.25, 1.14, 2.474 e+ 07, 10.98
1020   DATA 47, 10.29, 0.3, 14.25, 1.56, 2.499 e+ 07, 11
1030   DATA 48, 10.56, 8.4, 15.1, 0.53, 2.523 e+ 07, 11.03
1040   DATA 49, 8.44, 9.1, 13.55, 0.95, 2.547 e+ 07, 11.06
1050   DATA 50, 8.31, 9.8, 13.7, 0.84, 2.571 e+ 07, 11.09
1060   DATA 51, 9.5, 9.8, 16.35, 1.03, 2.596 e+ 07, 11.11
1070   DATA 52, 8.11, 6.8, 19.2, 0.7, 2.62 e+ 07, 11.14
1080   DATA 53, 8.05, 4.4 11.7, 2.63, 2.644 e+ 07, 11.17
1090   DATA 54, 8 44, 9.3, 12, 1.2, 2.669 e+ 07, 11.12
1100   DATA 55, 8.44, 8.6, 12, 1.09, 2.693 e+ 07, 11.22
1110   DATA 56, 9.63, 9.9, 13.25, 0.98, 2.717 e+ 07, 11.25
1120   DATA 57, 10.56, 9.4, 15.45, 1.09, 2.741 e+ 07, 11.28
1130   DATA 58, 11.08, 8.9, 16.6, 1.2, 2.765 e+ 07, 11.31
1140   DATA 59, 9.5, 9.8, 14.7, 1.34, 2.789 e+ 07, 11.34
1150   DATA 60, 7.52, 10.4, 13.35, 2.04, 2.813 e+ 07, 11.37
1160   DATA 61, 8.05, 9.6, 15.25, 1.42, 2.836 e+ 07, 11.39
1170   DATA 62, 8.58, 9.4, 20.55, 1.45, 2.86 e+ 07, 11.42
1180   DATA 63, 11.08, 8.2, 20.05, 2.15, 2.883 e+ 07, 11.45
1190   DATA 64, 10.69, 9.5, 18.15, 1.12, 2.906 e+ 07, 11.48
1200   DATA 65, 10.42, 10.3, 16.65, 1.23, 2.929 e+ 07, 11.51
1210   DATA 66, 10.29, 10.1, 16.75, 1.4, 2.952e+ 07, 11.54
1220   DATA 67, 10.56, 10, 18.2, 0.56, 2.974 e+ 07, 11.57
1230   DATA 68, 9.76, 9.3, 17.7, 0.81, 2.997 e+ 07, 11.6
1240   DATA 69, 10.56, 10.3, 18.45, 0.92, 3.019 e+ 07, 11.63
1250   DATA 70, 12.01, 9.8, 19.75, 0.75, 3.041 e+ 07, 11.66
1260   DATA 71, 13.72, 6.6, 20.6, 1.03, 3.062, e+ 07, 11.69
1270   DATA 72, 15.31, 8.3, 21.4, 0.98, 3.084 e+ 07, 11.72
1280   DATA 73, 16.89, 7.6, 20.9, 1.03, 3.105 e+ 07, 11.75
1290   DATA 74, 19.66, 4.1, 24.3, 1.09, 3.125 e+ 07, 11.78
```

1300   DATA 75, 17.02, 2.9, 22.15, 1.76, 3.146 e+ 07, 11.81

1310   DATA 76, 16.1, 8.6, 23, 1.14, 3.166 e+ 07, 11.84

1320   DATA 77, 17.29, 9.7, 22.6, 1.68, 3.186 e+ 07, 11.87

1330   DATA 78, 19, 9, 23.95, 1.76, 3.206 e+ 07, 11.9

1340   DATA 79, 17.29, 9, 23.75, 2.4, 3.225 e+ 07, 11.93

1350   DATA 80, 14.38, 8.4, 21.9, 0.64, 3.244 e+ 07, 11.96

1360   DATA 81, 13.06, 10, 23.7, 1.79, 3.263 e+ 07, 11.99

1370   DATA 82, 12.54, 10.5, 19.95, 1.62, 3.281 e+ 07, 12.02

1380   DATA 83, 14.52, 10.6, 20.8, 1.09, 3.299 e+ 07, 12.05

1390   DATA 84, 15.31, 10.1, 23.25, 1.09, 3.316 e+ 07, 12.08

1400   DATA 85, 14.12, 10.3, 23.15, 1.09, 3.334 e+ 07, 12.11

1410   DATA 86, 16.5, 10.5, 21.65, 1.68, 3.35 e+ 07, 12.14

1420   DATA 87, 14.65, 10.5, 22.45, 1.28, 3.367 e+ 07, 12.17

1430   DATA 88, 17.02, 9.1, 23.8, 0.7, 3.383 e+ 07, 12.2

1440   DATA 89, 15.04, 9.9, 26.95, 1.87, 3.399 e+ 07, 12.23

1450   DATA 90, 17.95, 8.5, 28.2, 1.34, 3.414 e+ 07, 12.26

1460   DATA 91, 16.56, 10.1, 23, 1.79, 3.429 e+ 07, 12.29

1470   DATA 92, 16.63, 10.3, 23.75, 1.09, 3.444 e+ 07, 12.32

1480   DATA 93, 15.31, 10.3, 24.4, 1.34, 3.458 e+ 07, 12.35

1490   DATA 94, 13.59, 6.1, 28.25, 2.38, 3.472 e+ 07, 12.38

1500   DATA 95, 20.19, 26.55, 3.27, 3.485 e+ 07, 12.41

1510   DATA 96, 20.59, 8.9, 24.7, 1.42, 3.498 e+ 07, 12.44

1600   REM Julian day, actual VP, min. temperture, max. temperature, wind velocity (m/s), Pot. Global rad. (Jm^{-1}d^{-1}), pot. Sunshine hr.

Computer Program of a Simple Model of Potential Crop Production in BASIC Language

10   REM   THE SCHEME OF PREDICTION OF POTENTIAL CROP PRODUCTION IS ADAPTED FROM H.V. KEULEN (1986)

20   REM   THE EXPERIMENTAL SITE IS PARAMARIBO, SURINAM, SOUTH AMERICA (5^{0}49' N, 55^0 09' W)

30   REM   THE VARIETY OF PADDY IS IR 8

40   REM   THE RICE WAS TRANSPLANTED ON 10th NOVEMBER, 1972 (VAN SLOBBE 1973)

50   REM   THE AIR TEMPERATURES USED IN THE CALCULATION WERE OBTAINED FROM REPORTED 10-DAY AVERAGES FOR THE EXPERIMENTAL PERIOD

60   REM   RADIATION WAS CALCULATED FROM MONTHLY

		AVERAGE OF SUNSHINE DURATION REPORTED

70 REM THE COMPUTER PROGRAM WAS PREPARED BY DR. PHOOL SINGH, BOTANY, CCS HAU, HISAR, INDIA, ON 20.05.2003.

90 REM PD (I) = PERIOD (I), TA (I) = DAILY AVERAGE TEMPERATURE IN DEGREE CENTIGRADE, DT (I) = TIME INTERVAL IN DAYS, MDVS (I) = HALFWAY OF THE DEVELOPMENT STAGE BETWEEN START AND END OF STAGE, FGS (I) = GROSS AS SIMULATION RATE OF CLOSED CANOPY IN CH_2O, KG ha^{-1} d^{-1}, FR (I) = FRACTION OF ASSIMILATES PARTITIONED TO THE ROOT, FL (I) = TO THE LEAF, FS (I) = TO THE STEM, FG(I) =TO THE GRAIN.

100 READ PD (I), TA (I), DT (I), MDVS (I), FGS (I), FR (I), FL (I), FS (I), FG (I)

110 DATA 1, 27.2, 10, 0.09, 336, 0.35, 0.395, 0.225, 0

111 DATA 2, 26.3, 10, 0.27, 318, 0.165, 0.445, 0.39, 0

112 DATA 3, 25.8. 10, 0.45, 300, 0.075, 0.48, 0.445, 0

113 DATA 4, 26.4, 10, 0.62, 283, 0.07, 0.40, 0.53, 0

114 DATA 5, 26.3, 10, 0.79, 301, 0.07, 0.265, 0.665, 0

115 DATA 6, 26.0, 7, 0.94, 319, 0.025, 0.06, 0.225, 0.69

116 DATA 7, 26.0, 3, 1.05, 319, 0, 0, 0, 1.0

117 DATA 8, 26.0, 10, 1.27, 335, 0, 0, 0, 1.0

118 DATA 9, 26.0, 10, 1.59, 346, 0, 0, 0, 1.0

119 DATA 10, 26.0, 8, 1.88, 340, 0, 0, 0, 1.0

200 TSUM (I) =0

210 FOR I = 1 to 10

220 TSUM (I)= TSUM + TA (I) * DT (I)

230 PRINT "TSUM=" TSUM (I)

240 NEXT (I)

300 DVS = 0

310 FOR I = 1 to 10

315 REM 1500 = DAY DEGREE REQUIRED FOR ANTHESIS FROM TRANSPLANTING DATE.

320 DVS (I) = DVS + (TSUM (I)/1500)

330 PRINT "DVS=" DVS (I)

340 NEXT I

400 WLV (0)=100: WRT (0)=40: WST (0)=0: WGR (0) =0: FH (0)=0.18 : TDWL (0)=140: LAI (0) = WLV (0) * 25 * 10^{-4}

450   FH (I) = $(1-e^{-0.8(LAI (I-1))})$

500   FOR I = 1 to 10

510   GASS (I) = FGS (I) * FH (I)

515   IF I $\leq$ 6 THEN 520 ELSE 525

520   MRES (I) = TDWL (I-1) * 0.015

525   MRES (I) = TDWL (I-1) * 0.01

530   ASAG (I) = GASS (I) - MRES (I)

535   IF I $\leq$ 6 THEN 540 ELSE 545

540   DMI (I) = 0.7 * ASAG (I)

545   DMI (I) = 0.8 * ASAG (I)

550   IWRI (I) = FR (I) * DMI (I)

560   WRI (I) = WRI (I-1)+IWRT(I) * DT (I)

570   IWLV (I) = FL (I) * DMI (I)

575   IF I $\leq$ 6, THEN 580 ELSE 585 ELSE WLV (I) = WL (I) * 0.02

580   WLV(I) = WLV (I-1) + IWLV (I) * DT (I)

585   WLV (I) = WLV (I) * 0.02

590   IWST (I) = FS (I) * DMI (I)

600   WST (I) = WST (I-1) + IWST (I) * DT (I)

610   LAI (I) = WLV (I) * 25 * 10^{-4}

615   IF I = 6, THEN 617 ELSE 620

617   IWGR (I) = FGR(I) * DMI (I) * 0.8/0.7

620   IWGR (I) = FGR (I) * DMI (I)

630   WGR (I) = WGR (I-1) + IWGR (I) * DT (I)

635   REM TADW = TOTAL ABOVE GROUND DRY WEIGHT

640   TADW (I) = WLV (I) + WST (I) + WGR (I)

645   REM TDW = TOTAL DRY WEIGHT OF THE VEGETATION

650   TDW (I) = TADW (I) + WRT (I)

655   REM TDWL = TOTAL DRY WEIGHT OF LIVE MATERIAL

660   TDWL (I) = TDW (I)

670   PRINT

I,   "WTR="WRT (I),

I,   "WST="WST (I),

I,   "WLV="WLV (I),

I,   "LAI=" LAI (I),

I,   "WGR=" WGR (I),

I,   "TADW=" TADW (I),

I,   "TDW = " TADW (I),

I,   "TDWL=" TDWL (I)

680   NEXT I

700   END

6.2.1.2 Computer Program for Predicting Water Limited Production

10   REM THE SCHEME IS ADAPTED FROM VAN KEULEN (1986)

20   REM THE EXPERIMENT WAS CARRIED OUT IN THE CENTRAL NAGAVE DESERT OF ISRAEL (30° N, 34° E)

30   REM THE SOIL TEXTURE WAS LOAMY SAND

40   REM THE FIELD CAPACITY IS 0.225 CM^3 CM^3

50   REM THE WILTING POINT IS 0.09 CM^3 CM^3

60   REM THE TOTAL PORE SPACE IS 0.40 CM^3 CM^3

70   REM SPRING WHEAT CV. LACHISH WAS SOWN ON 30 OCTOBER, 1977

80   REM GERMINATION WAS COMPLETED IN ABOUT 10 DAYS

90   REM THE CROP WAS AMPLY SUPPLIED WITH NUTRIENTS

100   REM SOIL-MOISTURE CONTENT IN THE DEEPER SOIL LAYER WAS 0.016 CM^3 CM^3, DUE TO RESIDUAL MOISTURE FROM A PREVIOUSLY IRRIGATED CROP

110   REM THE TOP SOIL (10 cm) WAS WETTED TO FIELD CAPACITY PRIOR TO SOWING TO ENSURE PROPER GERMINATION

120   REM THE CONTROL TREATMENT, SUFFICIENT MOISTURE WAS APPLIED FOR OPTIMUM SUPPLY, i.e. THE ROOT ZONE WITH IRRIGATED TO FIELD CAPACITY WHENEVER AVAILABLE MOISTURE AS DETERMINED BY GRAVIMETRIC SAMPLING FELL BELOW 30% OF ITS MAXIMUM VALUE

130   REM ROOTING DEPTH WAS DETERMINED AT TWO-WEEKLY INTERVALS BY EXAMINING THE SOIL CORE

140   REM WATER STRESS WAS ALLOWED TO DEVELOP BETWEEN MAXIMUM TILLERING AND ANTHESIS BY WITH-HOLDING IRRIGATION

150.   REM METEOROLOGICAL DATA WERE RECORDED AT A STANDARD METEOROLOGICAL STATION

160   REM RAINFALL WAS RECORDED AT THE SITE

170   REM METEOROLOGICAL DATA WERE USED TO CALCULATE TEN-DAY AVERAGES OF AIR, TEMPERATURE, RAINFALL AND POTENTIAL GROSS CH_2O ASSIMIL-ATION

180 REM TOTAL EVAPORATIVE DEMAND CALCULATED ON THE BASIS OF WEATHER DATA, APPLYING PENMAN'S EQUATION

200 REM THE COMPUTER PROGRAM WAS PREPARED BY DR. PHOOL SINGH, BOTANY, CCS HAU, HISAR, INDIA, ON 23.05.2003

210 REM PD(I) = NUMBER OF TEN-DAY PERIOD, TA (I)=DAILY AVERAGE TEMPERATURE IN DEGREE CENTRIGRADE, DT (I)=TIME INTERVAL IN DAYS, MDVS (I)=HALF WAY OF THE DEVELOPMENT STAGE BETWEEN START AND END OF STAGE, P(I)=RAINFALL IN MM, IR (I) = AMOUNT OF IRRIGATION IN MM, FGS (I) = GROSS ASSIMILATION RATE IN CH_2O, KG HA^{-1} DAY^{-1}, ETO (I) = EVAPORATIVE DEMAND IN MM DAY^{-1}, SMDF (I) = SOIL MOISTURE DEPLETION FACTOR, FR(I) = FRACTION OF ASSIMILATES TO THE ROOT, FL (I) = TO THE LEAVES, FS (I)= TO THE STEM, FGR (I) = TO THE GRAIN

300 READ PD (I), TA (I), DT (I), MDVS (I), P (I), IR (I), FGS (I), ETO (I), SMDF (I), FR (I), FL (I), FS (I), FGR (I)

310 DATA 1, 15.2, 10, 0.07, 0, 0, 251, 1.85, 0.86, 0.444, 0.556, 0, 0
2, 12.9, 10, .195, .1, 0, 257, 1.76, .86, .3345, .6655, 0, 0
3, 10.4, 10, .3, 7.4, 0, 225, 1.47, .86, .26, .6, .14, 0
4, 8.4, 10, .39, 19.6, 0, 263, 1.14, .86, .206, .528, .266, 0
5, 8.8, 10, .47, 1.6, 0, 257, 1.23, .86, .1418, .457, .378, 0
6, 7.9, 10, .545, .2, 0 , 258, 1.41, .86, .1275, .8895, .483, 0
7, 12.4, 10, .635, 0, 0, 260, 2.22, 83, .0895, 3.015, .609, 0
8, 11.5, 10, .74, 0, 0, 297, 2.29, .82, .058, .186, .756, 0
9, 12.3, 10, .85, 0, 0, 338, 2.87, .76, .03, .09, .63, .25
10, 11.7, 9, .955, 1.98, 72.9, 344, 2.72, .78, .009, .027, .189, .755
11, 11.7, 1, 1.01, .22, 8.1, 344, 2.72, .78, 0, 0 , 0, 1.0
12, 15.9, 10, 1.145, .1, 96, 402, 3.65, .68, 0, 0, 0, 1.0
13, 13.6, 10, 1.375, 3.9, 0, 374, 4.25, .62, 0, 0, 0, 1.0
14, 13.2, 10, 1.58, 0, 0, 437, 4.40, .61, 0, 0, 0, 1.0
15, 17.0, 10, 1.81, 4.2, 60, 464, 5.26, .54, 0, 0, 0, 1.0
16, 19.6, 2, 1.97, 0, 0, 457, 5.56, .52, 0, 0, 0, 1.0

400

470 REM SR is the metric head in the root zone, cm.

480	SR (0) $= EXP\left(\sqrt{\ln(0.4/0.225)/0.0189}\right)$
485	SR (0) = 250
490	REM KR = hydraulic conductivity in the rooting zone, cm day^{-1}
495	KR (0) = 22.6 * (250)$^{-1.4}$
500	REM KR (0) = 0.0099 cm day^{-1}
510	KR (0) = 0.01
520	REM 20 m^2 per kg of leaf of dry matter of leaf blade. If the leaf weight is 50 kg ha^{-1}, then leaf area per hectare = 50 x 20 = 1000 m^2/ha, so 1000 m^2/10000 m^2 = 0.1 m^2/m^2 = LA1=0.1.
530	LAI (0) = 50 * 20 * 10^{-4} = 0.1.
540	REM TADW = Total above ground dry weight in kg/ha
550	TADW (0) = WLV (0) = 50
555	REM TDW = Sum of the leaf blade and root weight (100 kg/ha)
560	WRT (0) = 50
570	TDW (0) = 50 + 50 =100
580	REM TDWL = Live plant tissue weight
590	REM TDWD = Dead plant tissue
600	TDNL (0) = 100
610	TDWD (0) = 100
614	TSUM =0 : DVS = 0 : SMA = 0.03 : SMFC = .225
615	REM SMA = Air dry soil moisture
616	REM SMFC = Soil moisture at field capacity
617	REM SMA = One-third of soil moisture at wilting point.
620	FH (0) = 1 - e$^{-0.6 \cdot LAI(0)}$
625	REM FH is the reduction factor in gross assimilation rate due to incomplete cover.
700	REM RM = maintenance respiration rate 0.015 kg CH$_2$O per kg dry matter day.
720	RM = 0.015
730	WR (0) = 22.5
740	REM WR(0) = 22.5 mm, soil's moisture content in the root zone of 100 mm depth at emergence.
745	DRM = 1500
750	REM DRM = Maximum rooting depth in mm.
760	REM WNR (0) = 225.5 = water content in non-rooted depth at emergence in mm.
770	WNR (0) = 222.5

780   SMA = 0.03

790   SMW = 0.09

800   SMFC = 0.225

900   REM SMA, SMW, SMFC IN cm^3/cm^3

1000   FOR I = 1 to 16

1010   TSUM (I) = TSUM + TA (I) * DT (I)

1020   IF I $\leq$ 10 THEN 1030 ELSE 1040

1030   DVS (I) = DVS + TSUM (I)/1100

1040   DVS (I) = (TSUM (I) - 1100/650+1

1050   PGASS (I) = FGS (I) * $(1-e^{-0.6 \cdot LAI (I-1)})$

1060   IF I $\leq$ 10 THEN 1070 ELSE 1180

1070   RD (I) = RD (0) * 12 * DT (I)

1180   RD (I) = RD (10)

1190   IM (I) = P (I) + IR (I)

1200   IF LAI (I) < 3.6 THEN 1210 ELSE 1220

1210   EM (I) = ETO (I) * $e^{-0.6 \cdot LAI (I-1)}$

1220   EM (I) = ETO (I) * 0.11

1230   EA (I) = EM (I) * (SMR (I-1)-0.03)/ (0.225 - 0.03)

1240   TM (I) = ETO (I) * $(1-e^{-0.6 \cdot CAI (I-1)})$

1250   SMCT (I) = (1-SMDF (I) * (SMFC - SMW) + SMW

1260   IF SMR (I) > SMCT (I) THEN 1270 ELSE 1280

1270   T (I) = TM (I)

1280   T (I) = TM (I) * (SMR (I) - SMW)/ (SMCT (I) - SMW (I)

1290   REM DMR = Amount of moisture added to the soil plant system
          by root growth (mm/day)

1300   REM WNR = Amount of moisture in the non-rooted part of the
          potential rooting zone (mm)

1310   REM RR = Growth rate of the roots (mm/day)

1320   REM DNR = Thickness of the non-rooted part of the potential
          rooting zone (mm), hence DNR (I) = DRM - RD (I-1)

1330   IF I $\leq$ 10 THEN 1340 ELSE 1350

1340   DMR (I) = WNR (I-1) * RR/(DRM-RD (I-1))

1350   DMR (I) =0

1360   REM DWR = The rate of change in soil moisture status of the root
          zone

1370   DWR (I) = IM (I) + DMR (I) - EA (I) - T (I)

1380   WR (I) = WR (I-1) + DWR (I) * DT (I)

1390   REM SMR (I) = Volumetric soil moisture content per unit soil

	depth for the present 10-day period (cm³/cm)
1400	SMR (I) = WR (I)/RD (I)
1410	REM SR (I) and KR (I) will not be treated, because unsaturated flow above the ground water table can be neglected in this situation

1400 SMR (I) = WR (I)/RD (I)
1410 REM SR (I) and KR (I) will not be treated, because unsaturated flow above the ground water table can be neglected in this situation
1420 IF I ≤ 10 THEN 1430 ELSE 1440
1430 WNR (I) = WNR (I-1) - DMR (I) * DT (I)
1440 WNR (I) = WNR (10)
1450 REM GASS = For the 10-day period, it equals actual gross assimilation
1460 IF T = TM THEN 1470 ELSE 1480
1470 GASS (I) = PGASS (I)
1480 GASS (I) = PGASS (I) * T (I)/TM (I)
1490 1F I ≤ 10 THEN 1500 ELSE 1510
1500 MRES (I) = 0.015 * TWDL (I-1)
1510 MRES (I) = 0.01 * TWDL (I-1)
1520 REM ASAG = The amount of assimilate available for increase in dry weight of vegetation is the difference between actual gross assimilation and maintenance respiration
1530 REM ASAG in kg/ha. day
1540 ASAG (I) = GASS (I) - MRES (I)
1550 1F I ≤ 10 THEN 1590 ELSE 1600
1590 DMI (I) = 0.7 * ASAG (I)
1600 DMI (I) = 0.8 * ASAG (I)
1605 REM DMI = The rate of increase in dry weight of the vegetation
1610 IWRT (I) = FR (I) * DMI (I)
1620 WRT (I) = WRT (I-1) + (IWRT (I) * DT (I))
1630 IWLV (I) = FL (I) * DMI (I)
1640 1F I ≤ 10 THEN 1650 ELSE 1740
1650 REM leaf blade deteriorates when the vegetation suffers from water shortage, in an attempt to reduce transpiration losses
1660 REM The rate of decline is proportional to the relative transpirational rate, T/TM
1670 REM severe moisture stress leaf blades are assumed to deteriorate at a maximum rate of 0.03 kg dry matter per kg leaf blade dry matter present per day
1680 REM DWLV = Rate of decline in leaf dry matter
1690 REM RDR = Rate of decline in leaf dry matter

```
1700    RDR (I) = 0.03 * (T (I)/TM (I))
1710    DWLV (I) = RDR (I) * IWLV (I)
1720    TDWD (I) = DWLV (I) * DT (I)
1730    WLV (I) = WLV (I-1) + (IWLV (I) - DWLV (I)) * DT (I))
1740    WLV (I) = WLV (I-1) + (IWLV (I) * 0.02) * DT (I))
1750    IWST (I) = FS (I) * DMI (I)
1760    WST (I) = WST (I-1) + (IWST (I)) * DT (I))
1790    IWGR (I) = FGR (I) * DMI (I)
1800    WGR (I) = WGR (I-1) + (IWGR (I) * DT (I))
1810    LAI (I) = WLV (I-1) * 20 * 10⁻⁴
1820    REM TADW = Total above ground dry weight
1830    TADW (I) = WLV (I) + WST (I) + WGR (I) + TDWD (I)
1840    REM TDW = Total dry weight of crop
1850    TDW (I) = TADW (I) + WRT(I)
1860    REM TDWD = Total dry weight of senesced leaf blade
1870    REM TDWL = Total dry weight of live material
1880    TDWL (I) = WLV (I) + WST(I) + WRT (I) + WGR (I)
1890    PRINT "TSUM =" TSUM (I),
                "DVS =" DVS (I),
                "PGASS =" PGASS (I)
                "RD =" RD (I),
                "EM =" EM (I),
                "EA =" EA (I),
                "TM =" TM (I),
                "T =" T (I),
                "DMR =" DMR (I),
                "DWR (I) =" DWR (I),
                "WR =" WR (I),
                " SMR =" SMR (I),
                "GASS =" GASS (I),
                "ASAG =" ASAG (I),
                "DMI =" DMI (I),
                "WRT =" WRT (I),
                "WLV =" WLV (I),
                "WST=" WST (I),
                "WGR=" WGR (I),
                "LAI =" LAI,
```

"TADW =" TADW (I),
"TWD =" TWD (I),
"TWDL =" TWDL (I),
"TDWD =" TDWD (I),

6.2.1.3 Sensitivity Analysis of Crop Model as a Help to Crop Production Research

Sensitivity analysis involves exploring the behavior of the model for different values of parameters. This is done to determine the extent to which a change in the value of a parameter influences the important outputs from the model.

The original objective may be related to research and the use of systems modeling to assist in a more thorough understanding of the system. In this case, the use of sensitivity analysis in guiding research efforts may be the intended applications (Jones and Luyten, (1998).

Singh (1995b) used a simulation approach to generate estimates of the senstivity of productivity of amphidiploid munbean, T1 - 18, to climate change, using modules PRODCL, POTTR, and model SMUNG. The simulation were performed using climatic data of Hisar, Haryana, India from Julian day 215-264. Weather data considered in this study were solar radiation, air temperature, actual vapour pressure and wind velocity. The results indicated that with decrease in either solar radiation, actual vapour pressure and wind velocity, there was a corresponding decrease in the crop productivity, except the vapour pressure, which caused an increase in the crop productivity. The most sensitive weather element appeared to be the solar radiation and the least, wind speed. When we increased the mean value of either solar radiation, vapour pressure, wind speed by 10, 20 and 30% and by 2, 4, and 6^0C of the ambient temperature, the wind velocity was the least sensitive weather element and productivity sensitivity to solar radiation, actual vapour pressure and temperature depended upon their magnitude of increments and the productivity components. Considering all these weather elements simultaneously and temperature separately, based on the change in weather elements due to doubling of atmospheric CO_2 predicted by General Circulation Models, it was found that the temperature alone was the most sensitive weather element which affected crop productivity of mungbean at Hisar. The effect of temperature alone was almost equivalent to the situation where all the weather elements simultaneously affected the productivity. The productivity response is the temperature change was related to the effect of increased temperature on shortening of phenology and, hence, decreasing crop productivity. This implies that the crop scientists should exploit homeostatic buffering, drought tolerance and yield potential together through breeding so that temporal and spatial shifts in ambient temperature as a consequence of global warming should not adversely effect crop phenology and the productivity.

Singh and Singh (2000) aimed at finding the magnitude and direction of the sensitivity of different morpho-physiological characters to grain and biomass productivity of wheat under global environmental change. The radiation use on efficiency and crop duration are found to be more sensitive as compared to other parameters studied. While the crop duration decreased due to global warming and was extremely sensitive to altering the wheat productivity, the combination of radiation use efficiency (+40%), stomatal resistance (+40%) and leaf area index (+15%) had more weight than the crop duration (–14%) in altering the productivity.

Singh (1995) aimed at finding the magnitude and direction of the sensibility of different morpho-physiological characters to grain and biomass productivity of mungbean under drought conditions. The radiation use efficiency and crop duration are found to be more sensitive as compared to other parameters studied.

6.2.1.4 Use of Crop Yield Models for Precision Farming

Precision farming is a tailoring of crop management, soil management and finance management to match different locations at global level through "Geographical Information System" and "Global Positioning System".

Singh and Ram (2000) studied the utilization of photoperiod and solar radiation resources in enhancement of genetic crop yield potential in south, south-east and far-east Asia . Though the study was confined to 22 countries of south, south-east and far-east Asia, it is equally applicable to other locations on the globe falling between these latitudes. With the main objective of presenting a meaningful comparison of potential biomass conversion in crop plants, solar radiation and effective photoperiod to phenology in 22 countries of south, south-east and far-east Asia, which could provide a basic framework for a successful collaboration on crop research programme and exchange of germplasm among these countries, output data have been generated for summer and winter solstices through a program written in Microsoft Basic-80 Rev; 5.2 and run on Apple 3 microcomputers for modeling daily effective photoperiod (EPP), potential solar radiation density (PSRD) and productivity of crops.

The EPP, PSRD and crop productivity on summer solstice increases from lower to higher latitudes while on winter solstice, the trend is the reverse. The values of EPP, PSRD and crop productivity are higher in summer as compared to the winter. The annual differences in the values of most of the variables increase from lower to higher latitudes, highlighting the lower latitude with no limiting energy resources for crop productivity throughout the years. The differences between the productivity of C_3 and C_4 crops on winter solstice increases from higher to lower latitudes and vice-versa on summer solstice, indicating thereby that C_4 crops should be preferred in lower latitudes during winters and in higher latitudes during the summer season. For wider adaptability, crop scientists should concentrate their research endeavour to develop photo-insenstive varieties of crops.

Singh et al. (1992) studied the potential crop biomass conversion, solar radiation and effective photoperiod to phenology at Hisar (India) and Reading (United Kingdom) with the main objective of presenting a meaningful comparison of potential biomass conversion in crop plants, solar radiation and effective photoperiod to phenology at Hisar (29.17 N) and Reading (51.45N). This may provide a basic framework for a successful collaboration on crop research program and exchange of crop germplasm between Haryana Agricultural University, Hisar and University of Reading. Output data have been generated for Julian days 13, 35, 57, 80 (80 = spring equinox). 103, 126, 149, 173, (173 = longest day); 196, 220, 243, 266 (266 = autumn equinox); 289, 312, 334, and 356 (356 = shortest day) through a program written in Microsoft Basic-80 Rev. 5.2 for modeling daily effective photoperiod, potential solar radiation density and potential biomass conversion. The interesting points that emerge from the study are: month of June has higher biomass conversion due to the higher solar radiation and longer effective photoperiod at Reading than at Hisar. An effective photoperiod exists for a longer duration at Reading than at Hisar. An effective photoperiod exists for a longer duration at Reading from March to September than at Hisar. Hisar, compared to Reading, receives more potential solar radiation density during autumn, winter and spring seasons and less during the summer. The potential productivity of both C_3 and C_4 species increases from day 13 to 173 and decreases thereafter. It is higher at Hisar than at Reading, except during summer.

Singh and Ram (1993) modeled the renewable resources for agricultural research and development in India. Through the present study was confined to India (8-36 degree N latitude), it is equally applicable to other locations on the globe falling between these latitudes. The effective photoperiod (EPP), inclusive of the twilight period, and the potential solar radiation density (PSRD), were modeled based on the astronomical and crop physiological principles. Apple 3 computer was used for running the program written in microsoft BASIC-80 Rev. 5.2 for modeling EPP and PSRD. At all the latitudes, EPP happened to be of a longer duration from the beginning of the year upto the summer solstice (Julian day 173) and shorter thereafter, upto the winter solstice (Julian day 356). At 16-36 degree N latitude, PSRD was greater in degree from the onset of the year upto the summer solstice and less thereafter upto the winter solstice. But at latitude 8 and 12 degrees N, this trend was not noticed. The values of EPP and PSRD were superimposed on the map of India. The presented information may be of great use to farmers, planners, plant breeders, and seed producers in improving the crop research activities.

Singh et al. (1996) studied the potential and water limited productivity of mungbean in south, south-east and far-east Asia. The study dealt with the modeling of daily potential gross biomass productivity (Goudriaan and van Laar 1978) in the countries of south, south-east and far-east Asia and actual water limited the productivity of mungbean "T1-18" at Hisar (29-17⁰N). The steps in computer implementation were programmed in Microsoft BASIC-80 Rev. 5.2 and run on Apple 3 Micro-Computer. Mungbean cultivation favours summer season at higher latitudes and autumn season at lower latitudes. Drought- and heat-tolerant and disease-and insect-resistant varieties of mungbean ought to be developed for its successful cultivation during summer season to lower latitudes, whereas cold-tolerant cultivars of mungbean may be introduced during the winter season in middle latitudes. The information derived on the basis of grouping of countries, for coverage and annual variation in potential productivity during clear and overcast days may be helpful in collaborating crop research activities and reducing the regional imbalances in productivity of mungbean. The initial validation of the mungbean model for predicting the water-limited productivity adapted from van Keulen (1986) using crop, weather and soil data of the environmental site, indicates that the model could simulate the various components of mungbean growth and yield with reasonable accuracy.

6.2.1.5 Crop Yield Models Using Remote Sensing DATA

Dadhwal and Ray (2000) reviewed the crop assessment using remote sensing-crop condition and yield assessment. Satellite-based remote sensing data can be suitably used to get pre-harvest crop yield estimate and early crop yield condition assessment. However, there are many constraints in using RS-based models. Those are: (a) long-time series of uniform spectral data being not available; (b) data gaps due to cloud cover; (c) spectral information modified by atmospheric effects, soil background viewing geometry; and (d) low correlation of yield with canopy vigour for some crops (e.g. groundnut).

Space technology in India has developed remarkably in the last few years. Sensors are now available with better spatial resolution (LISS-III on board IRS IC and ID), high temporal resolution (WiFS) and additional spectral bands (MIR for stress-related studies). To use this available technology for crop-related studies, further research work must be carried out in the following lines to develop a complete multi-stage regional level crop forecasting systems.

1. Use of satellite-based agrometeorological information, available from NOAA-AVHRR and INSAT VHRR sensors.
2. Use of WiFS and INSAT data for regular crop condition monitoring.
3. Development of simple, fast atmospheric correlation methology.
4. Development of physical spectral models from field experiments.
5. Integral use of GIS (geographical information system) and GPS (Global Positioning System) with RS data for development of crop yield models.

Dadhwal et al., 1989, developed an approach for the development of remote sensing data–based crop yield forecasting methodology.

6.2.1.6 Status of Research and Development on Crop Simulation Models

A list of some published crop simulation models is given as below:

Table 6.9. A list of some published crop simulation models.
(around data based crop models)

Crop	Model Name	Reference
Alfalfa	ALSIM (Level2)	Fick (1981)
	ALFACOLD	Kanneganti et al. (1998)
Barley	CERES-barley	Ritchie et al. (1989)
Cotton	GOSSYM	Baker et al. (1983)
	COTCROP	Brown et al. (1985)
	COTTAM	Jackson et al. (1988)
Dry bean	BEAN GRO	Hoogenboom et al. (1989)
	SMUNG	Singh (1995)
Maize	CERES-Maize V.1	Jones and Kiniry (1986)
	CERES-Maize V.2.0	Ritchie et al. (1989)
	CORN F	Stapper and Arkin (1980)
	VT-Maize	NewKirk et al. (1989)
	GAPS	Buttler (1989)
	CUPID	Norman and Campbell (1983)
Peanut	PNUT GRO	Boote et al. (1989)
Pearl Millet	CERES-Millete	Ritchie and Algarswami (1989)
	RESCAP	Monteith et al. (1989)
Potato	POTATO	Ng and Loomis (1984)
	POTATO (Modified)	Ewing et al. (1990)
	SUBSTOR	Hodges et al. (1989)
	(Unnamed)	Fishman et al. (1985)

contd....

Table 6.9 contd.

Rice	CERES-Rice (upland)	Ritchie et al. (1986)
	IRRIMOD	Angus and Zandstra (1980)
	(Unnamed)	Horrie et al. (1986)
	RICE MOD	Mc Megnnamy and O' Toole (1983)
Sorghum	CERES-Sorghum	Algarswamy and Ritchie (1990)
	SORGF	Arkin et al. (1976)
	SOR KAM	Rosenthal et al. (1989)
	RESCAP	Monteith et al. (1989)
Soybean	SOYGRO	Wilkerson et al. (1983)
	SOY GRO V5.0	Wilkerson et al. (1985)
	GLCIM	Acock et al. (1983)
	SOY MOD	Curry et al. (1975)
	SOY MOD/DARDC	Meyer et al. (1979)
Sugarcane	AUSCANE	Jones et al. (1989)
Wheat	CERES-Wheat	Ritchie et al. (1985)
	CERES-Wheat (Nitrogen)	Godwin and Vlek (1985)
	Unnamed	van Keulen and Saligman (1987)
	SMITFAG	Stapper (1984)
	TAMW	Mass and Arkin (1980)
	APSIM Nwheat	Mc Cown et al. (1996)
	SIM Wheat	Singh et al. (1989) Singh et al. (1982)
General Model	EPIC	Williams et al. (1984)
	MARCOS	Penning de Vries et al. (1989)
	SUCROS	van Keulen et al; (1982)
	Info Crop	Kalra (2003), Personal communication

6.2.1.7 Models Used for Directing Agricultural Research

Table 6.10. List of models used for directing research.

Crop	Model name	Reference	Important Parameter
Mung	SMUNG	Singh, (1995)	Daily tempertature, air water vapour concentration, crop duration, light use efficiency
Wheat	SIMWHEAT	Singh et al. (2000)	Light use efficiency
Barley	BARSIM	Teng et al. (1978) Teng et al. (1977)	The length of the latent period
Soybean	CROPGRO	Lal et al. (1999)	Cumulative heat units

In addition to these, Boote et al. (1996) have reviewed the models used for directing the agriculture research.

6.2.1.8 Models Used in Resource Management

Boote et al. (1996) reviewed the work on potential uses and limitation of crop models. Besides this, a list of models used in resource management is given in Table 6.11.

Table 6.11. List of models used in resource management.

Crop	Model/approach	Reference
Wheat	APSIM (Mc Cown et al., 1996)	Asseng et al. (2001a)
Wheat	APSIM (Agricultural production systems simulator)	Asseng et al. (2001b)
Wheat	APSIM	Asseng et al. (2001c)
Wheat	APSIM Nwheat	Asseng et al. (2000)
Wheat	APSIM wheat	Asseng et al. (1998a)
Wheat	APSIM wheat	Asseng et al. (1998b)
Agriculture crops	EPP (Effective photoperiod) and PSRD (Potential solar radiation density)	Singh and Ram (2000)
Cool season crops	Risk analysis	Ram and Singh (1995)
Mungbean	SMUNG	Singh et al. (1996)
Agricultural crops	EPP, PSRD	Singh and Ram (1995)
C_3, C_4 crops	EPP, PSRD, Potential biomass conversion	Singh et al. (1992)

6.2.1.9 Models of Crop Growth (Empirical and Mechanistic Models)

6.2.1.9.1 Empirical

France and Thornley (1984) gave a detailed review of crop responses and models. They have also tabulated some empirical crop models. The environmental or management factors are: I = irradiance; W = water status (rainfall, humidity, wind); T = temperature; F = fertilizer, nutrient status; D = day length; C = CO_2 concentration. (Table 6.12 and Table 6.14).

6.2.1.9.2 Mechanistic Models

France and Thornley (1984) listed the processes that may be important in mechanistic models of plant and crop growth.

Table 6.12. Some empirical crop models.

Crop	Factors	Method	Time interval	Reference
Grass	I,W,T,D	Growth and development rates are modified	7 days	Angus et al. (1980)
Rice	I, T	Regression of climatic factors on yield	Phase of development	Murata (1975)
Wheat	W, T	Stepwise multiple regression on yield	Day	Bridge (1976)
Wheat	W	Linear regression on seasonal rainfall	Season	Seif and Pederson (1978)
Wheat	T, W	Regression of yield on a derived index	Month	Sakamoto (1978)
Wheat	T, W	Multiple curvilinear regression	Month	Pitter (1977)
Wheat	T, W	Regression equation for daily growth rate	Day	Haun (1974)

Table 6.13. Important plant processes.

1. Phot	Light interception and photosynthesis: Canopy architecture, radiation characteristics; leaf characteristics.
2. Nutr	Root activity and nutrient uptake: root system architecture; soil nutrient status; root status and characteristics.
3. Part	Partitioning: substrate pools of carbon compounds and nutrients replenished by 1 and 2; transport between pools; utilization of pool substances for growth priorities.
4. Transp	Transpiration: Water balance of plant and soil; water status of plant.
5. Gr & R	Growth of structural dry matter and the recycling of structural components, respiration.
6. LA exp	leaf area expansion.
7. Dev	Devlopment and morphogenesis: initiation; growth and development of new organs (stems, leaves, flowers, fruits, storage organs, etc.)
8. Sen	Senescence.

Table 6.14. Some published mechanistic crop growth models: emp = *empirical, mech = **mechanistic.

Crop	Submodels								Environmental factors	Reference
	Photo.	Nutr.	Part	Transp.	Gr & R	LA exp	Dev	Sen		
Alfalfa	emp		mech		emp	emp			I, W, T, D	Holt et al. (1975)
Barley	mech				mech				I, W, T	Legg et al. (1979)
Chrysanthemum	mech				emp	emp			I, C	Charles-Edward and Acock (1977)
Clover (sub)	emp				emp	emp			I, T	Fukai and Salisbury (1978)
Cotton	emp	emp			emp				I, T, D	Mc Kinion et al. (1975)
Grass	mech		emp			emp		emp	I, T	Johnson et al. (1983)
Grass	emp		emp		emp	emp		emp	I, T	Sheehy et al. (1980)
Lettuce	emp				mech				I, C, T	Sweeney et al. (1981)
Maize	mech	emp	emp	mech	mech	emp		emp	I, W, T	de Wit et al. (1978)
Sorghum	mech	emp?	emp?	mech	mech	emp	emp		I, W, T	Arkin et al. (1976)
Soybean	mech	emp	mech	mech	mech	emp	emp	emp	I, W, T, F, D	Meyer et al. (1979)
Sugarbeet	mech		emp	mech	mech	emp	emp			Fick et al. (1975)
Tobacco	mech		mech	mech	mech	emp		emp	I, T	Wann et al. (1978)

*Empirical model: The objectives are to account for observed yield variation, to discover which factor affects yield greatly, and finally to consider whether management can manipulate some of these factors so as to increase yields or decrease costs.
*Mechanistic model: These models are constructed by assuming that the system has a certain structure, and assigning to the components of the system properties and processes which can be assembled within a mathematical model.

France and Thornley (1984) reviewed some published mechanistic crop growth models.

6.2.1.9.2.1 Status of Research and Development on Crop Simulation Model in India

A list of some published crop simulation models in India is given as below:

Table 6.15. A list of some published crop simulation models in India.

Crop/Factor	Model used	Reference
Wheat	CERES-Wheat	Attri et al. (2001)
Rice	CERES-Rice	Rathore et al. (2001)
Wheat	SPAW	Singh et al. (2000)
Rice	CERES-Rice	Saseendran et al. (1998a, 1998b)
Maize	SPAW	Rathore et al. (1998)
Wheat	SPAW	Rathore et al. (1994)
Wheat	CERES Wheat	Lal et al. (1998)
Rice	CERES Rice	Lal et al. (1998)
Potato		Rao et al. (1999)
Rainfall		Saseendran et al. (1996)
Weather forecast		Rao et al. (1996)
Weather forecast		Singh et al. (1999)
Soybean	CROPGRO	Lal et al. (1999)
C_3 and C_4 plants	Unnamed	Singh and Ram (2000)
Wheat	SIMWHEAT	Singh and Singh (2000)
C_3 and C_4 plants	Unnamed	Singh et al. (1992)
Peanut	PNUTGRO	Singh et al. (1994)
Wheat	Unnamed	Singh et al. (1982)
Mungbean	SMUNG	Singh (1995)
Mungbean	SMUNG	Singh (1995a)
Photoperiod	EPP	Singh and Ram (1993)
Potential solar radiation density	PSRD	Singh and Ram (1993)
Mungbean	SMUNG	Singh et al. (1996)
Temperature		Ram and Singh (1995)
Vapour pressure		Ram and Singh (1995)
Wheat	Unnamed	Singh et al. (1980)
Agricultural crops	Info crop	Kalra (2003)

6.2.1.9.2.2 Status of Research and Development on Crop Simulation Model in India and Abroad Through Remote Sensing Technology.

A list of some published crop simulation models through remote sensing technology is given below:

Table 6.16. List of models using remote sensing technology.
(remote sensing data based crop models)

Crop	Data used/Approach	Reference
Wheat	Landsat MSS and IRS-1A Data	Sharma et al. (1993)
Rice	IRS-1A Data	Patel et al. (1991)
Rice	Infrared	Rao et al. (1985)
Wheat		Sridhar et al. (1994)
Wheat	Spectral indices	Dubey et al. (1994)
Wheat	Spectral model	Kaluberme et al. (1995)
Wheat	Remotedly sensed data	Medhavy et al. (1993)
Wheat	Temperature, humidity, Sunshine hours, wind-speed, Potential evaporation.	Chaurasia and Minakshi, (1997)
Wheat	Landsat MSS and IRS LISS-I data	Patel et al. (1982) Dubey et al. (1985) Sharma et al. (1992) Kalebarma et al. (1992)
Agri. Crops	Remote sensing techniques	Sharma et al. (1998)
Wheat	Regression (Linear) yield to NDVI (Normalized difference vegetation Index)	Tucker et al. (1980)
Wheat	Regression (non-linear) VI (vegetation index) to grain yield	Dadhwal and Sridhar (1997)
Agri. Crops	Physical/Physiological approach	Monteith (1981)
Agri. Crops	RS technique can be used to estimate LAI	Dadhwal and Ray (1998)
Sugar beet	Linking of RS information with SUCROS (Simplified and universal crop growth simulator)	Clevers and van Leeuwen (1995)
Agri. Crops	Crop water stress index using RS and ground weather data	Jackson et al. (1981)
Agri. Crops	Temperature-stress-day	Gardner et al. (1981)
Agri. Crops	Satellite derived stress index (SDSI)	Dadhwal and Ray (1998)
Agri. Crops	Marriage between remote sensing based models with radiation interception models and canopy temperature models	Horie et al. (1992)
Agri. Crops	Satellite-based RS data on rainfall, temperature, soil moisture, etc.	Dadhwal and Ray (1998)
Wheat	Only RS parameter	Medhavy et al. (1995)
Rice	Only RS parameter	Patel (1996)
Mustard	Only RS parameter	Pokharna et al. (1995)
Sorghum	IRS LISS-1 data	Potdar et al. (1995)
Wheat	Agromet-spectral yield models	Kalubarme et al. (1995)

contd....

Table 6.16 Contd..

Crop	Data used/Approach	Reference
Wheat	Autoregressive models and RS based models combination	Pandey et al. (1992)
Sorghum	Integrated logarithmic senescence rate, derived from spectral profile using NOAA-AVH RR data.	Potdar (1993)
Wheat	Area under the profile	Kalubarme et al. (1997)
Cotton	Multi-data LISS-1 data	Ray et al. (1999)
Wheat	Crop cutting experiment estimate using TM (Thematic mapper) band derived VI.	Singh et al. (1992)
Agri. crops	NDVI images and CCE estimates	Murthy et al. (1996)
Peanut	Landsat MSS (multiple spectral scanner) band data used to detect moisture stress.	Dadhwal and Ray (1998)
Wheat	Landsat-TM- derived 6-band indices used to generate the time profile.	Dadhwal and Ray (1998)
Agri. crops	Hybrid approach, comparatively high resolution earth resources satellite data together with the coarse resolution meteorological satellite data from NO AA-AVHRR (National Oceanic and atmospheric administration-advanced very high resolution radiometer)	Ajai and Sahai (1986)
Agri. crops	NOAA data used to remove cloud contaminated pixels to develop a database of vegetation indices.	Boatwright and Whitehead, (1986)
Agri. crops	NDVI images from NOAA-AVHRR GVI (global vegetation index) to monitor large-scale drought pattern and their climatic pattern on vegetation.	Liu and Kogan (1996)
Agri. crops	VCI (Vegetation index) and TCI (temperature condition index) for monitoring drought	Kogan (1994)
Agri. crops	RS data from the NOAA satellite and ground observations of rainfall and agricultural conditions to monitor national agricultural drought.	NRSA, (1990)
Watershed	DEM (Digital Elevation Model)	Tripathi et al. (2002)
Phyto-plankton	IRS-P4 OCM sensor	Chauhan et al. (2002)
Phyto-plankton	IRS-P3 MOS-B Data	Sarangi et al. (2002)

contd....

Table 6.16 Contd..

Crop	Data used/Approach	Reference
Wheat	CROPWAT (FAO) Agrometeorology and Remote sensing	Raut et al. (2001)
Rice	Area estimation by remote sensing data; regression yield model using meteorological data	Dutta et al. (2001)
Watershed	IRS-B-LISS satellite imagery, SCS method	Durbude et al. (2001)
Agri. crops	CROPWAT (FAO) LANDSAT TM 5 digital data	Rao et al. (2001)
Wheat	RVI, NDVI, DVI, TVI, PVI, GVI	Singh et al. (2001)
Phyto-plankton	Optical sensor MOS-B, IRS-P3	Sathe and Jadhav (2001)
Wheat	WiFS (wide field sensors), IRS-P3, MOS-B bands, TD (Transformed divergence)	Singh et al. (1999)
Drought (Vegetation condition)	VCI (Vegetation condition index) TCI (Temperature condition index), NOAA operational polar orbiting satellite	Singh and Kogan (2002)
Temperature	Artificial neural network	George et al. (2001)
Soil moisture	Microwave radiometer data (remotely sensed data), IRS P4-MSMR, NIMBUS-SMMR	Rao et al. (2001)
Dew point (daily)	Maximum and minimum temperature.	Hubbard et al. (2003)
Wheat, sorghum and soybean	Seasonal reflection patterns	Kanemasu (1974)
Soybean and sorghum	Evapotranspiration model	Kanemasu et al. (1976)
Wheat	Price model (Leaf Area Index)	Rastogi et al. (2000)

6.2.1.9.2.3 Models Used as Policy Analysis Tools

Boote et al. (1996) reviewed the models used as policy analysis tools in the following heads:

1. Assist in best management decisions to reduce fertilizer and pesticide leaching and soil erosion.
2. Yield forecasting
3. Evaluate climate change effects.

6.2.1.9.2.4 Advantages and Limitations of Crop Models

Crop growth modeling is a potentially important tool in research, crop management technology, in policy decisions and teaching technology. Along with the considerable potential of crop modeling, there will also be misrepresentation, misuse and misunderstanding of the tools. Boote et al. (1996) encouraged crop model developers and users to be aware of the limitations and their possible misuse. Despite concerns about the difficulty of validating models, crop models proposed for broader crop management applications should be tested widely in diverse field environments.

As suggested by Boote et al. (1996), researchers can make use of crop growth models to conduct hypothetical studies and compare the outcomes with their own experiments. The widespread use of computers and the release of documented, comprehensive crop models make it relatively easy for crop physiologists and other discipline scientists to test the models, use them, and modify them. There is less cost in reinventing the wheel. Crop model improvement will be facilitated by such studies, through insight gained from testing the process relationships and through ideas for model simplifications. Simplifications can help one understand the basic responses but those using simplification in models should be aware of the limitations. Models will be dynamic in more than one way; not only do they respond dynamically to daily weather, but the models themselves will continue to evolve and change (Boote et al., 1996).

6.2.2 PLANT DISEASE PREDICTION

The barley-leaf-rust model (Teng, Blackie and Close, 1978), will be considered here. A plant-disease epidemic (widespread among many plants in the same place for a time) is a complex parasite-host system where growth of parasite and host and their interaction, are affected by a large number of factors. The system in this example may be defined as the growing crop and is represented in pictures in Figure 6.15. The conceptual boundary is drawn around the crop and the system consists of two major subsystems:

1.   The fungus *Puccinia hordei* Utth

2.   The crop *Hordeum vulgare* L. (barley)

Since the disease is foliar, the crop subsystem in the model is represented by plant leaf area and progress of the disease is calculated from the percentage of leaf area affected by disease. In the fungus subsystem, the organism assumes different forms during its life cycle, creating a multicomponent system in which the condition of each component is measured by the number of fungus bodies in each lifecycle

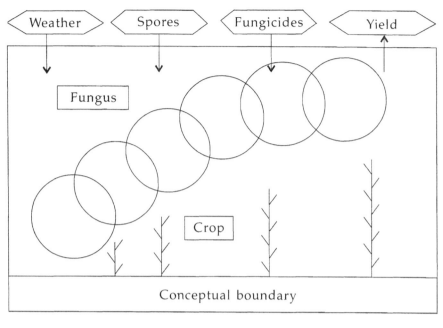

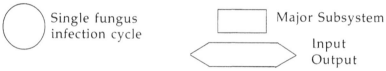

Fig. 6.15 The leaf-rust-barley crop agro-ecosystem
(After Teng, Blackie and Close, 1978)

form. Exogenous variables are inputs of weather, fungicide and external inoculum and these generate outputs of spores and crop yield across the boundary.

In the barley-leaf-rust model, a controllable exogenous variable is fungicide, whereas weather and outside inoculum represent uncontrollable exogenous variables. The crop will almost certainly be grown with the application of such other controllable inputs as fertilizer, but for the purpose of the model, this type of input is assumed to be constant. Effectively, this means that the model-builder has assumed normal application of fertilizers to the crop subsystem of the model.

In the barley-leaf-rust model, an important status variable is the number of spore-producing bodies present in the crop. Key elements are normally termed status variables.

6.2.2.1 Event Stepping Module in the Barley-Leaf-Rust Model

The barley-leaf-rust model has event-stepping modules. A model may be regarded as a part of the model concerned with some specific task. In the case of leaf rust, a spore which successfully germinates on a leaf enters a latent period before forming an infectious pustule. Field experiments conducted along side the modeling exercise showed that the length of the latent period was a function of ambient temperature and the density of infective pustules already on the leaf. Ambient temperature and pustule density were both related in some fashion to the time of year. The event of pustule eruption takes place not on a fixed day of the year but between 4 and 15 days after successful spore germination. Therefore, once the leaf-rust model has simulated the successful germination of any spores on a given day of the crop-growing season, those spores then enter an event-stepping module. The simplified example below in FORTRAN discusses the operation of this module.

The array LATENT, which has 15 elements, contains groups of latent pustules of different ages. LATENT (1) has that group of spores which germinated in this day of the simulation, while LATENT (8) has that group which germinated 8 days earlier. The variable IRUPT is the endogenously calculated length of the latent period or germination delay (some value between 4 and 15 days). The spores germinate and are placed in LATENT (1) by the model. In the next time step of the model (one day in this case), this group is moved to LATENT (2) and the new day's group of spores placed in LATENT(1). Similarly, all other groups of spores are moved up one position in LATENT—the model thus simulating the ageing of latent pustules by 1 day. The group of spores in LATENT (IRUPT) become infectious and are transferred into position 1 of the array INFECT. The action of this module is controlled by the irregular event step IRUPT and is only indirectly influenced by the main week time clock of the model. Figure 6.16 illustrates the operation of this event-stepping module.

6.2.2.1.1 The Barley-Leaf-Rust-Model Construction

The barley-leaf-rust model structure is depicted in Figures 6.16, 6.17, 6.18 and 6.19.

The rust model is conceived as two main segments—one which simulates the disease epidemic and the other which estimates the yield loss due to the simulated epidemic. The main biological and physical components in the epidemic segment are illustrated in Figure 6.19 (Teng, Blackie and Close 1977) and are an infectious period (which is the time between pustule eruption and pustule death), spore production,

Time 1 - First, the spores germinate and enter position 1. Germination delay calculated as 10 days - no spores present in position 10, so no infectious pustules appear.

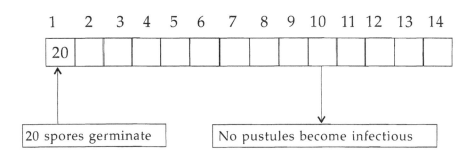

Time t+5 -Original spores are now 5 days old. Germination delay still 10 days and still no infectious pustules

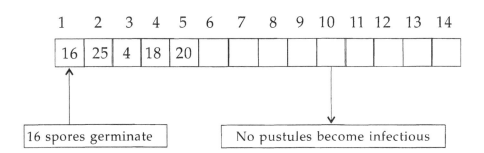

Time t+8 - Original 20 spores are now 8 days old. Germination delay now 8 days and first 20 pustules erupt.

Fig. 6.16 *Contd.*

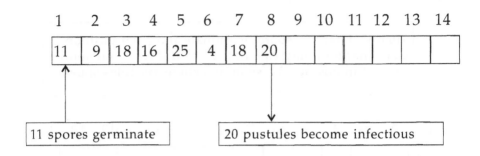

Time t+9 - Next 18 spores now 8 days old. Germination delay still 8 days so these now become infectious pustules.

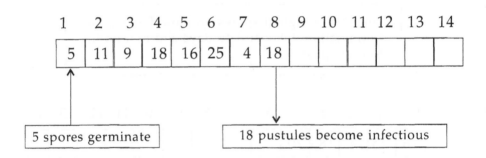

Fig. 6.16 Operation of event-stepping module.

liberation, survival, deposition, germination, penetration of host tissue by germ tube and, finally, a latent period (which is the time between penetration and pustule eruption). These components were established from field and laboratory studies of the leaf-rust pathogen. Having defined the lifecycle of leaf-rust fungus, the next task was to estimate the loss in yield in a barley crop affected by the disease. A measurement of the progress of the disease had to be established and, following the work of James (1971), the area of infected tissue expressed as a percentage of total tissue area was used. This concept is expressed in the term "percentage severity". Yield loss was then estimated from a function relating the percentage yield reduction to the percentage severity of disease on leaf 1 (flag) and leaf 2 of tillers at certain crop growth stage as defined within the decimal code of Zodoks, Chang and Konzak (1974). The relationship between disease progress and projected crop loss was derived from field experiments at the following crop growth stages; early boot, beginning of flowering, and milk dough on the Feekes' scale.

The model has been established to operate on a daily time interval ($\partial t = 1$ day) and the starting period of any simulation is the day on which the disease is first observed in the field. Data input to the model include percentage severity of initial observation of infection, number of days to be simulated and weather data.

The liberation of spores by an infectious pustule rises to a peak early in the infectious period and then declines gradually through the life of the pustule. In the model, therefore, the potential number of spores produced on any day is a function of the age distribution of pustules and the number of infectious pustules. The modeling of several events in the fungus lifecycle involved a stochastic element and the model was designed so that such elements could be either stochastic or deterministic, depending on the requirement for a given run of the model. In a deterministic run, the actual number of spores liberated equal the specific potential, while in a stochastic run, the actual spore number liberated is the result of multiplying the number of spores produced by a daily correction factor. This correction factor was drawn from a uniform random-number generator (0 to 1) for each day of the simulation. The deterministic or stochastic nature of the survival and deposition phases are modeled in the same manner. Spore germination is regarded as a binary activity since it is dependent on leaf wetness. A binary activity is one which can have only two values: in this case, germination either does or does not occur. If dew is present then all the spores deposited on that day germinate; if dew is absent, then no spores germinate.

The number of germinated spores that penetrate to form pustules is dependent upon the rust-race-host-cultivar combination. While the expected rate of penetration is included in the model as a 'penetration ratio', this relationship is modified by the amount of disease already present. Rusts are obligate parasites and can successfully infect healthy tissue. Therefore, the model includes an important negative feedback with respect to penetration success as the amount of unaffected tissue decreases. This relationship coresponds to the mathematical notation ($b - y$) in Verhulst's logistic equation

$$\frac{dy}{dt} = ay\,(b - y)$$

where dy/dt is the rate of growth, a is a rate parameter, y is the amount of growth at time t and b is the maximum attainable growth.

Output from the leaf-rust model includes a daily estimate of disease severity and an estimate of percentage yield at the end of the simulation run. Disease severity may be plotted to assist visual appraisal of the results of each run.

Model-building involves two main phases: an initial system model (Phase 1 model), which represents the epidemic caused by the urediniospore stage of rust disese, was constructed after a general literature review of the rust. Preliminary experimentation with this model enabled location of critical control points in the system and suggested the degree of emphasis to be placed on various system components (Teng, Blackid and Close, 1977) in subsequent development of the model. Empirical data for the structure of the detailed system model (Phase 2 model) was derived experimentally from controlled environment and field experiments. This involved determination of the effects of weather parameters on the development of the fungus on barley, the effect of the fungus on barley yield and the effects of fungicide on the progress of the disease epidemic. The detailed system model was assembled on a modular basis, with component subsystems linked by a main executive routine responsible for the time-keeping and event advancement (Dent, 1974). This model might have been incorporated into an agricultural-information system along the lines outlined by Blackie (1976). Farmers adopting the system would be able to use it to perform cost-benefit analysis on the economics of using fungicides to control barley-leaf-rust in the field during the growing season. The system would provide the individual farmer with a rational basis on which to base fungicide applications that will play a positive role in reducing the unnecessary use of pesticides.

6.2.2.1.2 *Validation of the Barley-Leaf-Rust Model*

6.2.2.1.2.1 *Statistical Assessment*

Teng, Blackie and Close (1978) conducted a field trial from the five sites. Field trial data from the five sites were used in regression analysis against model output under identical conditions. Separate analysis were carried out for each field on the tiller, leaf 1 and leaf 2 over the 15 different epidemics. Table 6.17 provides the regression coefficients and their associated standard errors, the value of R^2, the value of t for each coefficient (testing the intercept coefficient against zero and the slope coefficient against unity) and the F value for the null hypothesis that simultaneously intercept is equal to zero and slope is equal to unity for field number 1 only. High values for R^2 are a feature of the results as well as general support for the null hypothesis.

Model Structure

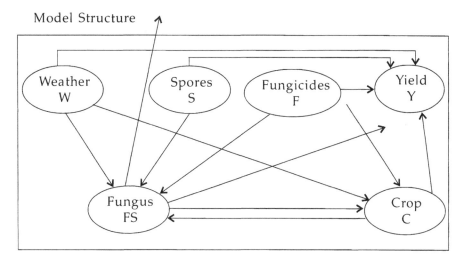

Fig. 6.17 The barley-leaf-rust model structure diagram.

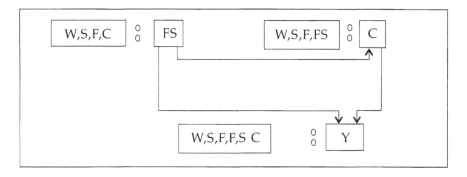

Fig. 6.18 The barley-leaf-rust model block diagram.

Goodness-of-fit testing proceeded using the Smirnov test, details of which can be found in Conover (1971). For each field and each of the three leaves separately, the distribution of per cent severity of infection over the 15 epidemics was established. Similar data from the model-results provided an analogous distribution. These density functions were converted to cumulative probabilty function (CDF) before being compared using the Smirnov test. The test assumes two independent random samples with CDFs represented by F(x) and G(x). Using a two-sided test, the hypotheses are:

H_0: $F(x) = G(x)$ for all x from $-\infty$ to $+\infty$

H_1: $F(x) \neq G(x)$ for at least one value of x.

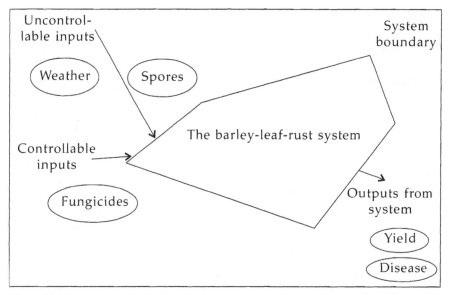

Fig. 6.19 The concept of a barley-leaf-rust system (*Source:* Dent and Blackie, 1979).

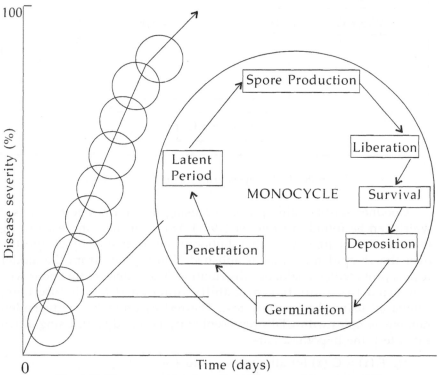

Fig. 6.20 The rust epidemic and components of its monocycle
(*Source*: Teng, Blackie and Close, 1977).

Table 6.17. Results of linear regression analysis of observed disease severity on model-predicted disease severity. (After Teng, Blackie and Close, 1978)

Field	Leaf	a	Standard error (a)	t_a	b	Standard error (b)	$t_{1\,b}$	F	R^2
1	Tiller	0.1460	0.0982	1.487	1.000	0.0063	–	0.18	0.999
	Leaf 1	0.1354	0.1347	1.005	1.0028	0.0051	0.549	0.71	0.999
	Leaf 2	-0.1852	0.0585	3.166**	1.0076	0.0039	1.949	10.99**	0.999

If the null hypothesis H_0 is accepted at a certain probability level, then the model-output for the situation described does not differ significantly from field data at this level. The test statistic is T, the greatest vertical distance between the two distributions and is compared with tabulated values at a specified confidence level. Results from the Smirnov test are presented in Table 6.18 and the null hypothesis is accepted in all the cases.

The validation described, however, must be seen in perspective. First, the model had only simulated epidemics occurring on five fields located within a radius of 1km. Secondly, validation data were collected over only two cropping seasons, and thirdly, the biological functional relationships built into the model-structure were for a 'virulent-race-susceptible host' combination found in Lincoln College, New Zealand where the model was built. The significance of these three factors cannot be judged until data from a wider area are available for validation.

Table 6.18. Results of Smirnov test on disease severity
(After Teng, Blackie and Close, 1978).

Field	Leaf	T	Two-tailed quantiles	
			$W_{0.95}$	$W_{0.99}$
1.	Tiller	0.1066	0.4737	0.5678
	Leaf 1	0.1212	0.6113	0.7326
	Leaf 2	0.1636	0.5942	0.7122
2.	Tiller	0.1103	0.4737	0.5678
	Leaf 1	0.1786	0.7039	0.8436
	Leaf 2	0.1515	0.8786	1.0531
3.	Tiller	0.1538	0.5444	0.6525
	Leaf 1	0.2222	0.6608	0.7926
	Leaf 2	0.0833	0.5677	0.6804
4.	Tiller	0.1333	0.5053	0.6057
	Leaf 1	0.2500	0.7039	0.8436
	Leaf 2	0.1727	0.5942	0.7122
5.	Tiller	0.1176	0.4737	0.5678
	Leaf 1	0.2000	0.6249	0.7489
	Leaf 2	0.1889	0.6249	0.7489

6.2.2.1.3 *Computer-based Experimentation of the Barley-Leaf-Rust Model*

The procedure for experimentation with a computer model of a bio-economic system is given in Figure 6.21.

In a disease-control experiment using the barley-leaf-rust model, the two alternatives of spraying versus no spraying may be considered. Let $x = 0$ represent   'no spray',

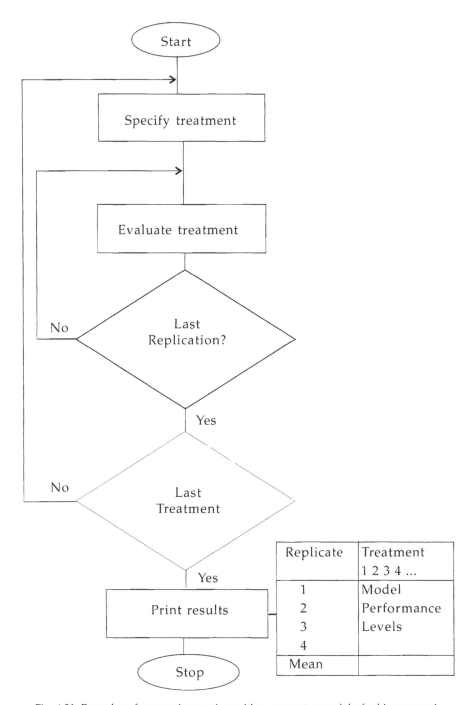

Fig. 6.21 Procedure for experimentation with a computer model of a bio-economic system (After Dent and Blackie, 1979).

$x = 1$ represent   'spray'

There is no need to replicate these treatments since identical results would be obtained for each replicate when using a deterministic model. Each treatment is evaluated once, performance being predicted in terms of, say, total cost of labour value Z_1 and Z_2 for 'no spray' and 'spray', respectively. No statistical analysis of the results is needed; the decision rule is simple.

If $Z_1 < Z_2$ then treat $x = 0$ is superior and no spraying is recommended;

If $Z_1 > Z_2$ then treatment $x = 1$ is superior and spraying is recommended.

However, while such an experiment is simple to conduct and interpret, the information provided is extremely limited. The assumption here is that spraying is uniformly effective over the entire range of conditions. While this may be true in some circumstances, more frequently it would be expected that exogeneous variables such as weather will affect the outcome Z to some significant extent. Therefore, it is needed to consider experimentation with models in which these stochastic or uncertain variables are explicitly included.

6.2.2.1.4 Application of the Barley-Leaf-Rust Model

6.2.2.1.4.1 Sensitivity Analysis of the Parameters of the Barley-Leaf-Rust Model

The barley-leaf-rust model was tested (Teng, Blackie and Close, 1977) for sensitivity to three parameters:

1.   Length of latent period;
2.   Length of infectious period; and
3.   Rate of spore production.

In this case, the output criterion was defined as percentage of disease severity. The experiment used a factorial design on the three parameters with two levels of spore production, three levels of infectious period and three levels of latent period. Values for the variables were within the range known for cereal rusts. The simulated epidemic assumed optimum dew conditions for germination and the activities of liberation, survival and deposition were treated as stochastic events. Each epidemic was replicated three times.

Table 6.19 provides an analysis of the variance results, indicating that spore production rate and latent period was significant in affecting the simulated epidemics, while the infectious period did not have any significant influence. The length of latent period appears to be the dominant factor in determining the rate of build up of rust-disease epidemics for the conditions specified.

Table 6.19. Abbreviated analysis of variance of a simulated factorial experiment to determine sensitivity (After Teng, Blackie and Close, 1977).

Source of variation	df	Sum of squares	Mean squares	F-value
Latent period	02	8.834	4.417	605.88***
Infectious period	02	0.012	0.006	0.859
Rate of spore production	01	0.126	0.126	17.30***

The most significant factor in determining the build-up of epidemic was the length of the latent period. Since this period is the result of host-parasite interaction, a recommendation from this sensitivity exploration was that breeding for rust resistance in cereals could be most effective if breeders concentrate on developing varieties which result in a long latent period.

An alternative approach is to feed the different input values of the parametres and simulate the output values in order to decide the most sensitive parameter. This is shown in Figure 6.22, where it is clear that parameter P_1 is more sensitive than the parameter P_2.

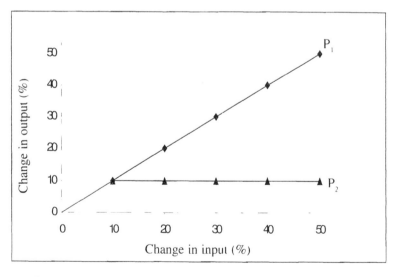

Fig. 6.22 Sensitivity results from varying two parameters of a model

6.2.2.2 Other Relevant Literature on Disease Prediction

France and Thornley (1984), reviewed the literature on disease prediction. Van der Plank (1963) is one of the pioneers of the mathem-atical approach to plant disease epidemics and their control; more accounts and surveys are given by Kranz (1974), Krause and Massie (1975) and James (1974).

6.2.2.2.1 Estimation of Crop Losses

This problem is invariably attacked using empirical methods (James, 1974). A disease assessment method is developed to measure the amount of disease at the different growth stages of plant (James, 1971).

Burleigh, Roelfs and Eversmeyer (1972), estimated the losses in wheat caused by leaf rust as:

$$Y = 5.3788 + 5.5260 \ X2 - 0.3308 \ X5 + 0.5019 \ X7 \qquad (6.47)$$

where $X2$, $X5$ and $X7$ are the percentage leaf-rust severity at the boot, early berry and early dough stages, respectively. This equation accounted for 79% of the variation. Similar work has been done on potato late blight and loss of tuber yield (James et al., 1972) and many other crops.

Equation (6.47) is sometimes referred to as a "multiple point" model, since it incorporates the disease at several points during the crop growth period. The so-called "critical point" models may be applicable for disease of short duration, and which affect the physiology of the growing plant at a particular and crucial phase. Katsube and Koshimizu (1970) used

$$Y = 0.57 \ X \qquad (6.48)$$

for estimating percentage loss of yield in rice blast, where X is the percentage of blasted neck nodes 30 days after heading. Manibhusanrao (1988) developed a computer model for epidemiology of blast disease of rice (Proceedings of the annual rice workshop of Directorate of Rice Research, April 26-29 held in TNAU, Coimbatore). Two sampling sites: at Nellore (AP)—as the place is considered one of the hot-spots for blast— and at Magali (Chengalpattu Dt., Tamil Nadu), where the breakdown of resistance in high-yielding IR 50 was reported, are identified. A linear prediction model is envisaged which will be quite handy if a few parameters are to be handled which are easily measurable. The subject of linear models has been reviewed by Kranz and Hau (1980) and Nagarajan (1983). In the case of blast incidence, the most common parameters that influence the epidemic spread will be temperature, humidity and leaf wetness. In the field study, effective spore trapping (Hirst, 1953) will be of immense use to develop any linear model, i.e. the number of spores trapped and correlate to the blast incidence. It has been found that the number of airborne spores present and the outbreaks of blast are closely correlated (Ono, 1965), and can be used to evolve one of the most reliable forecasting tools to study blast epidemics.

For stem rust in wheat, Romig and Calpouzos (1970) found that

$$Y = -25.33 + \ln X \qquad (6.49)$$

where X is the disease severity when the developing caryopsis has reached three-quarters of its final size, was the best predictor of yield loss.

The multiple-point and critical-point approaches of equations 6.47 and 6.48 may be generalized and combined in a single equation.

$$Y = \int w(t) \, X \, (t) \, dt \qquad (6.50)$$

where w is a weighting function that depends on the variable t which may denote the chronological time or developmental time, and X is disease severity at time t. Equation (6.50) is still a linear model, but this equation can be rewritten as

$$Y = \int w \, (t) \, f \, [X(t)] \, dt \qquad (6.51)$$

where f denotes some function; this would then encompass non-linear models such as equation 6.49. Effectively, equation 6.50 and 6.51 sum the area under the disease-progression curve.

The yield loss simulation model was used by Chander et al. (2002) to simulate the effect of bacterial leaf blight (BLB) on growth and yield of rice cultivar, IR-64.

6.2.2.2.2 Disease Prediction and Control

A disease is only able to progress if the conditions provided by the host plants and environment are favourable, and if some inoculum is present. The relevant components of the micro-environment largely depend upon weather, and weather cannot be forecasted reliably for more than very short periods in advance. It is, therefore, usual to make predictions only after certain biological and meteorological conditions favourable to the disease have been fullfilled. If warranted, disease control measures may then be applied. In fact, only a few disease prediction systems take account of the amount of inoculum of disease present at a given stage in the plant's development (Eversmeyer and Burleigh, 1970); it is more common to assume that there is always enough inoculum present to initiate an epidemic, to monitor the weather or microclimate within the crop, and to base disease predictions upon these measurements.

Many methods have been developed to predict the occurrence of potato late blight: A simple system due to Hyre (1954) is described to illustrate the general principles.

A day is "blight favourable" if

the 5-day temperature average <25.5°C

the total rainfall for the last 10 days $\geq$ 3.0cm

the minimum temperature on that day $\geq$ 7.2°C   (6.52)

when the consecutive blight-favourable days are first recorded, late blight is forecast to appear 7 to 14 days later. Appropriate control measures are recommended. This system works well in the north-eastern United States, but less well in the mid-western states, where different relationships between rainfall, relative humidity and temperature prevail. For different climatic conditions, alternative systems may be needed. Krause, Massie and Hyre (1975) describe a combination of Hyre's system with one of these alternatives.

Multiple regression equations have been used to predict disease-leaf-rust of wheat (Eversmeyer and Burleigh, 1970) and stem rust of wheat (Eversmeyer, Burleigh and Roellfs, 1973).These authors define about 15 meterological and biological parameters that predict the future rate of disease increase.

Gadre, Joshi and Mandokhot (2002) studied the effect of weather factor on the incidence of Alternaria leaf blight, white rust and powdery mildew of mustard and developed the empirical equations to predict the disease incidence as affected by weather factors.

Automatic data collection, good communciations, and the use of computers allow these schemes for disease prediction to make accurate, dependable and timely recommendations for disease control measures (France and Thornley, 1984).

6.2.2.2.3 *Plant Disease Deductive Model*

The methods of disease prediction described in the above section are essentially inductive in nature, although elements of understanding and mechanism are sometimes present. Deductive models are also known as simulators. The simulators of plant disease attempt to understand the system thoroughly, in terms of its parts and how they fit together (Waggoner, 1974). In Figures 6.20 and 6.23 the main features of the progress of the fungal disease epidemic are shown. The disease microcycle of Figure 6.23 is defined at the detailed level, largely that of microbiology, and using laboratory experiments. It is the activity of this disease microcycle, modified or stimulated by weather and host-plant status that gives rise to the epidemic shown in Figure 6.23, and hypothetical data given in Table 6.20.

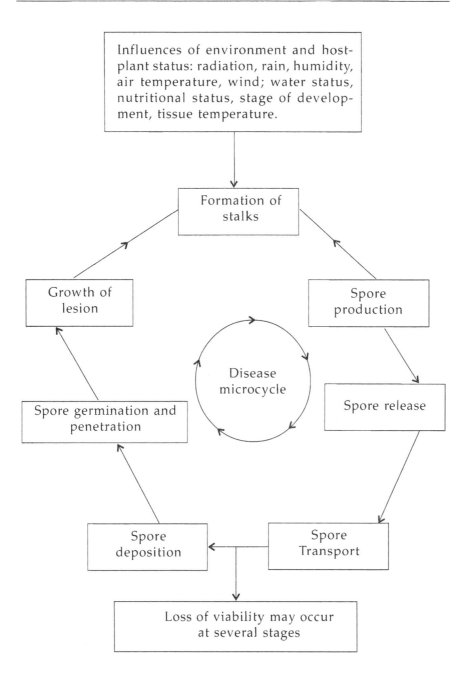

Fig. 6.23 Microcycle of a typical fungal disease
(*source:* France and Thornley, 1984).

Table 6.20. Hypothetical data on epidemic of a fungal disease in the field.

	$|x|$	$|y|$	$|ln(A/y -1)|$	$|lnx|$
1.	10	0.99	4.61	2.30
2	20	7.52	2.51	3.00
3.	30	21.23	1.31	3.40
4.	40	39.22	0.44	3.69
5.	50	55.55	−0.22	3.91
6.	60	68.49	−0.78	4.09
7.	70	77.52	−1.24	4.25
8.	80	84.34	−1.68	4.38
9.	90	87.22	−1.97	4.50
10	100	90.99	−2.31	4.61

x = time, y = disease severity and A = maximum value of severity.

Epidemic models for disease control are fairly new. In fact, an epidemic model or simulator aims to run like a real epidemic. It should mimic the epidemic of disease of the crop. It was Waggoner and his coworkers in 1972, who explored the response of the disease southern corn leaf blight caused by *Helminthosporium maydis* to weather in controlled experiments and suggested a computer program designed simulator, EPIMAY. The simulator was assembled from three parts: (i) weather as observations; (ii) The pathogen in the guise of knowledge of how all its stages react to weather; and (iii) the computer program for presenting the current weather to the pathogen and asking how it would react. With these three components of the simulator, the computer predicts whether the epidemic rises, falls or remains level. This was similar to the earlier disease simulator, EPIDEM by Waggoner and Horsfall, 1969, for the early blight of tomato and potato, caused by *Alternaria solani*. Shaner et al. (1971), evolved a plant disease display model as EPIMAY. Further, Bloomberg (1979) developed a model of damping off and root rot of Douglas fir seedlings caused by *fusarium oxysporum*. Excellent computer programs have been developed to simulate epidemics of grape powdery mildew (Sall, 1980), barley leaf rust (Teng, 1980) and late blight of potato-simulator (Bruhn and Fry, 1981). Further, EPIPRE-Pest and disease control in wheat (Zadoks, 1984) and a comparison of the epidemic modeling through simulation approaches (Teng, 1985) are some of the computer-based schemes that have evolved. Manibhushanrao (1988), developed a simulator, EPIBLA, which is the proposed code name for blast disease for warning system in rice (*Oryzae sativa* L.). EPIBLA is the abbreviation for Epidemology of BLAst. EPIBLA, in a way, is a new approach to control the blast disease through effective forecasting and thus, to minimize the blast epidemic threat and manage it by recommending the minimum protective sprayers. It has been aimed to benefit in planning the economics through the use of fungicides and crop loss so that it can enable the farmer to take the necessary protection measures. Krupinsky et al. (2004) found that crop sequencing and crop diversification reduced the plant disease risk in the cropping system.

These are large models, and it is not possible to give a complete description of one of them here. The best accounts are in some research reports and the serious reader cannot do better than study EPIMAY, which simulates southern corn leaf blight (Waggoner, Horsfall and Lukens, 1972).

A brief outline of the procedure:

1.   Choose state variables—quantities such as number of lesions, possibly with an age-size distribution; number of spores on stalks; number of spores on leaves; number of stalks, possibly with an age-size distribution.

2.   Definite rates of processes that connect these state variables, and particularly how these depend on environmental and host-plant variables. Much of this is based on microbiological experiment-ation, and may be in terms of analytic functions or tables of numbers.

3.   Specific initial conditions.

4.   Read weather reports and update the state variables.

Validation of plant disease simulators present similar problems to those encountered when validating most other large models—many parameters and often few data. The disease simulators are primarily research tools, aimed at increasing and organizing our understanding of the relevant processes. Practical disease prediction is still carried out by the inductive models described earlier (France and Thornley, 1984).

6.2.3 Insect Phenology

In population dynamics, the simplest modeling approach is in terms of total population number N. However, total population models are unrealistically simple, and have been found to be generally too inaccurate. For accurate and realistic models, prediction of insect phenology is necessary. Phenology deals with the age-specific distribution of population of insects. The specific knowledge of insect phenology helps in developing the biological control model. The developmental rates, which determine the time spent in each instar, are dependent on temperature, crowding and plant (host) status. Development rates of each instar are very important for integrated insect-pest management. Readers are advised to refer the sections 6.1.1.2.13 and 6.1.1.2.14 for models with age structure, Leslie matrix model and an example of a biological control model.

6.2.3.1 Development of an Organism

An organism may be a human being, plant, insect and pathogen. The way in which an organism develops can greatly affect the output of that organism, and this is, therefore, a topic with which physiologists and agronomists are much concerned (Landsberg, 1977). Although efforts to quantify growth have had considerable success over the years, attempts to similarly quantify development have not made the same progress, and we still largely fall back on a qualitative characterization.

The lifecycle of a plant is interrupted by events which take place effectively at points in time (in reality, over periods of time short compared with the length of the lifecycle); these events are used as markers to separate phases of the organism's development which occur over longer periods of time. A typical scheme for plant development is shown in Table 6.21. The marker should correspond with readily observable changes in the organism's morphology; for different organism species, different set of markers might be appropriate.

The four phases of organism (plant) development in Table 6.21 are denoted by G, V, R and S. A variable giving a measure of development, D_i, with $i = G$, V, R or S, may be notionally associated with each phase. The D_i have arbitrary units, and it is convenient to normalize D_i to lie between 0 and 1, so they are 0 at the beginning of a phase and 1 when the phase is complete, and the organism (plant) is about to move into the next phase. D is used to denote any of the D_i.

The rate of development is the rate at which D changes, and by definition this is equal to the developmental rate constant k, so that

$$\frac{dD}{dt} = k \qquad (6.53)$$

Table 6.21. Scheme for plant development

Markers (events occurring at points in time)	Processes (these occur over a period of time)
Sowing	
Germination, emergence	Processes of germination, G
Floral initiation	Vegetative growth, V
Maturity	Reproductive growth, R
Death	Senescence, S

Source: France and Thornley, 1984.

If the phase begins when time $t = t_l$ and $D = 0$, then

$$D = \int_{t_1}^{t} k dt \qquad (6.54)$$

gives the value of D at time t. At time t_l, $D = 1$, and this signifies the end of the current phase and the beginning of the next phase. It has been assumed that development can be measured by a single-valued quantity. The topic is not yet well enough understood to know if this is a reasonable assumption.

The essence of many agronomic studies of an organism (crop or plant and insect) development is how the developmental rate constant k depends on environment. The environmental quantities that affect development are temperature, day length, radiation, nutritional status and water status; however, interest is usually centred on the first two of these, with temperature having a dominant role.

6.2.3.1.1 Heat Sums or the Day-Degree Rule

If it is assumed that the development rate constant k is proportional to temperature T above a threshold temperature T_c, then

$$k = a \, H \, (T - T_c) \, (T - T_c) \qquad (6.55)$$

where a is a constant, and H is the unit step function:

$$H \, (T - T_c) = 0 \text{ for } T < T_c \qquad (6.56)$$
$$H \, (T - T_c) = 1 \text{ for } T > T_c$$

It is customary to form a heat sum, h_{-sum} [day °C], ignoring the normalization constant a, and using mean tempertures, to give

$$h_{-sum} = \sum_i H \, (\, Tm_i - T_c \,) \, (\, Tm_i - T_c) \qquad (6.57)$$

where Tm_i is the mean temperature on the ith day. To illustrate how this works, consider a sequence of 20 days, with a threshold temperature of $T_c = 10°C$, give

i	1	2	3	4	5	6	7	8	9	10	11	12	13	14	15	16	17	18	19	20
Tm_i	8	8	9	11	10	12	9	10	11	12	9	10	14	13	12	14	13	15	14	14
h sum	0	0	0	1	1	3	3	3	4	6	6	6	10	13	15	19	22	27	31	35

For a particular organism (crop), and a particular phase of development, it is found that the attainment of a certain value for the heat sum (50 day °C, say) corresponds quite well with the end of that phase of development. Thus, using the heat-sum method predictively, one can examine whether the phenology of a crop is suited by the average temperature experienced at a location.

6.2.3.1.2 Day Length and Other Environmental Factors

Day length is an environmental variable that is often next in importance after temperature in its effect on development. Some crop plants are "day neutral" or unaffected by day length, but for many organisms (plants), day length is critical, especially for the reproductive markers. Because of the dominant effect of temperature, it may be convenient to write the influence of daylength in terms of day-degree equivalents, for instance by replacing Equation (6.57) by

$$h_{sum} = \sum_i [H(Tm_i - T_c)(Tm_i - T_c) + bH(g_i - g_c)(g_i - g_c)] \quad (6.58)$$

where b is a constant, g the day length on ith day, and g_c is the critical value of the day length for development. Equation (6.58) is written for a 'long-day' organism (plant), where days of length greater than g_c promote development. For a "short day" plant, the subscripts i and c in the daylength term are reversed.

Other environmental factors may be added to equation (6.58) in a similar manner. Equation 6.58 illustrates linear addition of the environmental factors. Some authors have found it useful to indicate high-order terms, and combine these differently. For example, Robertson (1968) considered the developmental rate of a cereal crop (organism), and its dependence on day and night temperature and photoperiod. He obtained good results using an equation equivalent to

$$K = \sum_i \{H(g_i - g_c)[a_1(g_i - g_c) + a_2(g_i - g_c)^2]$$

$$+ H(T_{il} - T_c)[b_1(T_{il} - T_c) + b_2(T_{il} - T_c)^2]$$

$$+ H(T_{id} - T_c)[d_1(T_{id} - T_c) + d_2(T_{id} - T_c)^2]\} \quad (6.59)$$

where a_1, a_2, b_1, b_2 and d_1 and d_2 are constants; g_i, g_c and T_c are as before; and T_{il} and T_{id} are the daily maximum and minimum temperature on the ith day. It will be clear that many alternatives are possible when formulating empirical relations between developmental rate and environmental factors.

6.2.3.1.3 Calender Days as a Factor

Vishwar Dhar (2000) quantified the phenology of eriophyid mite (*Aceria cajani channabasavanna*) which is host specific on pigeonpea and a few of its wild relatives. The disease of sterility mosaic of pigeonpea is transmitted through this mite. This mite acts as a vector for transmit-ting this disease in pigeonpea.

Although mite is a not an insect, it is a non-insect pest. The life cycle of *A. Cajani* which includes an egg stage, two nymphal and adult stages is completed within 15 days (Reddy et al., 1990). The eggs hatch in 4-5 days and produce first nymphal and subsequently, the second nymphal stage. The nymphal stages last from 2-5 days after which they develop into adults. After 24 hours of attaining the adult stage, the female starts laying eggs at the rate of 1-3 per day (Oldfield et al., 1981)

Chander et al. (2002), assessed the yield losses due to stem borer, *scripophaga incertulas* in rice using simulation model. Chander et al. (2002) also studied the changes in pest profile in rice-wheat cropping system in Indo-Gangetic plains.

Sehgal et al. (2001) studied the management aspect of insect, disease and nemotode pests of rice and wheat in the Indo Gangetic plains.

6.2.3.1.4 Work Done on Insect Phenology in India and Abroad

Students and Readers are suggested to consult the reviewed work on the aspect by Karel et al. (1996).

Singh and Verma (1988) studied the antibiosis mechanism of resistance to stem borer, *Chilo partellus* (Swinhoe) in sorhgum. The biology of *Chilo partellus* (swinhoe) was studied on two resistant cultivars IS2205, IS5489 and two susceptible HC136, ICVS, sorghum genotypes. Significant differences were observed between resistant and susceptible genotypes in regard to larval mortality, including, larval period, larval weight, larval length, per cent pupation, pupal weight, fecundity per female and total life span of this pest. Growth index was also influenced by the resistant genotypes. It has been concluded that high mortality of larvae on resistant lines is due to antibiosis. The correlation between the characters of the pest biology have been worked out and the larval stage was the most important parameter influencing the total pest survival and population build up in the succeeding generations.

Table 6.22. List of work done on insect phenology.

Insect/Predator/ Parasite	Crop	Reference
Mustard Aphid	rapeseed and mustard	Arora and Sindhu (1992)
Phetella Maculipennis Curtis	-do-	Batra (1961)
Painted bug, *Bagrada cruciferarum*	-do-	Bhai and Singh (1961)
Syrphidae	-do-	Bhatia and Shaffi (1932)
Mustard sawfly	-do-	Bogawat (1967) Bogawat (1968)
Phetella xylostella (L.)	-do-	Dube and Chand (1977)
Indian insects	-do-	Ghosh (1914)
Mustard aphid, *Lipaphis erysimi* (Kalt.)	-do-	Kalra et al. (1987)
Neochrysocharis sp. (Parasite)	Rapeseed and Mustard	Kaurava et al. (1969)
Phytomyza atricornis Mg	-do-	Kaurava et al. (1970)
Mustard aphid *Lipaphis erysimi*	-do-	Kurl and Mishra (1979)
Myzus persicae sulzer	Potato	Lal (1950)
Mustard aphid *Lipaphis erysimi*	Rapeseed and Mustard	Landin (1982)
Parasitic chalcidoidea	-do-	Mani (1940)
Chilomenes sexmaculata Fabr.	-do-	Modawal (1941)
Parasites of Pea leaf minor *Phytomyza atricornis* Mg.	-do-	Narayanan et al. (1956)
Microtonus indicus a parasite of *Phyllotreta eruciferae* goeze	-do-	Narayanan et al. (1960)
Chrysopa scelesis Bank	-do-	Nasir (1947)
Bagrada cruciferarum Kirk	-do-	Rakshpal (1949)
Cabbage butterfly	-do-	Rataul (1959)
Mustard aphid, *Lipaphis erysimi*	-do-	Rout and Senapati (1968)

(Contd.)

Table 6.22. Contd.

Insect/Predator/ Parasite	Crop	Reference
Egg parasites of *Bagrada picta* Fabr.	-do-	Samuel (1942)
Ladybird beetles, *Coccinella septempunctata* L.	-do-	Sethi and Atwal (1963)
Aphididae	-do-	Sharma and Subba Rao (1964)
Mustard aphid, *Lipaphis erysimi* Kalt.	-do-	Sidhu and Singh (1964)
Mustard aphid, *Lipaphis erysimi* (Kalt.)	Rapeseed and mustard	Singh et al. (1990)
Painted bug, *Bagrada cruciferarum*	-do-	Singh and Malik (1993)
Pea-leaf-miner *Phytomyza horticola* Goureau	-do-	Singh (1980)
Coccinella undecimpunctata memetriesi muls, predator of mustard aphid	-do-	Singh and Malhotra(1979)
Tanymecus indicus	-do-	Srivastava et al. (1965)
Mustard aphid, *Lipaphis erysimi* (kalt.)	-do-	Srivastava and Srivastava (1961)
Mustard sawfly, *Athalia proxima* Klug.	-do-	Verma (1945)
Stem borer, *Chilo partellus* (Swinhoe)	Sorghum	Singh and Verma (1988)

6.2.4. SYMBIOSIS BETWEEN CROP MODELING AND GENOMICS

The paper presented at the symposium "Crop modeling and genomics" held at the 2000 ASA-CSSA-SSSA annual meetings in Minneapolis, MN, represents a variety of ways to address the incorporation of genomics into crop simulation modeling. Weiss (2003) gave an unbiased introduction of the symposium. The diversity in approaches ranges from modifying existing algorithms to simulate new responses to addressing this topic at the most basic level of current understanding. Having a detailed knowledge of a crop simulation model provided Ritchie and Algarswamy (2003) a way to modify and expand the ability of CERES-maize to simulate bareness and prolificacy. They relate kernel number per plant, with new genetic coefficients, to cumulative interrupted photosynthetically active radiation around silking. With the goal of assisting plant breeders gain information about the response of spring wheat to early vigour, simulated by increasing specific leaf area, Asseng et al. (2003), evaluated the spring wheat responses in several environments in Australia using APSIM-N wheat. Yields could increase or decrease depending on the climate, soil type, and nutrient management. Using the simulation model Cropsim to study wheat responses in different environments, Hunt et al. (2003) conclude that unknown factors may influence yield, so that incorporating a genetic response to growth will be difficult compared with simulating crop development. Rather than trying to deal with individual gene responses, they suggest the current focus of research should be on "grouping of genes into packages that can be characterized in terms of discrete coefficients". Boote et al. (2003) investigate a wide range of simulated soybean responses by varying the genetic coefficients in CROPGRO-Soybean to produce changes in phenological development, crop assimilation, vegetative vigour, leaf area expansion of determinate and indeterminate cultivars, seed-fill rates, and traits to improve production under drought. They attempt to relate the changes in these coefficients to the physiology of the plant and, where possible, extend their interpretation of the results to a lower level of organization.

The paper by White and Hoogenboom (2003) begins with an introduction to basic functional genomics, then discusses six levels of genetic detail that can be incorporated into crop simulation models, and concludes by looking at potential outcomes, what species and traits to simulate, and the importance of the necessary collaboration between modelers and molecular biologists. Stewart et al. (2003) focus on the development of a model that calculates the effects of alleles at six maturity loci on the photothermal response of soybean. They conclude that if the genetic makeup of a cultivar is known, the time from planting to first flower can be predicted using daily temperature, sowing date, and latitude. Welch et al. (2003) take a very different approach to predict flowering; they use a neural network model to predict flowering in contrasting strains of *Arabidopsis thalian*. This approach was able to demonstrate an important characteristics that will be necessary to quantify genetic inputs in future crop simulation models, the ability to predict a temperature-dependent change in transition order. Hoogenboom and White (2003) point out the promise and frustration associated with attempting to incorporate gene responses into Genegro, which was based on a simulation model of common bean (*Phaseolus vulgaris* L.). While there was a slight improvement in predicting days to flowering, the days to maturity and growth related variables were not improved with the incorporation of a gene that inhibits cold temperature response to photoperiod. A limitation to their approach was access to data that represents a wide range of environments where these genes are active.

A sought after but elusive goal of crop simulation modeling has been applications to plant breeding. Yin et al. (2003) suggest a complementary framework using quantitative trait loci (QTL) to represent input parameters in crop simulation model while the models may assist in QTL mapping by "dissecting yield into physiological components that are more likely related directly to gene expression". Chapman et al. (2003) take a different approach to address issues associated with plant breeding by combining the crop simulation model APSIM-Sorg with QTL-GENE, a genetical-based simulation model for developing plant breeding strategies. With this approach, they studied four traits (stay green, phenology, osmotic adjustment and transpiration efficiency) that look into account epistatis and genotype × environment interaction for yield.

Many interesting ideas were presented in the symposium. What is the next step? How do we proceed to pursue these ideas, assuming that these ideas are of sufficient merit that they should be pursued? How do two groups of scientists, crop modeler and plant geneticist, that work on different scales of plant organization using different methodologies corporate to achieve mutually beneficial results of the general answer is a combination of altruism and money. As one always goes from general to the specific, the devil is in the detail (Weiss, 2003).

Some other recent work somewhat related to this aspect are Welch et al.(2000a), Welch et al.(2000b), Welch et al. (2002), Gungula et al. (2003), Carbone et al. (2003), Miyasaka et al. (2003) and Bannayan et al. (2003).

APPENDIX A

Exercises on modeling crop production systems

1.   Choose one or more examples from your field of agrobiology and construct the appropriate symbol-arrow graph.

2.   You are asked to begin construction of a model of a system in your own field of agrobiology. What are the dimensions of the variables for that system?

3.   Does the system you have begun modeling have feedback loop? If not construct another model of another system that gives rise to negative or positive feedback?

4.   Cite one or more examples from your own field of agrobiology of pairs of scales that do not belong to the same dimension.

5.   Select one or more equations, more or less at random, from a recent issue of a journal in your field of agrobiology and analyze each for dimensional correctness.

6.   Suppose you have a chamber with 90 flies in it and drop in 10 additional flies. You wait for a few minutes and select a random sample of 10 flies from the chamber. What is a probability that you get back exactly the same 10 flies you added?

7.   In the system you are modeling, which dynamic processes can be described by linear rate laws?

8.   Cite from your own field of agrobiology, a system for which a compartmental description would not be acceptable. Why not?

9.   As a mathematical equation, the statement $x = x + 1$ is absurd. As a programming instructor, however, it might make perfectly good sense. What sequence of elementary operation instructions might it involve?

10.   Pick any convenient number, and find its square root by iterative procedures.

11.   Use finite difference methods to obtain a graph y versus t from the equation

$$dx/dt = -ky \quad k > 0$$

Hints: Use a convenient value of y_0 and a small value for k, say $k = 0.01$. Compute as many points as you think you need to get the feel of the procedure. Repeat the procedures with a different choice of Δt. After a certain number of time periods, compare with the exact solution, $y(t) = y_0 e^{-kt}$. Observe that if Δt is too large, very peculiar results emerge. Why?

12.   Write a $C^{..}$/ BASIC Program for simple exponential growth $dx/dt = kx$ with $x(t = 0) = 1$ and $k = 1$, integrating by Euler's method.

13. Using SI units (kg, m, day) what are the units of the parameters a, n, b and k in the equation?

$$dw/dt = at^n e^{-bt} - kt$$

14. Construct formulae converting the decimal point of the day, $0 < t_d < 1$, to hours, minutes and seconds and vice-versa.

    Hint: 0.01 day = 14.4 minutes = 864 sec. = 0.24 hr.

15. Use the equation

    N = integer part $(30. 6M + D - 91.3)$

    to find the climatological day number N, of the first day of each calender month, and tabulate the same.

    Hints: If $M \leq 2$, then $M = M + 12$. If $M \geq 13$, then $M = M - 12$.

    Remarks: M = month number of the year and D = day of the month.

16. Write an outline computer program on:

    A day is "blight favourable" if 5-day temperature average <25.5°C, the total rainfall for the last 10 days is 30 cm, the minimum temperature on that day ≥ 7.2°C, to give blight forecasts. The input data provided are daily value of mean temperature = [°C], minimum temperature T min [°C] and rainfall h [mm].

17. Write a possible sequence of calculations for equations: $dv/dt = \alpha_v - \beta_v$ VP, $dP/dt = \alpha_p$ VP $- \beta_p$ Pπ and $d\pi/dt = -\alpha_\pi \pi + \beta_\pi$ Pπ.

    Remarks: V = vegetation, P = Prey, π = Predators, α_v, α_p and α_π = intrinsic rate of increase of vegetation, prey and predators, respectively (time^{-1}) and β_v, β_p and β_π = predation coefficients for vegetation (mass of prey^{-1} time^{-1}, for prey (number of (predator)$^{-1}$ time^{-1} and for predators (number of prey^{-1} time^{-1}, respectively.

18. A quadratic equation $y = a + bx + cx^2$ is given. The parameter a is always the value of y for $x = 0$. If $x > x^2$ for very small x, equation is dominated by the $a + bx$ parts of very small x. For very large x, it is dominated by the cx^2 term. If you choose b to be positive, draw the curve. Choose the c as positive and draw the curve. Choose the $c < 0$ and draw the curve. Try it with some actual numbers. Note the qualitative appearance of the curves drawn according to the quadratic equation.

19. Describe an experiment in your own field of agrobiology that you have either done or read about

    (a) What were the things measured?

    (b) What were the things controlled?

(c) Which things may have been relevant but were neither measured nor controlled?

20.   Choose a system in your own field of agrobiology and carry the process through the following steps from 1 to 5:

1. formulation of the problem

2. qualitative description of the system.

3. definition of relevant components, subsystems, and interactions.

4. definition of relevant variables.

5. representation of the relations between the variables.

21.   You are asked to begin construction of a model of some system from your own field of agrobiology. Draw a symbol - arrow graph for this system.

22.   Convert

(a) 60 mi /hr to ft/sec.

(b) 60 mi/hr to km/hr

(c) 15 lb/m^2 to kg/cm^2

(d) 15 lb/m^2 to dyn/cm^2

(e) 32 ft/sec^2 to cm/min^2

23.   Given four different types of nucleotides, how many different sequences can be formed of length two? Length three? Why does this show that if the codes of all amino acids are non-overlapping and of the same length, then that length must be three?

24.   Suppose we have a collection of N amino acid molecules with n_1 of type one, n_2 of type two, upto n_{20} of type 20, with $\Sigma n_i = N$. Derive an expression for the number of different types of protein that can be made with the N molecules (use all N for each type).

25.   An experiment was performed with pea plant, in which two parents were crossed to get an F_1. Parent P_1 had round yellow seeds and parent P_2 had wrinkled green seeds. The F_1 plants all had round yellow seeds. Let R = allele for round seeds, Y = allele for yellow colour, r = allele for wrinkled seeds and y = allele for green colour. Assume that gene for shape and colour are independent. What can you say about dominance of R versus r, of Y versus y? In the F_2 generation, what are probabilities of P (round, yellow), P (round, green), P (wrinkled, yellow) and P (wrinkled green)?

26.   One estimate of the probability of a mutation at each nucleotide position in a single reproductive cycle is 10^{-8}. In an organism with 10^7 nucleotides, what is the probability that no mutation takes place? Assume that the probability for mutation at each

nucleotide is the same and that they are all independent. What is the probability of at least one mutation?

Hints: (a) remember that $0! = 1$;

(b) if x is a number whose absolute value is very small, then $\log_e (1 + x)$ is approximately equal to x. Therefore, for very small positive x, $\log_e (1 - x)$ is approximately $- x$.

27.   On what kind of time scale would it be reasonable to use a derivative formulation to describe the following:

(a) molecular changes

(b) microorganism growth

(c) change of size in population of rabbits.

(d) change in size of a population of elk.

(e) density of stars in the milky way.

(f) change of the number of dollars in a bank.

28.   For the system you were asked to begin modeling, which relationships involve dynamic processes? If your model system does not involve any dynamic processes, you should at this point begin work on a new model that does. For each of these dynamic processes is the process :

(a) continuous or does it involve discrete events?

(b) stochastic or deterministic? Which types of description would be appropriate: that of detailed decomposition into stochastic sub-processes, of stochastic formulation as a simple process, of stochastic formulation considering only expectations and variance or of continuous deterministic formulations? What assumptions would you make in each case? If a continuous deterministic description is appropriate, on what time scale would it be appropriate?

29.   Cite, from your own field of agrobiology, a system that can be described using a compartmental model. Formulate and justify a description of the individual components. Formulate (and justify) the designation of the allowable transfer pathways. Which of these allowable transfers may be assumed to follow linear rate laws?

30.   Perform a dimensional analysis of equation

$$J(x) = - D \frac{dc}{dx} \text{ and}$$

$$\frac{dc}{dt} = D \frac{d^2 c (x)}{dx^2} + q(x)$$

In particular, what are the dimensions of J and D?

Remarks: $J(x)$ = the flow per unit time flowing across a rectangle of unit area (That is, it is amount per unit time per unit of area). The negative sign is completely in accord with the idea that material should be flowing from high concentration to low concentration. If dc/dx is positive, it means that concentration is increasing as we go to the right and the net flow would be to the left opposite to the concentration gradient. Similarly, a negative value of dc/dx means concentration decreasing to the right and, therefore, a possible flow to the right. If $dc/dx = 0$ then there is no net flow.

D = diffusion coefficient; $q(x)$ = amount at position x.

31.   Make a physical analysis of equation

$$J(x) = -D\frac{dc}{dx} \text{ and}$$

$$\frac{dc(x)}{dt} = D\frac{d^2c(x)}{dx^2} + q(x)$$

for all combinations of dc/dx and d^2c/dx^2 positive, negative and zero. Assume $q(x) = 0$. Make liberal use of drawing and graphical representation.

Remarks: $J(x)$ = amount per unit time per unit area, D = diffusion coefficient, dc/dx = concentration gradient, $q(x)$ = net production.

32.   Construct an example in your own field of agrobiology that meets the assumptions (individuals or particles are identical in a stochastic sense and behave independently of each other, number of things and of events can be used instead of expectations of these numbers; things are changing slowly; change with respect to position have to be taking place slowly enough and the concentration gradient must not be so steep) so that flow or migration can be described by equation.

$$J = (J_x, J_y, J_z) = -\left(D_x\frac{\partial c}{\partial x}, D_y\frac{\partial c}{\partial y}, D_z\frac{\partial c}{\partial z}\right)$$

and change in concentrations or density can be described by equation

$$\frac{dc(p)}{dt} = D_x\frac{\partial^2 c}{\partial x^2}, D_y\frac{\partial^2 c}{\partial y^2}, D_z\frac{\partial^2 c}{\partial z^2} + q(p)$$

In constructing the example:

(a) Define the quantity being considered.

(b) Define what is meant by its concentration or density

(c) What is the production term q (p)?

(d) Define the spatial scale and time scale on which the use of the derivation is justifiable.

(e) Can the coefficient D_y, $D_{y'}$ and D_z be considered to be equal?

Remarks: J = net flow in those coordinate directions, (J_x, J_y, J_z) = an ordered triple, D = diffusion coefficient, δc = change in concentration at some point p;

Hints: areas of agrobiology: random spread of population, disease spread and epidemiology, spread of water, air pollution, intracellular metabolite movement.

33.   Assume X to be error free and Y to be measured with error ε_y. For each of the following cases, find an expression ε_z. If X and Y both vary over a range of 100 fold, what is the range of ε_z in terms of ε_y?

(a) $z = 1/y$ (b) $z = xy$ (c) $z = y^2$

(d) $z = (y^2+y)^x$ (e) $z = \ln y$

34.   Assume that the following pairs (x_i, y_i) are observations with "no" error in the values of x_i. Plot the points and find by eyes, the line of best fit.

x_i	1.17	1.74	2.50	3.50	4.00	5.04	7.48	8.00	8.50	9.00
y_i	1.04	1.58	1.83	2.50	3.12	5.08	4.91	7.00	4.80	6.86

35.   For the points in Exercise 34, find the line of best fit by unweighted least square. Find the mean square error and correlation coefficient.

36.   Assume that the expected error in measuring y is 10% of y. For the data in Exercise 34, find the weighted least-square, best fitting line.

37.   For the data of Exercise 34, find the best fitting line by the method of dividing into regions and finding $(\bar{x}, \bar{y})$ within each region. Two regions will suffice.

38.   The equation of the rectangular hyperbola in equation.

$$G = G_1 \frac{F}{K+F} - G_2$$

can be written $y = ax/(b + x)$, with variables x and y, and constants a and b. Drive the expression for the slope dy/dx. Verify that the initial slope dy/dx ($x=0$) and the asymptote y ($x - \infty$) are a/b and a, respectively. Derive an expression for the rate of change of slope, d^2y / dx^2, and verify that for $x \geq 0$, the magnitude of this is maximum at $x = 0$.

Remarks: F is the rate at which feed is supplied to the animal, and G is a growth rate of the animal. The quantities of G_1 and G_2 and k are called parameters.

39.   The result of an experiment are well fitted by a growth equation.

$$W = at /(b + t)$$

where W denotes weight, t is time, and a and b are constants. Differentiate to obtain a dynamic growth model of form $dW/dt = g\ (t)$, where g denotes a function of t. Use the original growth equation to eliminate t in favour of W to obtain a dynamic growth model in the more useful form of $dW/dt = h\ (W)$, where h denotes a function of W.

40.   The exponential quadratic growth curve has the equation, $x = \exp\ (a_0 + a_1\ t + a_2\ t^2)$ where x is the dependent variable (typically the fresh or dry weight of an animal or plant), t denotes time and a, a_1 and a_2 are constants. Show that this empirical equation is equivalent to the two-state variable problem.

$$dw/dt = x\ (a + 2y)$$

$dy/dt = a_2$; with $\ln x = a_0$ and $y = 0$ at time $t = 0$, y is the second state variable. Suggest a biological interpretation of the reformulated problem.

41.   Solve the differential equation $dy/dt = -\ y$, with $y = 1$ at time $t = 0$, using Euler's formula

$$x\ (t + \Delta t) = x\ (t) + \Delta tf\ [x(t),\ t]$$

for a few time steps using $\Delta t = 0.1$. Integrate the differrential equation and use the analytical solution to check your result.

42.   Re-do Exercise 41 using the second-order trapezoidal method of equation

$$x\ (t + \Delta t) = x(t) + 1/2\ \Delta t\ [f(x_1,\ t + \Delta t) + f(x,\ t)]$$
$$x_1 = x(t) + \Delta tf\ (x,\ t) = x(t + \Delta t),$$

$$f(x, t) = \frac{dx}{dt} = \frac{x(t + \Delta t) - x(t)}{\Delta t}$$

$$\frac{d^2x}{dt^2} = \frac{\dot{x}(t + \Delta t) - \dot{x}(t)}{\Delta t} = \frac{f[x(t + \Delta t), t + \Delta t - f\ (x,t)]}{\Delta t}$$

$$\dot{x} = \frac{dx}{dt}$$

43.   A model is fitted to 84 data points by adjustment of four

parameters to minimize the log residual sum of squares R.

(equations $r_i = y_i - Y$ or $r_i = \ln (y_i/Y)$ and $R = \sum_{i=1}^{n} g_i r_i^2$), giving an

R (minimum) of 0.8. Estimate the mean residual sum of squares, and estimate the average relative error or lack of fit of prediction and experiment.

Suppose the error residual has been found to be 0.4 with 50 degrees of freedom. Is the model an exceptable fit to the data at the 10% probability level?

Remarks: R = residual sum of squares, r_i = residual r_i, y_i = experimental value, g_i = weighting factor.

44. Solve graphically $Z = x_1 + x_2$ subject to

$$-2x_1 + x_2 \le 6$$
$$x_1 + 5x_2 \le 65$$
$$x_1 \le 10$$
$$x_1, x_2 \ge 0$$

and shade in the region representing the feasible solution. Check your graphical solution by resolving the problem using a linear programming package.

45. If the fractional nitrogen content of tissue, f_N is defined by the equation

$W_N = f_N W$, where W_N (kg nitrogen)

and W (kg total matter) are the respective masses of nitrogen and total matter, derive the unit of f_N. Can those units be simplified?

What is the unit of leaf area index?

46. Economics often talks about demand elasticity and cost elasticity. Find out what these are, and compare the idea with the analysis of model sensitivity and parameter ranking in following equation

$$V(P) = \frac{R}{df} \frac{1}{\partial^2 R / \partial P^2} \quad \text{and}$$

$$S(Y, P_i) = \frac{\partial Y}{\partial P_i} \frac{P_i}{Y} = \frac{\delta Y}{Y} = \frac{P_i}{\delta P_i}$$

Remarks:

$V(P)$ = the variance of P;

P = adjustable parameter

R = Residual sum of squares

df = degree of freedom

δP_i = a small finite change in parameter P_i

δY = The small change this causes in Y

$S(Y, P_i)$ = the sensibility of the quantity Y to the parameter P_i.

A 5% parameter increment is usually sufficient ($\delta P_i / P_i = 0.5$).

If $S(Y, P_i) = 1$, a given fractional change in the parameter value produces the fractional change in yield Y. Parameters with $S(Y, P_i) > 1$ have a larger effect on yield and vice-versa.

47.   Using equation

$\cos h_0 = \tan \phi \tan \delta$

to eliminate h_0, derive an alternative equation

($\cos a_0 = -\sin\delta / \cos\phi$) to equation $\sin a_0 = \cos \delta \sin h_0$ for the azimuth angle a_0, at sunrise or sunset.

Remarks: h_0 = hour angle of the sun, when the zenith angle is 90°, $h = (t_d - t_n)360$, h = the hour angle of the sun, this being the difference between actual time of day t_d and the time of apparent noon t_n (i.e. when the sun is highest in the sky), $h = (t_d - t_n)360°$. Since the sun is highest at midday, then $t_n = 0.5$, and at midday $t_d = 0.5$ and $h = 0$. Let Z, the zenith angle, be the angular distance of the sun from the local vertical, Z is given by (Sellers, 1965):

$\cos Z = \sin \phi \sin \delta + \cos \phi \cos h$ where ϕ is the latitude. The sun is on the horizon when $Z = 90°$, which, from equation $\cos Z = \sin\phi \sin \delta + \cos \phi \cos h$, occurs at $h = h_0$ with $\cos h_0 = -\tan\phi \tan \delta$. Here δ = solar declination angle in degrees between the line joining the sun and the earth, and the equatorial plane.

48.   For your location (i.e. latitude) write a computer program to calculate, at 10-day intervals throughout the year starting on 1st March: the day of the month D and the calendar month M using equations

N = Climatological day number $1 \leq N \leq 366$

N = integer part ($30.6M + D - 91.3$)

M = integer part [($N + 91.3$)/30.6],

D = integer part ($N - 30.6N + 92.3$) and

if $M \geq 13$, then $M = M - 12$; the declination δ using equation

$$Y = \frac{N - 21}{365} 360° \text{ and}$$

$\delta = 0.38092 - 0.76996 \cos y + 23.26500 \sin y + 0.369.58 \cos 2y + 0.10868 \sin2y + 0.01834 \cos 3y - 0.16650 \sin 3y - 0.00392 \cos 4y + 0.00072 \sin 4y - 0.00051 \cos 5y + 0.00250 \sin 5y + 0.00442 \cos 6y$;

the zenith angle Zn at solar noon (use equation $\cos Z = \sin \phi \sin$

$\delta + \cos \phi \cos \delta \cos h$ with $h = 0$ and ϕ equals to your latitude); the day length g_N corresponding to the zenith angle definition of $Z = 90°$, $90.83°$, $96°$, $102°$ and $108°$

$$\left(\text{equation } g_N = \frac{2}{360} \left(\cos^{-1} \frac{\cos Z}{\cos \phi \cos \delta} - \tan \phi \tan \delta \right) \right)$$

Remarks: M = month of the year, N = climatological day number; and D = day of the month, y = the year angle; $360°$ a period of one year, $21 = y$ is zero at the vernal equinox on 21st March, $t = 0$ falls on 1 March, climatological day number N is obtained from t by equations t_i = integer part (t) and N = integer part $(t) + 1 = t_i + 1$, Z_n = zenith angle at solar noon, ϕ = latitude of the location.

49. For your location and with a zenith angle of $96°$ (civil twilight), use your program from Exercise 48 to work out the day length on 21 June ($N = 113$) and on 21 December ($N = 296$): g_{113} and g_{296}. Construct a Sine-wave approximation to the variations in day length using.

$$g_N = (g_{113} + g_{296}) + (g_{113} - g_{296})$$

$$\sin \left[\frac{(N - 21)}{396} 360 \right]$$

Tabulate the results and compare these with the more accurate day lengths obtained in Exercise 48.

50. For Kew, London ($51.47°$ N, $0.32°$ W), a 10-year average of the monthly means of the daily receipt of photosynthetically active radiation (using the conversion factor 110 kilolux – hours = 400 $\times$ 3600 = 1.44 MJ m^{-2} PAR and for the period 1966-77 (omitting 1973 for which record is missing) is

	Jan	Feb	Mar	Apr	May	Jun	Jul	Aug	Sep	Oct	Nov	Dec
10 years Avg.	1.05	1.98	3.78	5.65	7.71	9.09	8.44	7.02	5.11	2.93	1.48	0.87
SD	0.13	0.36	0.64	0.85	0.89	1.31	1.10	0.89	0.61	0.45	0.27	0.14
SD/Avg.	0.12	0.18	0.17	0.15	0.12	0.14	0.13	0.13	0.12	0.15	0.18	0.16

The unit for the mean and standard deviation (S.D.) are MJ m^{-2} of PAR. Note that the coefficient of variation (SD/average) is fairly constant throughout the year at 0.15, with a slight suggestion that it might be lower in the summer month than in the winter. This supports the assumption in equations

$$\ln J_N - \ln \bar{J}_N = \epsilon_N \text{ and}$$

$$P(\epsilon)=\frac{1}{\sigma\sqrt[]{2\pi}}\exp\left(-\epsilon^2/2\sigma_1^2\right)$$

that in (daily light receipt) is normally distributed.

Assuming a relationship type of equation

$$J_N =a+b\sin\left[\left(\frac{N-21}{365}\right)360\right]$$

where N = the climatological day number (N = 1 on 1 March), estimate approximately the constant a and b from the average monthly means above, and plot a graph showing the equation and the average monthly means. This method only makes use of J_{113} and J_{296}.

Remarks: ϵ_N = a random variable with zero mean and standard deviation σ_1 obeying the normal disribution, $P(\epsilon)$ = the probability density function for ϵ.

51.   At Kew, London (51.47°N, 0.32°W), for the 21 days beginning on 11 June 1976 (N = 103), the daily illumination (in kilolux-hours) took the following values:

711, 513, 947, 442, 807, 184, 389, 837, 303, 513, 496, 938, 957, 992, 944, 842, 794, 716, 955, 1038, 1004. Assume that at this time of year J_N is constant. So, equation

$$\bar{J}_N = (J_{113} + J_{296}) + \dot{J}_{113} + J_{296}) \sin\left[\left(\frac{n-21}{365}\right)360\right]$$

does not vary much about N = 113 (21 June), Assuming equation

$$\ln J_N - \ln \bar{J}_N = t_N$$

$$P(\epsilon)=\frac{1}{\sigma_1 2\pi}\exp\left(-\epsilon^2/2\sigma_1^2\right)$$

$$\ln J_N - \ln \dot{J}_N = t_N + \rho_1 \epsilon_{N-1}$$

estimate the two parameters σ_1 and ρ_1 which are used to characterize the daily radiation receipt.

Remarks: It is assumed that $\ln J_N$ is normally distributed. ϵ_N = random variable with zero mean and standard deviation σ_1 obeying the normal distribution, $P(\epsilon)$ = the probability density function for ϵ. ρ_1 is an autocorelation coefficient, $\bar{J}_N$ = PAR receipt over the year.

52.   At Little Hampton, Sussex (50.82°N, 0.52°W), 10-year averages of the monthly mean of the air temperature [1/2 (maximum +

minimum), T_a, its standard deviation, SD, the soil temperature at a depth of 100cm, T_s and its standard deviation, SD, are as follows:

	Jan	Feb	Mar	Apr	May	Jun	Jul	Aug	Sep	Oct	Nov	Dec
T_a	5.6	5.0	6.3	8.2	11.6	14.7	16.6	17.0	14.5	11.9	7.7	5.6
SD	1.1	1.3	1.1	0.8	0.9	1.5	1.1	1.0	0.9	1.7	0.8	1.7
T_s	7.4	7.0	7.1	8.8	11.4	14.1	16.3	17.0	16.1	14.3	11.3	8.6
SD	0.7	0.5	0.6	0.7	0.4	0.7	1.1	1.0	0.7	0.7	0.7	0.7

For both T_a and T_s, calculate the parameters a_y, b_y and N_0 of the first harmonic Fourier approximations of equation

$$T_N = a_y + b_y \sin\left[\left(\frac{N - N_0}{365}\right)360\right]$$

where a_y = mean over the year, b_y is the amplitude of the Sine wave modulation, and N_0 gives the phase of the Sine wave.

53. The total October rainfall for the 10 years (1968-1977) at Little Hampton and the number of days on which the rainfall was above 0.2mm are:

95	1	35	59	18	55	86	35	145	76	mm
15	2	12	5	6	8	13	6	19	14	days

Calculate the arithmetic means and standard deviations of both, using the binomial rainfall model, find the parameters q and n_p, where n_p is the length (number of days) of good or bad weather, and q is the probability of rain falling during the given period of n_p days. What is the probability of the month of October passing without rain?

Remarks: Binomial model: p = probability of no rain, q = probability of h mm rain, mean monthly rainfall, $\bar{r} = qh$, $p + q = 1$. The probability of ih mm of rain during the n-day period is (n_i) $p^{n-i} q^i$. The mean monthly rainfall over the month of n days is

$h \sum_{i=0}^{i=n} i\,(n_i)\,p^{n-i}q^i = nqh$ in agreement with $r = qh$. The standard

deviation may be obtained by $(SD)^2 = h^2 \sum_{i=0}^{i=n} (i - nq)^2 = (n_i)$

giving $\qquad SD = h\,(npq) = h\,[n_q\,(1 - q)]$

$$CV = \frac{SD}{\text{mean}} = \left(\frac{1-q}{nq}\right)$$

The mean number of rainy days, n_r is $n_r = nq$

$$q = \frac{1}{n_{eff} \, (CV)^2 + 1}$$

$$n_{eff} = \frac{1-q}{q} = /CV^2 \,.$$

Number of days in month = $n_{eff} \, n_p = n$

54.   Assuming a canopy extinction coefficient $k = -0.16$ (typical of grasses) in equation $I = I_o e^{-kl}$, what is the leaf area index of a canopy which only allows 10% of the incident light to reach the ground.

If the intercepted light is assumed to be uniformly distributed over the leaves in the canopy, what is the mean light level (incident on the leaves as a fraction of that above the ground)?

Remarks: k = a constant (extinction coefficient L = leaf area index measured downwards from the canopy, I_o = downward light flux density (Wm^{-2}) on a horizontal plane above the crop canopy, I = the downward flux density on a horizontal plane within the canopy at a depth L from the upper canopy surface.

55.   Monteith (1965) proposed an equation for light interception

$$I \, (L) = [s + (1 - s)m]^L \, I_0$$

where $I \, (L)$ is the downward light flux density under leaf area index L, I_0 is that above the canopy, s = the fraction of radiation that passes through a layer by unit leaf area index without interception by foliage, and m is the transmission coefficient of foliage. This binomial expression assumes that the unit layers of foliage act independently.

Derive by using $(1 - x)^n \longrightarrow \exp \, (- nx)$ as $x \longrightarrow 0$, $n \longrightarrow \infty$ and $nx = $ constant, the Monsi-Saeki result of equation

$$I = I_0 e^{-kL}$$

56.   Assuming that equation $P_1 = \frac{\alpha I_1 P_{max}}{\alpha I_1 + P_{max}}$

describes leaf photosynthetic response to light I_1 with $P_{max} = 1.2 \times 10^{-6}$ kg CO_2m^{-2} s^{-1} and $\alpha = 13 \times 10^{-9}$ kg CO_2 (JPAR)$^{-1}$, calculate the light level I_1 for 90% saturation of leaf response, and the light level where the linear approximations $P_1 = \alpha \, I_1$ is 10% in error.

57.   Write a computer program to calculate canopy photosynthetic

rate by equation $P_c = \dfrac{P_{max}}{k} \ln_0 \left[\dfrac{k\,\alpha\,I_0 + P_{max}\,(1-m)}{k\,I_0\,\alpha e^{-kl} + P_{max}\,(1-m)} \right]$

for $P_{max} = 1.2 \times 10^{-6}$ kg $CO_2 m^{-2}$ s^{-1}

$\alpha = 13 \times 10^{-9}$ kg CO_2 (JPAR)$^{-1}$,

$m = 0.1$, examining the values when $I_0 = 10,100, 1000$ J m^{-2} s^{-1}, $k = 0.4, 1.2$, and $L = 1.6$. In particular, consider which values of the canopy extinction coefficient k give high canopy photosynthetic rates for various light: leaf area index combinations.

Remarks: P_{max} = maximum gross photosynthetic rate, k = extinction coefficient, α = initial slope, I_0 = PAR flux density at the surface of the canopy, m = maintenance coefficient.

58.   By comparing of equation

$R = kp - cW$ and equation

$$R = \left(1 - y_g\right) \dfrac{\Delta s}{\Delta t} = y_g m \, W$$

From the white clover parameters, measured by McCree and given in $k = 0.25$ and $c = 0.0125$ day^{-1}, obtain numerical values for the conversion efficiency y_g and the maintenance coefficient m.

Remarks: R = respiration in kg CO_2 day^{-1} plant^{-1}, P = gross photosynthesis during the light period, W = plant dry weight in kg CO_2 plant^{-1}, $k = 0.25$ and $c = 0.0125$ day^{-1}, Δs = gross supply of substrate, y_g= growth conversion yield, m = maintenance coefficient.

59.   In summer in the U.K., a typical total daily radiation receipt is 18 MJ m^{-2}. Calculate (a) the amount of water evaporated per unit area per day using equation.

$$E_e = \dfrac{sA}{\left[s + \gamma\left(1 + \dfrac{g_a}{g_c}\right) \right]}$$

and making the approximation $s >> \gamma$; (b) the quantity of dry matter produced per unit area per day assuming a water use efficiency of 0.0015 kg dry matter (kg water)$^{-1}$.

Remarks: E_e = energy driven evaporation rate, s = rate of change of saturated vapour pressure with temperature at air temperature T_a [PaK^{-1}], A = available energy = net radiation – soil heat flux ($JM^{-2} S^{-1}$), λ = latent heat of evaporation of water (J kg^{-1}), γ = psychometric constant [Pa K^{-1}], g = boundary layer conductance [m s^{-1}], g_e = canopy conductance [m s^{-1}].

60.   Using equation

$$\gamma = \frac{C_p P}{\lambda \epsilon}$$

and C_p (20°C, atmospheric pressure) = 1006 J (kg air)$^{-1}$ $^{0}c^{-1}$, P = 1013 × 10^5 Pa, λ = (20°C) = 2.453 × 10^6 J (kg water)$^{-1}$ and ϵ = 0.622, calculate the psychometric constant r.

Using equation

$$E_e = \frac{sA}{\lambda\left[s + \gamma\left(1 + g_c / g_o\right)\right]}$$

and s (20°C) = 145 PaK^{-1}, evaluate the correction factor to be applied to the estimate of Exercise 59 assuming g_a/g_c = 0.

61.   Assuming that the response of crop yield Y to nitrogen level N and phosphate level P is given by the quadratic equation
$Y = a_0 + a_1N + a_2N^2 + b_0 + b_1P + b_2P^2 + hNP$, where a_0, a_1, a_2, b_1, b_2 and h are constants, apply equations

$$\frac{\partial y}{\partial x_1} = 0, \frac{\partial y}{\partial x_2} = 0, ... \frac{\partial y}{\partial x_n} = 0$$

$$\frac{\partial Y}{\partial x_i} = \frac{c_i}{p} \ (i = 1, ...n)$$

to obtain equations for N and P in terms of the constants, giving the maximum crop yield and the economic crop yield.

Remarks: Y = crop yield (kg m^{-2}), x_i = level of fertilizer i (kg m^{-2}), c_i = (£kg^{-1}) the unit cost of fertilizer x_i, including the cost of application, p (£ kg^{-1}) be price obtained for the product.

62.   Derive equation
l_{dev} (minimum) = $p_a + p_c + 2p$ by differentiating the equation

$$l_{dev} = l_a + t_{c'}$$

with respect to l_c (the length of the cold temperature treatment) and assuming that equation

$$(t_a - t_c)(t_c - p_a) = p^2$$

relating t_c and t_a (the time to anthesis at normal growing temperature) is valid.

Remarks:

T_{dev} = total development time, p_a and p_c = the asymptotes of hyperbolic relationship, p = parameter determines how closely t_a: t_c graph is to the two asymptotes — for example, the equation

t_{dev} (minimum) = $p_a + p_c + 2p$ is satisfied at $t_a = p + p_a$, $t_c = p + p_c$.

63.   Derive equations

T (inflexion) = $E_a/(2R)$ and

$$k = \frac{4AR}{E_a e^2} \left[T - E_a /(4R) \right]$$

from equation $k = A \exp\left[- E_a / (RT)\right]$

Remarks:

T (inflexion) = inflexion point, E_a = the activation of energy of the reaction and R = gas constant, k = rate constant of a chemical reaction, A = constant and T = the absolute temperature

64.   Write a computer program to examine the dynamic behaviour of Voltera's equations

$$\frac{dy}{dt} = \mu y \left(1 - y/y_m\right) - kxy$$

$$\frac{dx}{dt} = k' xy - mx$$

Remarks:

μ [day^{-1}], y_m [prey], k [predator^{-1} day^{-1}]' , k' = [prey^{-1} day^{-1}], m [day^{-1}], x = the number of predators and y = the number of prey.

65.   Modify your program of the Exercise 64 to take account of the alterations in Voltera's equations

$$\frac{dy}{dt} = \mu y \left(1 - y/y_m\right) - kxy$$

$$\frac{dx}{dt} = k' xy - mx$$

outlined in equations

$$\frac{dx}{dt} = k'xy + \mu_x x \left(1 - x/m\right),$$

$$\frac{dx}{dt} = k'x \left(\frac{y}{1+ay} \right) - mx,$$

$$\frac{dy}{dt} = \mu y \, (1 - y/y_m) - kx \, (y - y_0) \, H \, (y - y_0)$$

and
$$\frac{dx}{dt} = k'x \, (y - y_0) \, H \, (y - y_0) - mx$$

Check that your results are biologically plausible.

Remarks:

y = number of prey, x = the number of predators, μ [day^{-1}], y_m [prey], k [predator^{-1} day^{-1}], k' [prey^{-1} day^{-1}], m [day^{-1}], a = constant which causes the term to approach a maximum value of $k' \, x/a$ at high prey number, y_0 = the maximum number of prey that can use the cover, H = the unit step function which prevents the predator - prey interaction term from going negative, defined by $H \, (y - y_0) = 1$, for $y - y \geq 0$ and $H \, (y - y_0) = 0$ for $y - y_0 < 0$.

66.   Write a computer program to examine the behaviour of equation

$$\frac{dy}{dt} \, (t) = - ay \, (t - \tau) - hy \, (t),$$

which describes a system with a simple time delay. Draw a figure between population deviation y (y-axis) and time t (x axis). The equation is solved for $\tau = 10$, $h = 0$, y ($t \leq 0$) = 1 and the value of a the either side of $\pi/2\tau = 0.157$. Integration is by Euler's method with $\Delta t = 1$.

Remarks: y = the insect population, τ = incubation period, h = death rate, $a = \eta m$, η = efficiency, m = maintenance constant.

67.   Use your program for the Exercise 66 for equation

$$\frac{dy}{dt} \, (t) = - ay \, (t - \tau) - hy \, (t)$$

to examine the epidemic model of equation

$$\frac{dy}{dt} = - \mu y \, (t) + cy \, (t - \tau)$$

Remarks: $\frac{dy}{dt}$ derivative of the insect population with derivative of the time, τ = incubation period , y = infected population, μ = probability per unit time of a susceptible organism become infected, and c = natural recovery or cure rate.

68. Satisfy yourself, graphically or otherwise that equations

$$\frac{\partial n}{\partial t} + \frac{\partial n}{\partial a} = 0 \text{ and in the presence of death rate}$$

$$\frac{\partial n}{\partial t} + \frac{\partial n}{\partial a} = - D\,(t_1 a)n\,(t_1 a) \text{ are valid}$$

Remarks: $\frac{\partial n}{\partial t}$ = partial derivative of number of insects with partial

derivative of time , $\frac{\partial n}{\partial a}$ = partial derivation of insect number with

partial derivative of age of the insect, D = death rate.

69. A company produces four types of compound fertilizers by using nitrates, phosphates, potash and inert ingredients. Nitrates cost the company £225 t^{-1}, phosphates 220, potash 175 and inert ingredients 25. The amount of nitrates and phosphates that can be obtained are limited to 3000 and 3250 t month^{-1}, respectively. The other ingredients can be bought to unlimited amounts. The composition and price of the compounds produced are shown below:

Compound	Composition [g ingredient (100g compound)$^{-1}$]			Price £t^{-1}
	Nitrates	Phosphate	Potash	
1:1:1	15	15	15	150
High N	20	10	10	142
Low N	10	20	20	157
No N	0	20	20	132

It costs £35 t^{-1} to mix, bag and distribute each compound and the company can sell whatever it produces, but total production cannot exceed 20000t month^{-1}. Furthermore, the firm has contracts to supply 5000 tonnes of "high N", 3000 tonnes of "low N" and 2500 tonnes of "No N" next month. What amount of each compound should the company produce in the coming month so as to maximize the profit?

70. An arable farmer has 200 ha of high grade and 100 ha of low grade land available for the coming season and £ 150000 to spend on crops. He also has 5000 man-hour of free labour available (his own and his family's), additional labour can be bought at a cost

of £250 man-hours. The crops under consideration are spring wheat, spring barley, main crop potato, sugar beet and oilseed rape. The available information on each crop is as follows:

Crop	Expected yield (t ha⁻¹)		Labour requirement (man-hour ha⁻¹)	Other cost (e.g. seed, fertilizer, spray [£ha⁻¹]	Expected price £t⁻¹
	High grade land	Low grade land			
Spring wheat	4.6	3.4	12.4	135	112
Spring barley	4.8	3.5	12.4	117	107.5
Main crop potato	40	20	72.5	725	69
Sugarbeet	45	28	67.3	365	25
Oil seed rapes	2.5	1.5	14.0	185	22.5

Determine the crop plan that maximizes his profits.

71. In a hypothetical grazing system, pasture growth [kg herbage ha⁻¹ day⁻¹] is described by a Gompertz function, with initial specific growth rate parameter g_o and specific growth rate decay parameter g_D, and animal consumption [kg herbage animal⁻¹ day⁻¹] by a rectangular hyperbola, with maximum consumption rate parameter C_M and Michaelis–Menten constant C_k. Determine the first order differential equation describing the dynamics of this system and show that the stocking density S [animal ha⁻¹] necessary to maintain herbage biomass at a constant level H^* [kg herbage ha⁻¹] is given by $S(t) = g_0 (H^* + C_k) \exp(-g_D t)/C_M$.

72. Derive the bimolecular growth equation, by replacing the equation

$$\frac{dw}{dt} = kS$$

by $\sqrt{\dfrac{dw}{dt}} = kS^2 / W_f$, and integrating.

Sketch it when $W_0 = 1$, $W_f = 100$, and $k = 0.2$ for $0 \le t \le 30$, and briefly discuss its properties.

Remarks: $\dfrac{dw}{dt}$ = derivative of weight with the derivative of time,

k = a constant,

S = substrate level and W_f = final weight.

73. Derive a growth equation by simultaneous monomolecular and bimolecular growth rate, from the growth rate equation.

$$\frac{dw}{dt} = k_1 S + k_2 S^2 / W_f$$

Show that the resulting equation reduces to equation

$$W = W_f - (W_f - W_0) \, e^{-kt}$$

if $k_2 = 0$, and to solution of Exercise 72 if $k_1 = 0$. Find the initial growth rate.

Remarks: W_0 = initial weight, k_1 and k_2 are constants.

74. Derive an equation whose solution will give the inflexion point

of the Chanter growth function $\dfrac{dw}{dt} = \mu w \left(1 - \dfrac{W}{B} \right) e^{-Dt}$

Remarks: μ, B and D are constants, B = constant related to substrate level, D = constant related to senescence level, μ = a parameter known as the specific or relative growth rate.

75. The logistic equation is based on a linear response to substrate

$$\frac{dw}{dt} = k_1 WS,$$

whereas the diminishing-return type of response of a Michaelis equation is more commonly encountered in biology. Derive and discuss such a modified logistic.

Hint: Divide the right side of above equation by $(k + S)$

Remarks: k' = a constant,

$k' = \mu / W_f$, μ = a parameter known as the specific or relative growth rate,

$$k' = \frac{\mu}{W_f} = \frac{\text{relative grown rate}}{\text{Final weight}}$$

$$k = \text{Michaelis–Menten constant,}$$
$$S = \text{Substrate level}$$

76. The logistic equation

$$\frac{dw}{dt} = k'WS$$

describes growth that can continue indefinitely as long as substrate is available. For animals and many determinate plants and particularly with organs, growth ceased before the substrate is exhausted, due to differentiation, which may be expressed in the Gompertz equation as

$$\frac{dw}{dt} = \mu / We^{-Dt}$$

Combine the letter effect with that of Exercise 75 to give a new growth function.

Hints: Multiply the right side of logistic equation by

$$\exp [(-Dt)/k + S]$$

Note that this may be viewed as a Michaelis–Menten modification of the Chanter growth equation

$$\frac{dw}{dt} = \mu w \left(1 - \frac{W}{B} \right) e^{-Dt}$$

Remark: μ_0= the value of μ at time $t = 0$, D = parameter describing the decay in the specific growth rate. Specific growth rate or relative growth rate = $1/w$ (dw/dt). B = constant related to substrate.

77. In a paper on oxygen transport in tissue, the author developed a formula for the penetration of oxygen into metabolizing cells. Among the assumptions used in the development of the formula were the following:

(1) The cells are all spherical and of the same diameter.

(2) Oxygen concentration on the outside of each cell is uniform over the entire surface of the cell. It is assumed to be the same for all cells and is held as constant. In order to test the resulting equation, would you prefer to use data on the respiratory activity of:

(a) Whole animals? (b) Suspension of single cells? (c) Tissues slices? Justify your answer.

Most of aforementioned exercises have been selected from France and Thornley (1984), Gold (1977) and Keulen and Wolf (1986). (See references in Chapter 6). The book by Goudriaan and van Laar (1994) is the ideal textbook with exercises and their solutions.

Further reading:

Goudriaan, J, and H.H.van Laar (1994). Modeling potential crop growth processes. Textbook with exercises. Kluwer Academic Publishers, London pp.238.

APPENDIX B

Discussion and solutions of exercises

6.   Sampling without replacement gives

$\binom{100}{10}$ possible different combinations of 10 from 100 flies. If each

is equally likely, the probability of getting any given 10 is $1/\binom{100}{10}$

$$= \frac{90!\ 10!}{100!} = 5.89 \times 10^{-14}$$

9.   A position would have to be assigned that stores the constant 1 .
Then we might have

Fetch x

Add 1

Store x

10.   Suppose $x = 4$ and we choose

$Q = 1$, then

$$R_1 = \frac{4}{1} = 4$$

$$Q_2 = \frac{4+1}{2} = 2.5$$

$$R_2 = \frac{4}{2.5} = 1.60$$

$$Q_3 = \frac{2.5 + 1.60}{2} = 2.05$$

$$R_3 = \frac{4}{2.05} = 1.95$$

$$Q_4 = 2$$
$$R_4 = 2$$

12.   The possible sequence of calculations for equation

$$\frac{dx}{dt} = kx$$

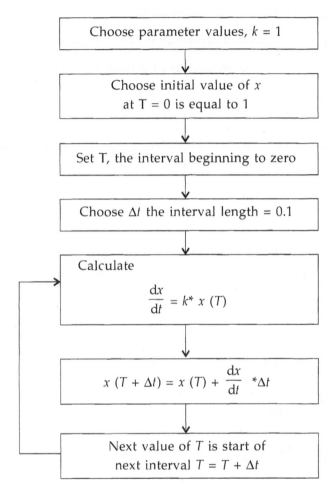

Choose parameter values, $k = 1$

Choose initial value of x
at T = 0 is equal to 1

Set T, the interval beginning to zero

Choose Δt the interval length = 0.1

Calculate

$$\frac{dx}{dt} = k^* \, x \, (T)$$

$$x \, (T + \Delta t) = x \, (T) + \frac{dx}{dt} \,^* \Delta t$$

Next value of T is start of
next interval $T = T + \Delta t$

Computer program in BASIC

10   Rem calculation for equation $\frac{dx}{dt} = k^*x$

20    $k = 1$
30    $x \, (0) = 1$: T (0) = 0 : DT = 0.1
40   Input N
50   For I = 0 to N step 0.1

52      $J = 1 + 0.1$

55      $T (J) = T (J - 1) + 0.1$: Print T (J), "RAM"

56   DN (J) = k * x (J - 1) * dt

58      $X(T) = x (J) + DN (J)$

60

70   Print "x =" x (J); "dn =" DN (J) : Next I

80   End

Output? 0.2, 0.1 RAM, x = 1.1 dn = .1, 0.2 RAM, x=1.21 dn = .11, 0.3 RAM, x=1.331 dn = 0.121

Computer Program in C++

```
// My first.cpp for calculation dx/dt = k*x
# include <iostream.h>
int main (Void)
{ const int k=1;
 float dx ;
 float dt = 0.1;
 float pop = 1.0
 float a
 cout << "Enter max value of Time" ;
 cin >> a;
 For (float T =0; T < a; T = T + dt)
 { dx = k*pop*dt;
 pop = pop + dx;
 cout << "At time =" <<T "dx is ="
 <<dx <<"population is = "
 <<pop << endl; }
 return 0;
```

Output :? 0.5, T=0 dx = 0.1 pop = 1.1 ....T = .4 dx = .1464 pop = 1.6105

13.   Each term must have the same units as dw/dt, i.e. kg day^{-1}, k, therefore, has units of kg day^{-2}. The argument of an exponential is unit free, and thus b has units of day^{-1}, n must be a unit-free number, and a has units of kg day^{-1-n}. Note that the argument of any function such as e^a or sin (x) can be expanded as a series (e.g. $1 + x +1/2 x^2 +.......$) must be unit free.

14.   Hours = integer part (60 * 24 t_d + 60 * hours) and
        seconds = integer part (60 * 60 * 24 t_d - 60 * 60 hours - 60 * minutes)

For conversion in the opposite direction

$t_d = 1/24$ (hours + minutes/60 + seconds/3600)

Remark t_d = decimal part of the day.

15.   First day of | *Jan* | *Feb* | *Mar* | *Apr* | *May* | *June*

First day of	*Jan*	*Feb*	*Mar*	*Apr*	*May*	*June*
N	307	338	1	32	62	93

First day of	*Jul*	*Aug*	*Sep*	*Oct*	*Nov*	*Dec*
N	123	154	185	215	246	276

16.   Define variables

T_5 = five-day average temperature [^{0}C]

S_{10} = rainfall totalled over the past 10 days [mm]

T_{min} = minimum daily temperature [^{0}C]

n = number of blight favourable days

$T_{1\ to\ 5}$ = array for five daily temperature means [^{0}C]

$h_{1\ to\ 10}$ = array for 10 daily rainfall values [mm]

Initialize all variables to zero

Input the last day's value of $\overline{T}$, T_{min} and h into

$\overline{T}_5$, T_{min}, h_{10}

Compute $\overline{T}_5$ and S_{10} by means of

(" : = " means " is assigned the value of ")

$$\overline{T}_5 = \left(\sum_{i=1}^{5} T_i \right) \Big/ 5, \quad S_{10} = \left(\sum_{i=1}^{10} h_i \right)$$

Test and accumulate blight favourable days

if $\overline{T}_5 < 25.5$ and $S_{10} \geq 30$ and $T_{min} \geq 7.2$ then $n : = n + 1$

if $n = 10$ then print "blight warning"

Move the arrays along one position

for $i = 1$ to 4, $T_i : = T_{i+1}$

for $i = 1$ to 9, $h_i : = h_{i+1}$

Go to "Input the last day's value ....".

17.   Possible sequence of calculations for equation:

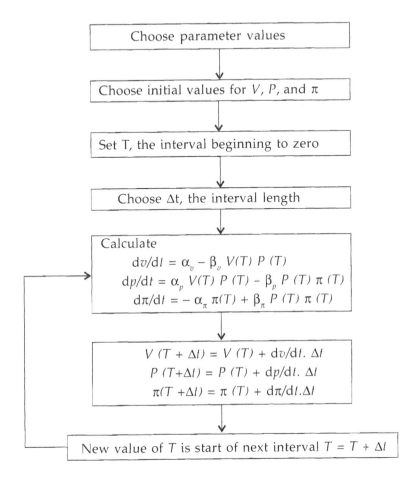

22.   (a) 60 miles/hr = 60 miles/hr $\dfrac{5280 \ ft/mile}{3600 \ sec \ /hr}$

= 60 x 1.4667

= 88 ft/sec

(b) 60 miles per hour = 60 miles/hr $\dfrac{1.6 \ km/mile}{hr}$

= 60 x 1.6

= 96 km/hr

(c) 15 lb/m² = (15 lb/m²) $\left[\dfrac{0.45 \ kg/lb}{(2.54 \ cm/in)^{2}}\right]$

$$= 15 \text{ lb/in}^2 \ \frac{0.45 \text{ kg/lb}}{0.45 \text{ cm}^2/\text{in.}^2}$$

$$= 15 \ (0.0698)$$

$$= 1.05 \text{ kg/cm}^2$$

(d) $15 \text{ lb/in.}^2 = \dfrac{4.4 \times 105 \text{ dyn/lb}}{6.45 \text{ cm}^2/\text{in.}^2}$

$$= 15 \times 0.6822 \times 10^5$$

$$= 1.02 \times 10^6 \text{ dyn /cm}^2$$

(e) $32 \text{ft/sec}^2 = 32 \text{ ft/sec}^2. \ \dfrac{30.48 \text{ cm./ft}}{[(1/60) \ (\text{min/sec})]^{\,2}}$

$$= 32 \times 1.097 \times 10^5$$

$$= 3.51 \times 10^6 \text{ cm/min}^2$$

23.   The "population" being sampled consists of four types. It is sampling with replacement; neither 4 nor 4^2 combinations is enough to code for 20 amino acids. The minimum would be $4^3 = 64$.

24.   If every rearrangment of the N molecules gave a different protein, we would have $N!$ different possibilities. Since for the Jth type of amino acid, the n_1 molecules may be permuted among themselves n_1 ! ways without changing the type of protein, we get $N!/n_1!n_2!........n_{20}!$ different protein types. Another way of getting the same answer is to ask the number of ways of dividing the "population" of N positions into 20 sub-populations; so that group1 has n_1 elements, group2 has n_2 elements, and so on, and then applying equation

$$= \frac{N!}{n_1!n_2!n_3!........n_{r-1}!n_r!}$$

25.   Since the F_1 all had round yellow seeds, then R must be dominant and Y must be dominant. For F_2, we get

  P (round) = 3/4, P (wrinkled) = 1/4
  P (yellow) = 3/4, P (green) = 1/4

Assuring shape and colour to be independent, P (round, yellow) = P (round) P (yellow) = (3/4) (3/4) = (9/16)
Similarly,

  P (round, green) = (3/4) (1/4) = 3/16
  P (wrinkled, yellow) = (1/4) (3/4) = 3/16
  P (wrinkled, green) = (1/4) (1/4) = 1/16

26. Since the probability of a mutation at a given site is 10^{-8}, the probability of no mutation of a given site is $(1 - 10^{-8})$. Since each site is independent, the probability of no mutation at any site $(1 - 10^{-8})$ raised to the power 10^7. This may be estimated using logs,

$$P \text{ (no mutation)} = (1 - 10^{-8})^{10^7}$$
$$\ln_e P \text{ (no mutation)} = 10^7 \ln_e (1 - 10^{-8})$$

using hint *b*, this becomes approximately

$$\ln P \text{ (no mutation)} = - (10^7) \, 10^{-8}$$
$$= - 0.1$$

Taking the antilog gives

$$P \text{ (no mutation)} = 0.9048$$

Since there must either be no mutation or at least one,

$$P \text{ (at least one mutation)} =$$
$$= 1 - 0.9048$$
$$= 0.0952$$

27. In each case, it is necessary to find a time scale that is long relative to the rate at which elementary events take place. These might be

a   minutes for many chemical reactions
b   hours, perhaps, it would depend in the organism and the conditions
c   months, perhaps, depending on condition
d   change in size of a population of elk.
e   billions of years
f   days if compounded daily, years if compounded yearly.

30. *J* was defined as an amount per unit time per unit area, so it would have dimensions.

$$\text{(amount) } T^{-1} \, L^{-2}$$

Since *c* is amount per unit volume (amount) L^{-3}, we get, for equation $J (x) = -D \cdot dc/dx$

$$\text{(amount) } T^{-1} \, L^{-2} = -D \times \frac{\text{(amount) } L^{-3}}{L}$$

D must, therefore, have dimensions of T^{-1} that is per unit time.

33. *a* $\varepsilon_z = (-1/y^2) \, (\varepsilon_y)$
  b $\varepsilon_z = x\varepsilon_y$
  c $\varepsilon_z = 2y \, \varepsilon_y$
  d $\varepsilon_z = (2y + 1) \, x\varepsilon_y$
  e $\Sigma_z = (-1/y) \, (\varepsilon_y)$

35.   For this exercise we need

$$\sum_i x_i = 50.9300$$

$$\sum_i y_i = 38.7200$$

$$\bar{x} = 5.093$$
$$\bar{y} = 3.872$$

$$\sum_i x_i y_i = 250.6400$$

$$\sum_i (x_i - \bar{x})^2 = 78.112$$

$$\sum_i (y_i - \bar{y})^2 = 42.0016$$

$$\sum_i (x_i - \bar{x})(y_i - \bar{y}) = 53.4400$$

These give
$$a = 0.3876$$
$$b = 0.6841$$
$$r = 0.9330$$

36.   Taking $\varepsilon_i = 0.1 \, y$, we get

$$\sum_{i=1} \frac{1}{\varepsilon_i^2} = 208.9789$$

Thus $q = 10 / \Sigma^{-2}\varepsilon_i = 0.0479$

Letting $W_i = q / \varepsilon_i^2$ gives the following set of weights

4.79, 1.87, 1.48, 0.76, 0.50, 0.18, 0.20, 0.10, 0.21, 0.10

Applying equation

$$b^*_w = \frac{\sum_i w_i x_i y_i - \bar{y}_w \sum_i w_i x_i}{\sum_i w_i x_i^2 - \bar{x}_w \sum_i w_i x_i}$$

$$a^* = \bar{y}_w - b^* \bar{x}_w$$

gives
$$a = 0.310$$
$$b = 0.662$$

37.   The lines obtained from Exercises 34, 35 and 36 should all be plotted and compared. Note that Exercise 36 gives more importance to accurately fitting the smaller values.

38. Differentiating $y = ax/(b + x)$ gives

$$\frac{dy}{dx} = \frac{a(b+x)-ax}{(b+x)^2} = \frac{ab}{(b+x)^2}$$

at $x = 0$, $dy/dx = a/b$, and as $x \longrightarrow \infty$

$Y \longrightarrow ax/x = a$. A second differentiation of equation

$$\frac{dy}{dx} = \frac{ab}{(a+b)^3} \text{ gives}$$

$$\frac{d^2y}{dx^2} = -\frac{2ab}{(b+x)^3}$$

The magnitude of this is maximum for $x \geq 0$ is when $x = 0$ with the value $- 2a /b^2$. Note that the least linear part of the rectangular hyperbola is in the region of the origin.

39. Differentiating

$$W = \frac{at}{b+t}$$

gives

$$\frac{dw}{dt} = \frac{ab}{(b+t)^2}$$

from equation $W = \dfrac{at}{b+t}$, therefore

$$t = \frac{bw}{a-w}$$

substituting $t = \dfrac{bW}{a-W}$ into $\dfrac{dW}{dt} = \dfrac{ab}{(a+t)^2}$

gives

$$\frac{dW}{dt} = \frac{(a-W)^2}{ab}$$

40. This problem can be worked either way; by differentiating the growth equation twice and eliminating the time variable t; or, more easily, by integrating the two differential equations with the given boundary conditions.

The exponential quadratic growth equation can be written as $\ln x = a_0 + a_1 t + a_2 t^2$

Differentiation gives

$$\frac{1}{x}\frac{dx}{dt} = a_1 + 2\,a_2 t \text{ with } \ln x = a_0 \text{ at } t = 0$$

We introduce a second state variable y by means of

$$y = a_2 t$$

and differentiation gives

$$\frac{dy}{dt} = a_2 \text{ with } y = 0 \text{ at } t = 0$$

The two differential equations specifying the problem are thus

$$\frac{1}{x}\frac{dx}{dt} = a_1 + 2y \text{ and } \frac{dy}{dt} = a_2$$

with at $t = 0$, $\ln x = a_0$ and $y = 0$

Integration of these two equations to recover equation $\ln x = a_0 + a_1 t + a_2 t^2$ is straightforward.

A biological view of equations

$$\frac{1}{x}\frac{dx}{dt} = a_1 + 2y \text{ and } \frac{dy}{dt} = a_2 \text{ is that at time } t = 0,$$

the specific growth rate $(1/x)\,(dx/dt)$ of the organism has the value a_1, and as time progresses, some factor (parameterized in a_2) causes the specific growth rate to change (usually decreasing) linearly with time. An exponential quadratic describes the cell numbers in meristematic tissue where the progeny of cell division have an exponentially diminishing probability of remaining meristematic. This function is not suitable for organ or organisms which show asymptote dry weight: time behaviour, with a non-zero asymptote. More commonly, a physiologist would use only a part of the response, stopping short somewhere in the region of the maximum.

41.   Euler's formula, given in equation $x\,(t + \Delta t) = x(t) + \Delta t f\,\{x(t),\ t\}$, may be written as

$$y\,(t + \Delta t) = y\,(t) + \Delta t\,\frac{dy}{dt}$$

Taking for example, $t = 0.1$, a table can be constructed, line by line, working from left to right:

t	y	dy/dt	$\Delta t\, dy/dt$	$\exp(-t)$
0	1.000	−1.000	−0.1000	1.000
0.1	0.900	−0.900	−0.0900	0.905
0.2	0.810	−0.810	−0.0810	0.819
0.3	0.729	−0.729	−0.0729	0.741
etc.				

Integration of $dy/dt = -y$ with $y = 1$ at $t = 0$ gives $y = e^{-t}$ which is given in the fifth column of the table.

42.   For $dy/dt = f(y) = -y$, we define

$$f_1 = f(y) = -y$$
$$y_1 = y + \Delta t\, (dy/dt)\,(y)$$
$$f_2 = (dy/dt)\,(y_1) = -y_1$$
$$g = 1/2\,(f_1 + f_2) \text{ and}$$
$$\Delta y = g\, \Delta t$$

Updating is achieved by

$$t \longrightarrow t + \Delta t \text{ and } y \longrightarrow y + \Delta y$$

As before, we construct a table, working line by line from left to right (with $\Delta t = 0.1$):

t	y	f_1	y_1	f_2	g	Δy
0	1.000	−1.000	+0.900	−0.900	−0.950	−0.095
0.1	0.905	−1.905	0.815	−0.815	−0.860	−0.086
0.2	0.819	etc.				

Comparing the second column with the exact solution in the last column of the solution to Exercise 41, it is seen that the accuracy is much improved.

43.   The degree of freedom $= 84 - 4 = 80$

The mean residual sum of square is $R = R_1 + R_e$

where $r_1 = $ the part of the residual due to the lack of fit on the model and $R_e = $ due to error in the experimental data. R_e has an expected value of

$$R_e = (m - n)\, \sigma^2$$

where $m = $ number of data point,

$n = $ number of parameters in P,

and $\sigma^2 = $ the error variance

$$R = R_1 + (m - n)\sigma^2$$
$$\sigma_t^2 = R/(n - m) = 0.8/80 = 0.01$$

Thus $\ln(y/Y) = (0.01)^{1/2} = \pm\, 0.1,$

and $y/Y = 0.9$ or 1.1. On an average, there is a 10% difference between predicted and experimental values.

The error variance is

$$\sigma^2 = 0.4/50 = 0.008$$

Hence

$$F = \sigma_t^2/\sigma^2 \text{ with } N_1 = 80 \text{ and } N_2 = 50$$

The 10% probability level is given by the 95% point of the F distribution. The table F is about 1.5. Therefore, the model gives an acceptable fit to the data at the 10% level.

44. The optimal solution is

$$z = 32,\ x_1 = 10,\ x_2 = 11$$

45. The unit of f_N are kg nitrogen (kg total matter)$^{-1}$. These units cannot be simplified; it is not, for instance, permissible to cancel out kg.

The unit of LAI are:

m^2 leaf (m^2 ground)$^{-1}$ and again m^2 cannot be cancelled out.

46. If the price of a goods is p, and x goods are sold (per unit time), then demand elasticity E may be defined as

$$E = \frac{\delta x/x}{\delta p/p} = \frac{\delta(\ln x)}{\delta(\ln p)}$$

where the prefix δ indicates a small increment in the variable. This is equivalent to the model sensitivity index, which measures the sensitivity of the predicted quantity Y to the parameters P_i.

47. Using equation

$\sin a_0 = \cos\delta \sin h_0$

$\cos^2 a_0 = 1 - \cos^2\delta \sin^2 h_0$

which, using

$\sin^2 h_0 = 1 - \cos^2 h_0 = 1 - \tan^2\phi \tan^2\delta$

from equation

$\cos h_0 = -\tan\phi \tan \delta$, becomes

$\cos^2 a_0 = 1 - \cos^2\delta (1 - \tan^2\phi \tan^2\delta)$

$\qquad = 1 - \cos^2\delta + \cos^2\delta \tan^2\phi \tan^2\delta$

$\qquad = 1 - \cos^2\delta + \cos^2\delta \tan^2\phi \dfrac{\sin^2\delta}{\cos^2\delta}$

$\qquad = 1 - \cos^2\delta + \tan^2\phi \sin^2\delta$

$\qquad = \sin^2\delta + \tan^2\phi \sin^2\delta$

$$=\sin^2\delta\,(1 + \tan^2\phi)$$

$$= \frac{\sin^2\delta}{\cos^2\phi} \text{ since } 1 + \tan^2\phi = \frac{1}{\cos^2\phi}$$

Taking the negative square root, therefore $\cos a_0 = -\sin\delta/\cos\phi$

48.   Calculations were performed for a latitude of 50.75^0 (Glasshouse Crop Research Institute, Little Hampton, West Sussex, UK) with the following results.

Climato-logical day, N	Day of the month D	Month of the Year, M	Decli-mation (degree)	Zenith-angle at solar noon z (degree)	Day length (fractional) for different definition of Zenith angle				
					90°	90.83°	96°	102°	108°
1	1	3	-7.8	58.6	0.45	0.45	0.50	0.55	0.61
11	11	3	-4.0	54.7	0.47	0.48	0.53	0.58	0.63
21	21	3	-0.0	50.8	0.50	0.51	0.55	0.61	0.66
31	31	3	6.9	46.8	0.53	0.53	0.58	0.64	0.69
41	10	4	7.7	43.0	0.55	0.56	0.61	0.67	0.73
51	20	4	11.3	39.4	0.58	0.59	0.64	0.70	0.77
61	30	4	14.6	36.2	0.60	0.61	0.66	0.73	0.81
71	10	5	17.5	33.3	0.63	0.63	0.69	0.76	0.85
81	20	5	19.9	30.9	0.65	0.65	0.71	0.79	0.91
91	30	5	21.7	20.1	0.66	0.67	0.73	0.82	1.00
101	09	6	22.9	27.9	0.67	0.68	0.75	0.84	1.00
111	19	6	23.4	27.3	0.68	0.69	0.75	0.85	1.00
121	29	6	23.3	27.5	0.68	0.69	0.75	0.85	1.00
131	09	7	22.4	28.3	0.67	0.68	0.74	0.83	1.00
141	19	7	21.0	29.8	0.66	0.66	0.72	0.81	0.96
151	29	7	18.9	31.9	0.64	0.65	0.70	0.78	0.88
161	08	8	16.3	34.4	0.62	0.62	0.68	0.75	0.83
171	18	8	13.3	37.5	0.59	0.60	0.65	0.72	0.79
181	28	8	9.9	40.8	0.57	0.58	0.62	0.68	0.75
191	07	9	6.3	44.5	0.54	0.55	0.60	0.65	0.72
201	17	9	2.5	48.3	0.52	0.52	0.57	0.62	0.68
211	27	9	-1.4	52.2	0.49	0.50	0.54	0.60	0.65
221	7	10	-5.3	56.1	0.46	0.47	0.52	0.57	0.62
231	17	10	-9.1	59.8	0.44	0.44	0.49	0.54	0.60
241	27	10	-12.6	63.4	0.41	0.42	0.47	0.52	0.57
251	06	11	-15.9	66.6	0.39	0.40	0.44	0.50	0.55
261	10	11	-18.6	60.4	0.36	0.37	0.42	0.48	0.53
271	26	11	-20.9	71.6	0.35	0.35	0.41	0.46	0.52
281	06	12	-22.5	73.2	0.33	0.34	0.39	0.45	0.51
291	16	12	-23.3	74.1	0.32	0.33	0.39	0.45	0.50
301	26	12	-23.4	74.1	0.32	0.33	0.39	0.45	0.50
311	05	01	-22.7	73.4	0.33	0.34	0.39	0.45	0.51
321	15	01	-21.2	72.0	0.34	0.35	0.40	0.46	0.52
331	25	01	-19.1	69.9	0.36	0.37	0.42	0.48	0.53
341	04	02	-16.4	67.2	0.38	0.39	0.44	0.49	0.55
351	14	02	-13.3	64.0	0.41	0.41	0.46	0.52	0.57
361	24	02	-9.7	60.5	0.43	0.44	0.49	0.54	0.59

49.   Application of equations

$$y = \left[\frac{N - 21}{3.65}\right] 360°;$$

$\delta = 0.38092 - (0.76996 \cos y) + 23.26500 \sin y$
    $+ 0.36958 \cos 2y + 0.10868 \sin 2y + 0.01834 \cos 3y$
    $- 0.16650 \sin 3y - 0.00392 \cos 4y + 0.00072 \sin 4y$
    $- 0.00051 \cos 5y + 0.00250 \sin 5y + 0.00442 \cos 6y$

and

$$g_N = \frac{2}{360} \cos^{-1} \frac{\cos z}{\cos \phi \cos \delta} - \tan \phi \tan \delta$$

for a latitutde $\phi = 50.75°$, gives
$g_{113} = 0.7517$ day and $g_{296} = 0.3859$ day
for the longest and shortest days.
Assuming a Sine wave, then the expressions for g_N becomes

$$g_N = 0.5688 + 0.1829 \sin \left[\left(\frac{N - 21}{365}\right) 360\right]$$

$g_{113} = 0.5688 + 0.1829 (0.99)$
    $= 0.5688 + 0.181071 = 0.7498$ day.
$g_{296} = 0.5688 + 0.1829 (-0.99)$
    $= 0.5688 - 0.181071 = 0.3877$ day

The more accurate results are 0.7517 day and 0.3859 day

for $g_N = 113$ and $g_N = 296$, respectively.

50.   The 10-year average of the mean daily light receipts for the month
of December and June are 0.87 and 9.09 MJ m^{-2} day^{-1}, respectively.
In December and June, the radiation is at the bottom and top of
the Sine curve, and changing least quickly.

One day is approximately equivalent to one degree (in fact, 360/
365°), and as $\cos 15° = 0.97$, to assume that the monthly means
apply to the values on 21 December and 21 June may produce
an error of about 3%. Since the coefficient of variation is 15%, this
is acceptable. An estimate of the coefficients of the equation.

$$\bar{J}_N = a + b \sin \left[\left(\frac{N - 21}{365}\right) 360\right] \text{ is obtained by taking}$$

    $a = \frac{1}{2} (0.87 + 9.09)$ and $b = \frac{1}{2} (9.09 - 0.87)$

for $\bar{J}_N = 4.98 + 4.11 \; sin\left[\left(\dfrac{N-21}{365}\right)360\right]$

and is compared directly with the average monthly means, which are placed at the middle of each calender month.

Beter agreement could be obtained by adding extra terms to the Fourier series, replacing equation

$\bar{J}_N = a + b\left[\left(\dfrac{N-21}{365}\right)360\right]$ by

$\bar{J}_N = a_0 + a_1\left[3\left(\dfrac{N-21}{365}\right)360\right] + a_2 \; cos\left[2\left(\dfrac{N-21}{365}\right)360\right]$

$= a_3 \; cos\left[3\left(\dfrac{N-21}{365}\right)360\right] \; ..... \; + b_1 \; sin\left[\left(\dfrac{N-21}{365}\right)360\right]$

$+ \; b_2 \; sin\left[2\left(\dfrac{N-21}{365}\right)360\right] + b_3 \; sin\left[3\left(\dfrac{N-21}{365}\right)360\right] ....$

With only 12 months to fit, one should not include too many terms in the series. Using the methods for the first three terms (a_0, a_1, b_1) in above equation, and simplifying, one obtains the equation.

$\bar{J}_N = 4.59 + 4.04\left[sin\left(\dfrac{N-21}{365}\right)360\right]$ instead of equation

$\bar{J}_N = 4.98 + 4.11\left[sin\left(\dfrac{N-21}{365}\right)360\right]$

The coefficient may be estimated by

$$a_0 = \frac{1}{n} \sum_{i=0}^{n} y_i$$

and for $J \geq 1$. J is the subscripts of coefficients.

$$a_1 = \frac{2}{n} \sum_{i=1}^{n} y_i \; cos\left[\bar{j}\left(\dfrac{N-21}{365}\right)360\right]$$

$$b_1 = \frac{2}{n} \sum_{t=1}^{n} y_t \sin\left[\bar{j}\left(\frac{N-21}{365}\right)360\right]$$

The last two terms are only approximations, although the approximation is reasonable if n is not too small.

If n is even, and if terms in the Fourier series upto $j = n/2$ are included, then the $n + 1$ independent measurements are exactly method by $n + 1$ Fourier coefficients, and the Fourier represent-ation will go exactly through the measured values. It is usually only appropriate to use a Fourier series if the first few terms of the series dominants. For example, when describing, light receipt or temperature throughout a day or a year, one might find that the terms a_0, a_1, and b_1 give a sufficiently accurate description of the environmental variable for most purposes.

51.   First work out logarithms to the base e:

ln J_N for $N = 103, 104, ..., 123$

Then estimate the mean $\bar{J}$, and the standard deviation σ, by

$$\bar{J} = \frac{1}{2} \sum_{N=103}^{123} \ln J_N$$

and

$$\bar{J} = \frac{1}{2} \sum_{N=103}^{123} \ln (J_N - \bar{J})^2$$

Finally evaluate ρ_1 by means of

$$\rho_1 \sigma_1^2 = \frac{1}{20} \sum_{N=104}^{123} \ln (J_N / \bar{J})(J_{N-1}/\bar{J})$$

The answers are $\sigma_1 = 0.45$ and $\rho_1 = 0.24$

52.   Application of the method of Fourier analysis the coefficient

$$a_0 = \frac{1}{n} \sum_{t=1}^{n} y_t = 10.4 \ \text{for } T_a$$

$$b_1 = \frac{2}{n} \sum_{t=0}^{n} y_t \sin\left[1\left(\frac{N-60}{365}360\right)\right] = 6.3 \ \text{for } T_a$$

$$a_0 = \frac{1}{n} \sum_{t=1}^{n} y_t = 11.6 \ \text{for } T_a$$

$$b_1 = \frac{2}{n} \sum_{i=1}^{n} y_i \sin\left[1\left(\frac{N-76}{365}\ 360\right)\right] = 5.1 \text{ for } T.$$

So, $\quad T_a = 10.4 + 6.3 \sin\left[\left(\frac{N-60}{365}\ 360\right)\right]$

and $\quad T_s = 11.6 + 5.1 \sin\left[\left(\frac{N-76}{365}\ 360\right)\right]$

These are best worked out by writing a short computer program.

53. The mean monthly rainfall for October is 60.5mm with a standard deviation of 42.03mm, giving a coefficient of variation of 0.69. The mean number of rainy days in the 31-day month is 10, with a standard deviation of 5.37, giving a coeffcient of variation of 0.537.

From equation

$$n_i = nq$$

where $n = n$ – day period; q = probability of h mm rain,

$$q = 10/31 = 0.323$$

From equation

$$n_{eff} = \left(\frac{1-q}{q}\right) / (cv)^2 \text{ with c.v.} = 0.69,\ n_{eff} = 4.41 \text{ day,}$$

which with equation,

number of days in month = $n_{eff}\ n_p = n$ gives

$n_p = 31/4.41 = 7.03$ day (n_{eff} periods of n_p days in a month). The probability of the month of October passing without rain is $(q)^{n_{eff}}$ $= 0.323^{4.41} = 0.007$, or rather less than 1% according to the rainfall model.

54. Applying equation $I = I_0 e^{-kL}$ with $I/I_0 = 0.1$
and $k = 0.6$, therefore, $0.1 = \exp(-0.6L)$, giving $L = 3.84$
If the intercepted light (0.9 of the total) is spread over a leaf area index of 3.84, then the mean light level is $0.9/3.84 = 0.23$.

55. s = the fraction of light reaching the ground and let x = the fraction of light that is intercepted by unit foliage layer, so that

$$s = 1 - x$$

Monteith equation becomes

$$I\,(L) = [1 - x\,(1 - m)]^L\,I_0$$

Now assume that all the foliage elements are independent (rather than Monteith's implicit assumption that each group of foliage elements within leaf area index layer acts as an independent unit, but within each unit layer there may or may not be independent action by the foliage elements). Let the elemental layer ΔL intercept $y\,\Delta\,L$ of the radiation incident on it, and take $n\Delta L = L$. Equation $I\,(L) = [1 - x\,(1 - m)]^L\,I_0$ becomes (replacing x by $y\,\Delta L$, and L by n on the right hand side).

$$I\,(L) = [1 - y\Delta L\,(1 - m)]^n\,I_0$$

In the limit $n \longrightarrow \infty$, $\Delta L \longrightarrow 0$, $n\,\Delta L = L$, therefore

$$I\,(L) = e^{-y\,(1-m)L}\,I_0$$

This is identical to equation $I\,(L) = e^{-kl}$, with $k = y\,(1 - m)$

56.   With $P = fP_{max}$ where f is the fraction of the maximum rate, equation

$$P_1 = \frac{\alpha I_1\,P_{max}}{\alpha I_1 + P_{max}}$$

where α is a constant and P_{max} is the value at saturated light level ($I_1 \longrightarrow \infty$).

The equation becomes

$$f\,(P_{max}) = \frac{\alpha I_1\,P_{max}}{\alpha I_1 + P_{max}}$$

which is rewritten to give

$$I_1 = \frac{f\,P_{max}}{\alpha}$$

with $f = 0.9$ and the given values of P_{max} and α, therefore

$$I_1\,(f = 0.9) = 831\,Jm^{-2}\,s^{-1}\,PAR$$

This unrealistically high value is a measure of the inadequacies of equation of P_1 estimation as a leaf response equation.

Defining the photosynthetic rate estimated using the approximation as

$$P \text{ (app)} = \alpha I_1$$

The error $\in$ is given by

$$\in = \frac{P \text{ (app)} - P_I}{P_I}$$

which, with equation

$$P_I = \frac{\alpha I_I P_{max}}{\alpha I_I + P_{max}}$$

and

$$P \text{ (app)} = \alpha I_I$$

becomes $\in = \dfrac{\alpha I_I}{P_{max}}$

For a 10% error, therefore

$$I_1 = 0.1 P_{max} / \alpha = 9 \text{ Jm}^2\text{s}^{-1} \text{ PAR}$$

Again, this low value results from the strong curvature of equation

$$P_I = \frac{\alpha I_I P_{max}}{\alpha I_I + P_{max}}$$

near the origin, which is not realistic.

57.   The calculated canopy photosynthetic rate (g CO_2 m^{-2}s^{-1}), as affected by extinction coefficients and PAR flux density, is given in tabular form as:

PAR flex density (Jm^{-2}s^{-1})	1 0	100	1000
k			
0.4	0.066×10^{-3}	0.5×10^{-3}	1.49×10^{-3}
1.2	0.11×10^{-3}	0.7×10^{-3}	1.6×10^{-3}

The main points are that at low light and low value of extinction coefficient with low L.A.I. do not intercept the available light efficiently. whereas at high light with a high value of extinction coefficient with low leaf area index distributes the available light more evenly over the canopy so that it is used more efficiently.

58.   Comparison of equations

$$R = k P - cW \text{ and}$$

$$R = (1 - y_g) \frac{\Delta s}{\Delta t} y_g \; mW$$

gives

$$k = 1 - y_g \text{ and } c = y_g m$$

with equations

$$k = 0.25 \text{ and } c = 0.0125 \text{ day}^{-1}$$

$$y_g = 0.75 \text{ and } m = 0.017 \text{ day}^{-1}$$

59.   Dividing the radiation receipt by the latent heat of water $\lambda = 2.5MJ$ kg^{-1}, the quantity of water evoporated is $18/2.5 = 7.2$ kg m^{-2} day^{-1}. This is equivalent to a depth of 7.2 mm day^{-1}.

7.2 kg water corresponds to dry matter production of $7.2 \times 0.0015 = 0.011$ kg dry matter m^{-2} day^{-1}.

60.   The psychrometric constant at 20^{0}C and atmospheric pressure is $\gamma = 66.8 \; Pa \; c^{-1}$.

From equation

$$E_e = \frac{sA}{\lambda[s+y(1+g_a/g_c)]},$$

the correction factor is simply

$$s \, (s + \gamma) = 0.68.$$

61.   Differentiating the yield response with respect to N and P give

$$\frac{\partial Y}{\partial N} = a_1 + 2a_2 \, N + hP$$

and

$$\frac{\partial Y}{\partial P} = b_1 + 2b_2 \, P + hN$$

Equating these to zero

$$\frac{\partial Y}{\partial X_1} = 0, \frac{\partial Y}{\partial X_1} = 0, ..., \frac{\partial Y}{\partial X_n} = 0,$$

gives

$$2a_2 N + hP = - a_1$$

and

$$hN + 2b_2P = -b_2$$

$$N_{max} = \frac{b_1h - 2b_2\,a_1}{4\,a_2b_2 - h^2}$$

and

$$P_{max} = \frac{a_1h - 2a_2\,b_1}{4\,a_2b_2 - h^2}$$

For the most economic levels of fertilizer, combining equations

$$\frac{\partial Y}{\partial X_i} = \frac{c_i}{p}\,(i = 1,......n) = 0$$

and

$$\frac{\partial Y}{\partial X_1} = a_1 + 2a_2N + hP,\ \text{therefore}$$

$$2a_2N + hP = -\left(a_1 - \frac{c_p}{p}\right)$$

and

$$hN + 2b_2P = -\left(b_1 - \frac{c_p}{p}\right)$$

where c_N and c_p are the unit costs of nitrogen and phosphate fertilizer, and p the price obtained for the harvested crop. Solving to obtain the economic fertilizers dressing gives

$$N_{ec} = \frac{(b_1 - c_p/p)\,h - 2b_2\,(a_1 - c_N/p)}{4a_2b_2 - h^2}$$

and

$$P_{ec} = \frac{(a_1 - c_N/p)\,h - 2a_2\,(b_1 - c_p/p)}{4a_2b_2 - h^2}$$

62.   The total development time t_{dev} is $t_{dev} = t_c + t_a$.
Substituting for t_a from equation
$$(t_a - p_a)(t_c - p_c) = p^2, \text{ therefore,}$$

$$t_{dev} = t_c + p_a + \frac{p^2}{t_c - p_c}$$

Differentiating gives

$$\frac{d\, t_{dev}}{dt} = 1 - \frac{p^2}{(t_c - p_c)^2}$$

and equating this to zero, hence

$$p^2 = (t_c - p_c)^2$$

and taking the positive square root (which is on the branch of the hyperbola that is of interest), thus

$$t_c = p_c + p$$

Substituting of equation
$$t_c = p_c + p \text{ into equation}$$
$$t_a - t_c\,(t_c - p_a) = p^2 \text{ now gives}$$
$$t_a = p_a + p$$
and finally, substituting equations
$$t_c = p_c + p \text{ and } t_a = p_a + p \text{ into equation}$$
$$t_{dev} = t_c + t_a \text{ leads to}$$
$$t_{dev} = p_a + p_c + 2p$$

63.   Since
$$k = A \exp\left[- E_a/(RT)\right],$$

therefore,

$$\frac{dk}{dT} = A \exp(-E_a / RT)\left(\frac{E_a}{RT^2}\right),$$

and

$$\frac{d^2 k}{dT^2} = A \exp\left(- E_a/RT\right)\left[\left(\frac{E_a}{RT^2}\right)^2 - \frac{2E_a}{RT^3}\right]$$

Equating equation

$$\frac{d^2 k}{dT^2} = A \exp\left(- E_a / RT\right)\left[\left(\frac{E_a}{RT^2}\right)^2 - \frac{2E_a}{RT^3}\right] = 0,$$

hence at the inflexion point (T^*, k^*)

$$T^* = \frac{E_a}{2R} \text{ or } \frac{E_a}{RT^*} = 2$$

AT $T = T^*$, from equation
$k = A \exp(-E_a/RT)$, therefore,
$k^* = A \exp(-2)$.
Substituting with equation

$$\frac{E_a}{RT^*} = 2$$

in equation $\dfrac{dk}{dT} = A \exp(-E_a/RT)\left(\dfrac{E_a}{RT^2}\right)$

gives $\dfrac{dk}{dT} = (T = T^*) = A \exp(-2)\left(\dfrac{4R}{E_a}\right)$,

and the straight line with gradient of equation

$$\frac{dk}{dT} = (T = T^*) = A \exp(-2)\left(\frac{4R}{E_a}\right) \text{ through}$$

the point (T^*, k^*) is $k - Ae^{-2} = A \exp(-2)\dfrac{4R}{E_a}\left(T - \dfrac{E_a}{2R}\right)$,

which simplifies to equation $k = \dfrac{4AR}{E_a \exp(2)}\left(T - \dfrac{E_a}{4R}\right)$.

68. Consider the fate of the cohort of organisms $n(t, a) \, \delta a$. After a period of time δt, this cohort is now of age $a + \delta t$, and the surviving organism belong to $n(t + \delta t, a + \delta t) \, \delta a$. The number that has died is, by the definition of the death rate $D(t, a)$, equal to $D(t, a) \, n(t, a)$. Thus, we can write $n(t + \delta t, a + \delta t) = n(t, a) - D(t, a) \, n(t, a) \, \delta t$ where the common factor of δa has been cancelled throughout. To first order, using two-dimensional Taylor series

$$n(t + \delta t, a + \delta t) = n(t, a) + \frac{\delta n}{\delta t} \delta t + \frac{\delta n}{\delta a} \delta t$$

substitution into equation

$$n (t + \delta t, a + \delta t) = n (t, a) - D (t, a) n (t, a) \delta t$$

gives the required result

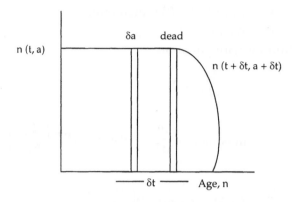

The fate of a cohort of organisms, n(t, a) δa

69.   The problem can be formulated as a mixed integer programme: minimize

$$Z = 300 x_1 + 295 x_2 + 296 x_3 + 9.9\delta_1 + 16.8\delta_2 + 16 \delta_3$$

subject to

$$\left.\begin{array}{l} x_1 - 2.2\delta_1 \le 0 \\ x_2 - 2.8\delta_2 \le 0 \\ x_3 - 3.2\delta_3 \le 0 \end{array}\right\} \text{ (Site capacity constraints)}$$

$\delta_1 + \delta_2 + \delta_3 = 1$ (Site selection constraint)

$x_1 + x_2 + x_3 \ge 2$ (Factory size constraint)

$x_1, x_2, x_3, \delta_1, \delta_2, \delta_3 \ge = 0$ (non-negativity conditions)

$\delta_1, \delta_2, \delta_3 =$ integer 0 – 1 (integer restrictions)

where Z is the cost of the development (£ x 10^3),

x_i is the size of the factory [Ha] to be built at site ($i = 1, 2, 3$), and δ_i is a zero-one variable which must be either zero or one ($i = 1, 2, 3$). The optimal solution is $Z = 606.8$, $\delta_2 = 1$, $x_2 = 2.0$, $\delta_1 = \delta_3 = x_1 = x_3 = 0$; thus the optimal solution to select site 2 and build a 2 ha factory there.

70.   The expected profit on a crop [£ ha⁻¹] = expected price – labour cost – other cost, e.g. profit on spring wheat grown on high-grade land = 112 x 4.6 – 12.4 x 2.5 – 135 = 349.27 £ ha⁻¹.

Let X_{HW} be the area [ha] of high-grade land allocated to spring wheat , X_{LW} low-grade to wheat, X_{HB} high grade to barley, X_{LB} low grade to barley, X_{HP} high-grade to potatoes, X_{LP} low grade to potatoes, X_{HS} high-grade to sugarbeet, X_{LS} low grade to sugarbeet, X_{HO} high-grade to oilseed rape, X_{LO} low-grade to oilseed rape, and Z be the profit [£]. The linear program is maximize

$Z = 349.2\ X_{HW} + 214.8\ X_{LW} + 368\ X_{HB} + 228.25\ X_{LB} + 1493.75\ X_{HP} + 293.75\ X_{LP} + 591.75\ X_{HS} + 166.75\ X_{LS} + 342.5\ X_{HO} + 117.5\ X_{LO}$ (objective function).

Subject to $X_{HW} + X_{HB} + X_{HP} + X_{HS} + X_{HO} \le 200$ [High-grade land constraint]

$X_{LW} + X_{LB} + X_{LP} + X_{LS} + X_{LO} \le 100$ [Low-grade land constraint]

$135\ (X_{HW} + X_{LW}) + 117\ (X_{HB} + X_{LB}) + 725\ (X_{HP} + X_{LP}) + 365\ (X_{HS} + X_{LS}) + 185\ (X_{HO} + X_{LO}) < 150000$ [Constraint on capital available for costs other than labour]

$166\ (X_{HW} + X_{LW}) + 148\ (X_{HB} + X_{LB}) + 906.25\ (X_{HP} + X_{LP}) + 533.25\ (X_{HS} + X_{LS}) + 220\ (X_{HO} + X_{LO}) \le 162500$ [Capital constraint]

The constraints on capital are arrived at as follows. The farmer has available 5000 man hours of free labour, which is equivalent to £ 12500 of extra capital that can only be spent on labour. The farmer, therefore, has £ 162500 of capital but at most £ 150000 of which can be spent on costs other than labour. The optimum crop plan is to allocate 155.75 ha of high grade land to potatoes, 44.25 ha of high-grade land to barley and 100 ha of low-grade land to barley. This yields a profit of £ 271764.36.

71. Let H [kg herbage ha^{-1}] denote herbage biomass at time t (days)], then

$$\frac{dH}{dt} = \text{growth} - \text{consumption}$$

$$= g_0\ He^{-gDt} - C_M\ \frac{HS}{H + C_k}$$

For equilibrium at H^*, $dH/dt = 0$ and

$$g_0\ H^*\ e^{-gDt} - C_M\ \frac{H^* S}{H^* + C_k} = 0$$

that is

$$S\ (t) = g_0\ (H^* + C_k\ e^{-gDt}\ /C_M)$$

72. Using $S = W_f - W$ from equation

$W + S$ = a constant = $W_0 + S_0 = W_f + S_f = C$
With $S_f = 0$, there

$$W_f \int_{W_0}^{W} \frac{dW}{(W_f - W)^2} = \int_0^t k \, dt$$

which gives

$$\frac{W}{W_f} = \frac{W_0 + kt(W_f - W_0)}{W_f + kt(W_f - W_0)}$$

This equation describes a rectangular hyperbola, similar to equation

$$G = G_1 \frac{F}{K + F} - G_2$$

where G = growth rate, F = the rate at which food is supplied to the animal, and G_1, G_2 and K are constants. It does not pass through the origin, but through the point $t = 0$ and $W = W_0$. It approaches an asymptote at $W = W_f$ as $t \rightarrow \infty$. There is no point of inflexion. It resembles the monomolecular equation in shape, but with a much slower approach to the asymptote.

73.   Using equation
$W + S$ = a constant = $W_0 + S_0 = W_f + S_f = C$, the growth rate equation becomes

$$\frac{dW}{dt} = (W_f - W)[k_1 + k_2(W_f - W)/W_f]$$

Which can be divided into partial fractions to give

$$\frac{k_1}{k_2} \int_0^t dt = \int_{W_0}^{W} \left[\frac{1}{k_2(W_f - W)} - \frac{1}{(k_1 + k_2)W_f - K_2 W} \right] dW$$

After integration and arrangement, therefore

$$\frac{W}{W_f} = \frac{[(k_1 + k_2)W_f - k_2 W_0]e^{k_1 t}(k_1 + k_2)(W_f - W_0)}{[(k_1 + k_2)W_f - k_2 W_0]e^{k_2 t} - k_2(W_f - W_0)}$$

This equation shares the qualitative properties of the rectangular hyperbola of equation

$$\frac{W}{W_f} = \frac{W_0 + kt(W_f - W_0)}{W_f + kt(W_f - W_0)}$$

and the monomolecular equation
$W = W_f - (W_f - W_0)\, e^{-kt}$ but lies between the two curves and, thus, can have either a fast or a slow approach to the asymptote with $k_2 = 0$, equation

$$\frac{W}{W_f} = \frac{[(k_1 + k_2)W_f - k_2 W_0]e^{k_1 t} - (k_1 + k_2)(W_f - W_0)}{[(k_1 + k_2)W_f - k_2 W_0]e^{k_1 t} - k_2(W_f - W_0)}$$

becomes identical to equation
$$W = W_f - (W_f - W_0)\, e^{-kt}$$
With $k = k_1$. Taking the limit $k_1 \to 0$
(write exp $(k_1 t) \approx 1 + k_1 t$), equation

$$\frac{W}{W_f} = \frac{[(k_1 + k_2)W_f - k_2 W_0]e^{k_1 t} - (k_1 + k_2)(W_f - W_0)}{[(k_1 + k_2)W_f - k_2 W_0]e^{k_2 t} - k_2(W_f - W_0)}$$

becomes identical to equation

$$\frac{W}{W_f} = \frac{W_0 + kt(W_f - W_0)}{W_f + kt(W_f - W_0)}$$

With $k = k_2$. The initial slope of the growth equation is (from equation

$$\frac{dW}{dt} = (W_f - W)\left[k_1 + k_2\left(W_f - W\right)/W_f\right]$$

with $W = W_0$

$$\frac{dW}{dt}(t = 0) = (W_f - W_0)\left[k_1 + k_2\left(W_f - W_0\right)/W_f\right]$$

74.   Differentiating equation

$$\frac{dW}{dt} = \mu W\left(1 - \frac{W}{B}\right)e^{-Dt}$$

(where μ, B and D are constants, W = weight and t = time) with respect to time gives

$$\frac{d^2W}{dt^2} = \mu e^{-Dt}\left[\frac{dW}{dt}\left(1 - \frac{2W}{B}\right) - DW\left(1 - \frac{W}{B}\right)\right]$$

which, with substitution for $\dfrac{dW}{dt}$ from equation

$$\frac{dW}{dt} = \mu W\left(1 - \frac{W}{B}\right)e^{-Dt}$$

becomes

$$\frac{d^2W}{dt^2} = \mu W e^{-Dt}\left[\mu\left(1 - \frac{W}{B}\right)\left(1 - \frac{2W}{B}\right)e^{-Dt} - D\left(1 - \frac{W}{B}\right)\right]$$

equating this to zero for a point of inflexion.
Therefore,

$$0 = \mu\left(1 - \frac{2W}{B}\right)e^{-Dt} - D$$

From equation

$$W = \frac{W_0 B}{W_0 + (B - W_0)\exp\{-[\mu(1 - e^{-Dt})/D]\}}$$

It may be shown that

$$e^{-Dt} = 1 - \frac{D}{\mu}\ln\left[\left(\frac{B - W_0}{B - W}\right)\left(\frac{W}{W_0}\right)\right]$$

which, with equation

$$0 = \mu\left(1 - \frac{2W}{B}\right)e^{-Dt} - D$$

gives

$$0 = \mu\left(1 - \frac{2W}{B}\right)\left\{1 - \frac{D}{\mu}\ln\left[\left(\frac{B-W_0}{B-W}\right)\left(\frac{W}{W_0}\right)\right]\right\} - D$$

Although an analytical solution to this equation cannot be obtained, the limiting logistic and Gompertz cases are easily verified.

For $D = 0$, $W = B/2 = W_1/2$

(equation $W = 1/2\ W_1$) with $W_1 = B$) . For $B \to \infty$,

$W = W_1 \exp[(\mu-D)/D]$, in agreement with equations

$$t^* = 1/D\ \ln\frac{\mu_0}{D} \text{ and } W\ (t = t^*) = \frac{W_f}{2}$$

(where $t^* =$ time at inflection, $\mu_0 =$ the value of constant of proportionality (μ) at time $t = 0$)

and

$$W_1 = W_0\ e^{\mu_0/D}$$

75. In place of equation

$$\frac{dW}{dt} = k^1 WS$$

from which the logistic is derived

$$\frac{dW}{dt} = \mu_0 W\left(\frac{S}{K+S}\right)$$

where μ_0 and K are constants, is taken. Since growth will continue until the substrate is exhausted and $S_1 = 0$ in equation

$$W + S = a \text{ constant} = W_0 + S_0 = W_1 + S_1 = C,$$

therefore,

$$W_1 = S + W$$

Where W_1 is the final weight.

Putting this into equation

$$\frac{dW}{dt} = \mu_0 W\left(\frac{S}{K+S}\right)$$

gives

$$\frac{dW}{dt} = \mu_0 W \left(\frac{W_f - W}{K + W_f - W} \right)$$

$$\frac{1}{W_f} \int_{W_0}^{W} \frac{K + W_f}{W} + \frac{K}{W_f - W} \, dW = \int_0^t \mu_0 dt$$

After integration, the equation connecting W and t is

$$\frac{1}{W_f}(K + W_f) \ln\left(\frac{W}{W_0}\right) + K \ln + \left(\frac{W_f + W_0}{W_f - W}\right) \mu_0 dt$$

To obtain the logistic, one takes the limit $\mu_0 \longrightarrow \infty$, $K \longrightarrow \infty$ and

$$\mu_0 / K \longrightarrow \mu / W_f$$

and equations

$$\frac{dW}{dt} = \mu_0 W \left(\frac{W_f - W}{K + W_f - W} \right) \text{and}$$

$$\frac{1}{W_f}(K + W_f) \ln\left(\frac{W}{W_0}\right) + K \ln + \left(\frac{W_f + W_0}{W_f - W}\right) \mu_0 t$$

become identical to equations

$$\frac{dW}{dt} = \mu W \left(1 - \frac{W}{W_f} \right) \text{and}$$

$$W = \frac{W_0 \, W_f \, e^{\mu t}}{W_f - W_0 + W_0 \, e^{\mu t}}$$

The point of inflexion of the curve is obtained by differentiating equation

$$\frac{dW}{dt} = \mu_0 W \left(\frac{W_f - W}{K + W_f - W} \right)$$

equating to zero, and taking the negative root of the resulting quadratic (the other root being $> W_f$ and therefore non-physiological) to give

$$W \text{ (inflexion)} = (K + W_f) [K (K + W_f)]^{1/2}$$

The time of inflexion is obtained by substituting this value into equation

$$\frac{1}{W_f}(K+W_f)\ln\left(\frac{W}{W_0}\right)+K\ln+\left(\frac{W_f+W_0}{W_f-W}\right)=\mu_0 t$$

76.   Instead of equation

$$\frac{dW}{dt}=\mu_0 W\left(\frac{S}{K+S}\right)$$

assume

$$\frac{dW}{dt}=\mu_0 W\left(\frac{S}{K+S}\right)e^{-Dt}$$

However, in contrast with the last exercise, growth need not now continue until the substrate is exhausted, so in place of equation.

$$W_f = S + W$$

Therefore

$$S + W = \text{constant} = A$$

so that equation

$$\frac{dW}{dt}=\mu_0 W\left(\frac{S}{K+S}\right)e^{-Dt}$$

becomes

$$\frac{dW}{dt}=\mu_0 W\left(\frac{A-W}{K+A-W}\right)e^{-Dt}$$

Proceeding as in the previous example, integration gives an equation very similar to equation

$$\frac{1}{W_f}\left[(K+W_f)\ln\left(\frac{W}{W_0}\right)+K\ln+\left(\frac{W_f-W_0}{W_f-W}\right)\right]=\mu_0 t,$$

namely

$$\frac{1}{W_f}\left[(K+A)\ln\left(\frac{W}{W_0}\right)+K\ln+\left(\frac{A+W_0}{A-W}\right)\right]=\frac{\mu_0}{D}\left(1-e^{-Dt}\right)$$

The final weight W_f may be found by substituting $W = W_f$ in the above equation and solving for W_f.

77.   It seems clear that of the three alternatives, *b* is the only one that has a chance of meeting assumption 2. Alternative *a* is least likely to meet either assumption. The author of the article, however, chooses alternative *a* and thus generates what is, in my opinion, a "textbook example" of a mismatch between experimental conditions and mathematical assumptions.

The most of aforementioned solutions of exercises have been selected from France and Thornley (1984), Gold (1977) and Keulen and Wolf (1986). (See references in chapters 6).The book by Goudriaan and van Laar (1994) is the best text book with exercises and their solutions.

Further reading:

Goudriaan, J, and H.H.van Laar (1994). Modeling potential crop growth processes. Text book with exercises. Kluwer Academic publishers, London pp.238.

REFERENCES

Acock, B., Reddy, V.R., Whisler, F.D., Baker, D.N., Mckinion, J.M., Hodges, H.F. and Boote, K.J. (1983). The Soybean Crop Simulator GLYCIM: Model Documentation 1982. Report No. 2, Washington, D.C.: United States Department of Energy, Carbon Dioxide Research Division, Office of Energy Research.

Aggarwal, P.K. and Mall, R.K. (2002). Climate and rice yields in diverse agro-environments of India. II. Effect of uncertainties in scenarios and crop models on impact assessment. Climate Change **52:** 331-343.

Aggarwal, P.K., Kalra, N., Bandopadhyay, S.K., Pathak, H., Sehgal, V.K., Kaur, R., Rajput, R.K., Joshi, H.C., Chaudhary, R. and Roetter, R. (1998). Exploring agricultural land use options for the state of Haryana: Biophysical modeling. In: Exchange of Methodologies in Land Use Planning. (eds. Roetter, R. et al.), pp.59-65.

Aggarwal, P.K., Kalra, N., Kumar, S., Bandopadhyay, S.K., Pathak, H., Vashisht, A.K., Hoanh, C.T. and Roetter, R.P. (2000a). Exploring land use options for a food grain production in Haryana: Methodological frame work. (eds. Roetter, R. et al.), SYSNET Research for Optimizing Future Land Use in South and Southeast Asia (SYSNET Research Paper) Series No.2, pp.57-69.

Aggarwal, P.K., Kalra, N., Pathak, H., Bandopadhyay, S.K., Vashisht, A.K., Roetter, R.P. and Hoanh, C.T. (2000b). Haryana case study: Trade off between cereal production and environmental impact. In: (eds.) Roetter et al., Synthesis of Methodology Development and Case Studies. SYSNET Research Paper Series No.3: IRRI, Los Banos, Phillipines, 11-18.

Aggarwal, P.K., Kumar, S., Vashisht, A.K., Hoanh, C.T., Keulen, H. Van, Kalra, N., Pathak, H. and Roetter, R.P. (2001a). Balancing food demand and supply. In: Land use analysis and planning for sustainable food security with an illustration for the state of Haryana, (eds) Aggarwal, P.K., Roetter, R.P., Kalra, N., Keulen, H.V., Hoanh, C.T. and LAAR, H.H.V., Published by: IARI , India, IRRI, Phillipines and WUR, The Netherlands, pp.137-151.

Aggarwal, P.K., Roetter, R.P., Kalra, N., Hoanh, C.T., Keulen, H.V. and Laar, H.H.V. (eds.) (2001b). Land use Analysis and Planning for Sustainable Food Security. IARI, New Delhi, India, IRRI, Los Banos, Phillipines, Wageningen University and Research Centre, Wageningen, The Netherlands.

Aggarwal, P.K., Singh, A.K., Kalra, N., Thyagrajan, T.M., Reddy, P.R. and Mohandas, S. (1999). Using simulation models to analyze rice weather relationships. In: Rice in a variable climate, (eds) Y.P. Abrol and S. Gadgil, Narosa Publishing House, New Delhi, pp. 151-166.

Ajai and Sahai, B. (1986). Drought detection and quantification by remote sensing. Jal Vigyan Sameksha **1**: 138-152.

Algarswamy, G. and Ritchie, J.T. (1990). Phasic development in CERES – Sorghum model. In: T. Hodges (ed.), Predicting crop phenology, Boca Raton: CRC Press (in press).

Angus, E.C. and Zandstra, H.G. (1980). Climatic factors and the modeling of rice growth and yield. In: Proceedings of a Symposium on the Agrometeorology of the rice crop. pp.189-199. Los Banos: IRRI.

Angus, I.F., Kornner, A. and Torsell, B.W.R. (1980). A system approach to estimation of Swedish ley production. Progress Report 1979/80. Uppsala Swedish University of Agricultural Sciences.

Arkin, G.F., Vanderlip, R.L. and Ritchie, J.T. (1976). A dynamic grain sorghum growth model. Transactions of the American Society of Agricultural Engineers **19**: 722-630.

Arkin, G.F., Richardson, C.W. and Mass, S.J. (1978). Forecasting grain sorghum yields using probability functions. Transactions of the American Society of Agricultural Engineers **21**: 874-880.

Arora, R. and Sindhu, H.S. (1992). Effect of some juvenile hormone analogues on the fecundity and longevity of mustard aphid, *Lipaphis erysimi* (Kalt.). J. Insect Sci. **4** (2): 138-140.

Asseng, S., Fillery, I.R.P., Anderson, G.C., Dolling, P.J., Dunin, F.X. and Keating, B.A. (1998a). Use of the APSIM wheat model to predict yield, drainage and NO_3^- leaching for a deep sand. Aust. J. Agric. Res. **49**: 363-377.

Asseng, S., Keating, B.A., Fillery, I.R.P., Gregory, P.J., Bowden, J.W., Turner, N.C., Palta, J.A. and Abrecht, D.J. (1998B). Performance of the APSIM – Wheat model in Western Australia. Field Crops Research **57**: 163-179.

Asseng, S., Keulen, H., Van and Stol, W. (2000). Performance and application of the APSIM N Wheat model in The Netherlands. European Journal of Agronomy. **12**: 37-54.

Asseng, S., Turner, N.C. and Keating, B.A. (2001a). Analysis of water and nitrogen use efficiency of wheat in a Mediterranean climate. Plant and Soil. **233**: 127-143.

Asseng, S., Fillery, I.R.P., Dunwin, F.X., Keating, B.A. and Meinka, H. (2001b). Potential deep drainage under wheat crops in a Mediterranean climate. I. Temporal and spatial variability. Aust. J. Agric. Res. **52**: 45-56.

Asseng, S., Dunwin, F.X., Fillery, I.R.P., Tennant, D. and Keating, B.A. (2001c). Potential deep drainage under wheat crops in a Mediterranean climate II. Management opportunities to control drainage. Aust. J. Agric. Res. **52**: 57-66.

Asseng, S., Turner, N.C., Botwright, T. and Condon, A.S. (2003). Evaluating the impact of a trait for increased specific leaf area on wheat yield, using a crop simulation model. Agron. J. **95**: 10-19.

Attri, S.D., Singh, K.K., Kaushik, A., Rathore, L.S., Mendiratta, N and Lal, B. (2001). Evaluation of dynamic model for wheat genotypes under diverse environments in India. Mausam **52** (3): 561-566.

Baker D,M., Lambert, J.R. and Mckinion, J.M. (1983). GOSSYM: A simulation of cotton crop growth and yield. Technical Bulletin 1089. Clemson: South Carolina Agricultural Experiment Station.

Bannayan, M., Crout, N.M.J. and Hoogenboom, G. (2003). Application of CERES - Wheat model for within-season production of winter wheat yield in the United Kingdom. Agron. J. **95**: 114-125.

Batra, H.N. (1961). Studies on the application of Dyar's law to larval stages of *Plutella maculipennis Curtis*. Indian J. Ent. **23**: 69-70.

Berry, G. (1967). Mathematical model relating plant yield with arrangement for regularly spaced crops. Biometrics **23**: 505-515.

Bhai, B.D. and Singh, S. (1961). On the biology of painted bug, *Bagrada cruciferarum* in the Punjab. Proc. 48th Indian Sci. Congr., Part 3 p.494.

Bhatia, H.L. and Shaffi, M. (1932). Life histories of some Indian Syrphidae. Indian J. Agric. Sci. **2**: 543-570.

Blackie, M.J. (1976). Management information systems for the individual farm firm. Agr. Systems **1**: 23-36.

Bloomberg, W.J. (1979). A model of damping-off and roof rot of Douglas fir seedlings caused by *Fusarium oxysporum*. Phytopathology **69**: 74-81.

Boatwright, G.O. and Whitehead, V.S. (1986). Warning and crop condition assessment research. IEEE Trans. Geosc. Remote Sens, GE **24**: 54-64.

Bogawat, J.K. (1967). Biology of mustard sawfly on different host plants. Indian J. Ent. **29**: 270-274

Bogawat, J.K. (1968). Effects of temperature on the development of mustard sawfly, *Athelia proxima* Klug (Hymenoptera: Tenthredinidae). Indian J. Ent. **30**: 80-82.

Boote, K.J., Jones, J.W. , Batchelor, W.D., Nafziger, E.D., and Meyers, O. (2003). Genetic coefficients in the CROPGRO-Soybean model; Links to yield performance and genomics. Agron. J. **95**: 32-51.

Boote, K.J., Jones, J.W. and Pickering, N.B. (1996). Potential uses and limitations of crop models. Agron. J. **88**: 704-716.

Boote, K.J., Jones, J.W., Hoogenboom, G., Wilkerson, G.G. and Jagtap, S.S. (1989). PNUTGRO V1.02: Peanut Crop Growth Simulation Model. User's guide. Florida Agricultural Experiment Station Journal No.8420. Gainesville: University of Florida.

Bridge, D.W. (1976). A simulation model approach for relating effective climate to winter wheat yields in the great plains. Agricultural Meteorology **17**: 185-194.

Brown, L.G., Jones, J.W., Hesketh, J.D., Hartsog, J.D., Whisler, F.D. and Harris, F.S. (1985). COTCROP: Computer Simulation of Growth and Yield.

Information Bulletin No.69. Mississippi state; Mississippi Agricultural and Forestry Experiment Station.

Bruhn, J.A. and Fry, W.E. (1981). Analysis of potato late blight epidemiology by simulation modeling. Phytopathology **71**: 612-616.

Burleigh, J.R., Roelfs, A.P. and Eversmeyer, M.G. (1972). Estimating damage to wheat caused by *Puccinia recondite tritici*. Phytopathology **62**: 944-946.

Buttler, I.W. (1989). Predicting water constraints in productivity of corn using plant-environmental simulation models. Ph.D. Thesis, Ithaca: Cornell University.

Carbone, G.J., Mearns, L.D., Mavromatis, T., Sadlar, E.J. and Stooksbury, D. (2003). Estimating CROPGRO - Soybean performance for use in climate impact studies. Agron. J. **95**: 537-544.

Chander, S., Aggarwal, P.K. and Reddy, P.R. (2002). Assessment of yield loss due to bacterial leaf blight using simulation models. Ann. Pl. Protect. Sci. **10** (2): 277-281.

Chander, S., Aggarwal, P.K., Kalra, N., Swarooparani, D.N. and Prasad, J.S. (2002a). Assessment of yield losses due to stem borer, *Scirpophaga incirtulas* in rice using simulation models. Journal of Entomological Research **26** (1): 23-28.

Chander, S., Aggarwal, P.K. and Swarooparani, D.N. (2002b). Changes in pest profiles in rice-wheat cropping system in Indo-Gangetic plains. Annals of Plant Protection Sciences (in press).

Charles-Edwards, D.A. and Acock, B. (1977). Growth response of a chrysanthemum crop to the environment. II. A mathematical analysis relating photosynthesis and growth. Annals of Botany **41**: 49-58.

Chauhan, P., Mohan, M., Nayak, S.R. and Navalgund, R.R. (2002). Comparison of ocean color chlorophyll algorithms for IRS –P4 OCM sensor using in-situ data. Journal of the Indian Society of Remote Sensing **30** (1, 2): 87-94.

Chaurasia, R. and Minakshi (1997). Wheat yield prediction using climate parameters. Indian Journal of Soil Conservation **25** (2): 147-150.

Clevers, J.G.P.W. and Leeuwen, H.J.C. Van. (1995). Linking remotely-sensed information with crop growth models for yield prediction. A case study of sugarbeet. In: Proceedings of seminar on yield forecasting. FAO, 24-27 October, 1994, France.

Conover, W.J. (1971). Practical Non-parametric Statistics, New York: Wiley.

Curry, R.B., Baker, C.H. and Streeter, J.G. (1975). SOYMOD I: A dynamic simulator of soybean growth and development. Transactions of the American Society of Agricultural Engineers **18**: 963-968.

Dadhwal, V.K., Pokharna, S.S., Ravi, N. and Oza, M.P. (1989). An approach for development of remote sensing data-based crop yield forecasting methodology - A Report of Space Application Centre, Ahmedabad - 380053, 43pp.

Dadhwal, V.K. and Sridhar, V.N. (1997). A non-linear regression form for VI-crop yield relation incorporating acquisition data normalization. International Journal of Remote Sensing **18** (6): 1403-1408.

Dadhwal, V.K. and Ray, S.S. (2000). Crop assessment using remote sensing - Part-II: Crop condition and yield assessment. In: Remote Sensing and Agricultural statistics: Rational, scope and aims, suppliment to Indian Journal of Agricultural Economics **55** (2): 55-67.

Dent, J.B. (1974). Application of Systems Concepts and Simulation in Agriculture. School of Agriculture, University of Aberdeen, Misc. Pub.

Dent, J.B. and Blackie, M.J. (1979). Systems Simulation in Agriculture. 180pp. London: Applied Science Publishers Ltd.

Dhar, V. (2000). Seven decades of research on sterility mosaic of pigeonpea. Indian J. Pulses Res. **13**: (2): 1-10.

Dube, R.B. and Chand, P. (1977). Effect of food plants on the development of *Plutella xylostella* (L.) (Lapidoptera: Plutellidae). Entomon. **2** (2): 139-140.

Dubey, R.P., Sharma, T., Garg, J.K., Malik, K.D. and Patel, J.R. (1985). Relationship of wheat yield with spectral and agrometeorological data. In: Proceedings of Sixth Asian Conference on Remote Sensing, pp. 406-411. November 21-26, Hyderabad, (A.P.).

Dubey, R.P., Mahey, R.K., Ajwami, N., Sindhu, S.S., Kalubarme, M.H., Sridhar, V.N., Jhorar, O.P., Cheema, S.S., Narang, R.S. and Navalgund, R.R. (1994). Pre-harvest wheat yield and production estimation for Punjab, India. International Journal of Remote Sensing **15**: 2137-2144.

Durbude, D., Purandas, B.K. and Sharma, A. (2001). Estimation of surface runoff potential of a watershed in semi-arid environment – A case study. Journal of the Indian Society of Remote Sensing **29** (1,2): 47-58.

Dutta, S., Patel, N.K. and Srivastava, S.K. (2001). District-wise yield models of rice in Bihar based on water requirement and meteorological data. Journal of the Indian Society of Remote Sensing **29** (3): 175-182.

Eversmeyer, M.G. and Burleigh, J.R. (1970). A method of predicting epidemic development of wheat leaf rust. Phytopathology **60**: 805-811.

Eversmeyer, M.G., Burleigh, J.R. and Roelfs, A.P. (1973). Equations for predicting wheat stem rust development. Phytopathology **63**: 348-351.

Ewing, E.E., Heym, W.D., Batutis, E.J., Snyder, R.G., Ben Khedher, M., Sandlan, K.P. and Turner, A.D. (1990). Modifications to the simulation model POTATO for use in New York. Agricultural System **33**: 173-192.

Fick, G.W.(1981). ALSIMI (Level 2) User's manual. Agronomy Mimeograph 81.35, Ithaca: Department of Agronomy, Cornell University.

Fick, G.W., Loomis, R.S. and Williams, W.A. (1975). Sugarbeet. In: Crop Physiology, (ed.) L.T. Evans, pp.259-295. Cambridge: Cambridge University Press.

Fishman, S., Talpaz, H., Dinar, M., Levy, M., Arazi, Y., Rozman, Y. and Varshavsky, S. (1985). A model for simulation of potato growth on the plant community level. Agricultural Systems. **18**: 115-128.

France, J. and Thornley, J.H.M. (1984). Mathematical Models in Agriculture. 335pp. London: Butterworths.

Fukai, S. and Salisbury, J.H. (1978). A growth model for *Trifolium subterraneum* swards. Australian Journal of Agricultural Research **29**: 51-65.

Gardner, B.R., Blad, B.L., Garrity, D.P. and Watts, D.G. (1981). Relationships between crop temperature, grain yield, evapotranspiration and phenological development of two hybrids of moisture-stressed sorghum. Irrigation Science **2**: 213-224.

George, R.K., Rammohan, S., Kulshrestha, M.S., Shekh, A.M. and Saita, H. (2001). Prediction of soil temperature using artificial neural network. Journal of Agrometeorology **3** (1, 2): 169-173.

Ghos, C.C. (1914). Life-histories of Indian insects. Mem. Dep. Agric. India. **5** (1): 1-72.

Godre, U.A., Joshi, M.S. and Mandakhot, A.M. (2002). Effect of weather factors on the incidence of alternaria leaf blight, white rust and Powdery mildew of mustard. Ann. Pl. Protect. Sci. **10** (2): 337-339.

Godwin, D.C. and Vlek, P.L.G. (1985). Simulation of nitrogen dynamics of wheat cropping system. In: W. Day and R.K. Arkin (eds.), Wheat Growth and Modeling. pp.311-332. New York.: Plenum Press.

Gold, H.J. (1977). Mathematical modeling of biological systems—An introductory guide book. 357p. New York: John Wiley and Sons.

Goudriaan, J. and Laar, H. Van. (1978). Calculation of daily totals of the grass CO_2 assimilation of leaf canopies. Netherlands Journal of Agricultural Science **26**: 416-425.

Gungula, D.T., King, J.G. and Togun, A.O. (2003). CERES - Maize predictions of maize phenology under nitrogen-stressed conditions in Nigeria. Agron. J. **95**: 892-899.

Haun, J.R. (1974). Prediction of Spring Wheat Yields from Temperature and Precipitation Data. Agronomy Journal **66**: 405-409.

Hirst, J.M. (1953). Changes in atmospheric spores content. Diurnal periodicity and the effect of weather. Trans. Br. Mycol. Soc. **36**: 376-393.

Hodges, T., Johnson, B.S. and Manrique, L.A. (1989). Substore: A Model for Potato Growth and Development. In: Agronomy Abstract, p.16. Madison: American Society of Agronomy.

Holt, D.A., Bula, R.J., Miles, G.E., Schreiber, M.M. and Peart, R.M. (1975). Environmental physiology, modeling and simulation. I. Conceptual development of SIMED. Research Bulletin of Purdue University Agricultural Experiment Station. No. 907, 26pp.

Hoogenboom, G., White, J.W. and Jones, J.W. (1989). A computer model for the simulation of bean growth and development. In: Advances in Bean (*Phaseolus vulgaris* L.) Research and Production. Cali: CIAT Publication No.23.

Hoogenboom, G. and White, J.W. (2003). Improving physiological assumptions of simulation models by using gene-based approaches. Agron.J. **95**: 82-89.

Hook, R.L.V. (1971). Energy and nutrient dynamics of spider and orthopteran populations in a grassland ecosystem. Eco. Monogr. **41**.

Horie, T., Nakagawa, H. and Kira, T. (1986). Studies on the modeling and prediction of rice development processes. Japanese Journal of Crop Science **55**: 214-215.

Horie, T., Yajima, M. and Nakagawa, H. (1992). "Yield forecasting", Agricultural Systems: 211-236.

Hubbard, K.B., Mehmood, R. and Carison, C. (2003). Estimating daily dew point temperature for the Northern Great Plains using maximum and minimum temperature. Agron J. **95**: 323-328.

Hunt, L.A. Reynolds, M.P., Sayre, K.D., Rajaram, S., White, J.W. and Yan W. (2003). Crop modeling and the identification of stable coefficients that may reflect significant groups of genes. Agron. J. **95**: 20-31.

Hyre, R.A. (1954). Progress in forecasting late blight of potato and tomato. Plant Disease Reporter **38**: 245-253.

Jackson, B.S., Arkin, G.F. and Heary, A.B. (1988). The cotton simulation model "COTTAM": Fruiting model callibration and testing. Transactions of the American Society of Agricultural Engineers **31**: 846-854.

Jackson, R.D., Idso, S.B., Raginato, R.J. and Pinier Jr, P.J. (1981). Canopy temperature as a crop water stress indicator. Water Resource Research **17**: 1133-1138.

James, W.C. (1971). An illustrated series of assessment keys for plant diseases, Their preparation and usage. Can. Plant Disease Survey **51**: 39-65.

James, W.C., Shih, C.S., Hodgson, W.A. and Callbeck, L.C. (1972). The quantitative relationship between late blight of potato and loss of tuber yield. Phytopathology **62**: 92-96.

James, W.C. (1974). Assessment of plant diseases and losses. Annual Review of phytopathology **12**: 27-48.

Johnson, I.R., Ameziane, T.E. and Thornley, J.H.M. (1983). A model of grass growth. Annals of Botany **51**: 599-609.

Jones, C.A. and Kiniry, J.R. (eds.) (1986). CERES – Maize. A Simulation Model of Maize Growth and Development. College Station; Texas A and M University Press.

Jones, C.A., Wegener, M.K., Russell, J.S., Mc Leod, I.M. and Williams, J.R. (1989). AUSCANE: Simulation of Australian sugarcane with EPIC. Technical paper No.29. Division of tropical crops and pastures. Australia; CSIRO.

Jones, J.W. and Luyten, J.C. (1998). Simulation of biological process. In: Agricultural Systems Modeling and Simulation. (eds.) Peart, R.M. and Curry, R.B. Chapter 2. New York: Marcel Dekker, Inc.

Jorgensen, S.E. (ed.) (1984). Modeling the fate and effect of toxic substances in The environment. Proceedings of a symposium held from 6 to 10 June, 1983, In Copenhagen, Denmark, sponsored by the International Society for Ecological Modeling (ISEM). 342p. Amsterdam: Elsevier.

Kalra, N. (2003). Info Crop. Personal Communication.

Kalra, V.K., Singh, H. and Rohilla, H.R. (1987). Influence of various genotypes of Brassica on biology of mustard aphid, *Lipaphis erysimi* (Kalt). Indian J. Agric. Sci. **57** (4): 277-279.

Kalubarme, M.H., Mahey, R.K., Dhaliwal, S.S., Sidhu, S.S., Singh, R., Mahajan, A. and Sharma, P.K. (1995). Agromet spectral wheat yield modeling in Punjab. In: Proceedings of National Symposium of Indian Society of Remote Sensing, pp.11-17. November 22-24, Ludhiana. .

Kalubarme, M.H., Potdar, M.B., Manjunath, K.R., Mahey, R.K. and Sidhu, S.S. (1997). Spectral Wheat Yield Modelling based on Growth Profile parameter derived from NOAA-11 AVHRR data, scientific note: RSAM / SAC / CAPE – II / SN / 69 / 67, Space Application Centre, ISRO, Ahmedabad.

Kalubarme, M.H., Vyas, S.P., Manjunath, K.R., Bhagia, N., Sharma, R., Gupta, P.C., Zadoo, S. and Prasad, D.V.V. (1992). Pre-harvest wheat production forecast for rabi 1990-91 and 1991-92 in Western Uttar Pradesh using IRS LISS –1 digital data. Proceedings of National Symposium on Remote Sensing for Sustainable Development, November 17-19, Lucknow.

Kanemasu, E.T. (1974). Seasonal canopy reflectance pattern of wheat, sorghum and soybean. Rem. Sens. Environ. **3**: 43-47.

Kanemasu, E.T., Stone, L.R. and Power, W.L. (1976). Evapotranspiration model tested for soybean and sorghum. Agron J. **68**: 569-572.

Kanneganti, Y.R., Rotz, C.A. and Walgenbach, R.P. (1998). Modeling freezing injury in Alfalfa to calculate forage yield.: I. Model development and sensitivity analysis. Agron. J. **90**: 687-697.

Karel, A.K., Singh, S., Kumar, S., Ahmad, N. and Singh, S.P. (1996). Integrated Pest Management in Mustard: Bibliography on Insect Pests, Diseases, Nematodes and Weeds in Rapeseed-Mustard. New Delhi: NCIPM (ICAR).

Katsube, T. and Koshimizu, Y. (1970). Influence of blast disease on harvest in rice plants. I. Effect of panicle infection on yield components and quality. Bulletin Tohoku National Agricultural Experiment Station **39**: 55-96.

Kaurava, A.S., Odak, S.C. and Dhamdhere, S.V. (1969). Preliminary studies on the biology of *Neochrysocharis* sp. (Eulophidae, hymenoptera) parasite of *Phytomyza atricornis* Meigen. J. Bombay Nat. Hist. Soc. **66** (2): 396-398.

Kaurava, A.S., Odak, S.C. and Dhamdhere, S.V. (1970). Studies on the biology of *Phytomyza atricornis* Mg. (Agromyzidae: Diptera). J. Bombay Nat. Hist. Soc. **67**: 597-601.

Keulen, H. Van and Seligman, N.G. (1987). Simulation of Water Use, Nitrogen Nutrition and Growth of a Spring Wheat Crop. Wageningen: PUDOC.

Keulen, H. Van. (1986). A simple model of potential crop production. In: Modeling of Agricultural Production: weather, soils and crops, (eds.) H. Van Keulen and J. Wolf), Chapter 2.3. Wageningen: PUDOC.

Keulen, H. Van, Penning De Vries, F.W.T. and Drees, E.M. (1982). A summary model for crop growth. In: F.W.T. Penning De Vries and H.H. Van Laar (eds). Simulation of Plant Growth and Crop Production. Simulation Monograph Series. pp.464. Wageningen: PUDOC.

Kogan, F.N. (1994). Application of vegetation index and brightness temperature for drought detection. Adv. Space Research **15** (11): 91-100.

Kowal, N.E. (1971). A rationale for modeling dynamic ecological sytems. In: B.C. Patten (ed.). System Analysis and Simulations in Ecology, 1., New York: Academic press.

Krahz, J. and Hau, B. (1980). Systems analysis in epidemiology. Ann. Rev. Phytopath. **18**: 67-83.

Kranz, J. (ed.) (1974). Epidemics of Plant Diseases: Mathematical Analysis and Modeling. London: Chapman and Hall.

Krause, R.A. and Massie, L.B. (1975). Predictive systems: modern approaches to disease control. Annual Review of Phytopathology **13**: 31-47.

Krause, R.A., Massie, L.B. and Hyre, R.A. (1975). Blitecast: A computerized forecast of potato late blight. Plant Disease Reporter **59**: 95-98.

Kropff, M.J. and Laar, H.H. Van. (1993). Modeling crop-weed interactions. 274p. Wallingford: CAB International.

Krupinsky, J.M., Tanaka, D.L., Lares, M.L. and Merrill, S.D. (2004). Leaf spot diseases of barley and spring wheat as influenced by preceeding crops. Agron. J. **96** (1): 259-266.

Kurl, S.P. and Mishra, S.D. (1979). Biology of laboratory reared mustard aphid, *Lipaphis erysimi*. GEOBIOS (Jodhpur), **6**: 81-84.

Lal, M., Singh, K.K., Rathore, L.S., Srinivasan, G. and Saseendran, S.A. (1998). Vulnerability of rice and wheat yields in North West India to future changes in climate. Agricultural and Forest Meteorology **89**: 101-114.

Lal, M., Singh, K.K., Srinivasan, G., Rathore, L.S., Naidu, C.N. and Tripathi, C.N. (1999). Growth and yield responses of soybean in Madhya Pradesh, India to climate variability and changes. Agricultural and Forest Meteorology **93**: 53-70.

Lal, R. (1950). Biology of *Myzus persicae* Sulzer as a pest of potato at Delhi. Indian J. agric. Sci. **20**: 87-100.

Landin, J. (1982). Life history, population dynamics and dispersal in the mustard aphid, *Lipaphis erysimi*: A. literature study, Swedish Univ. Agric. Res., Research Information Centre, Uppsala (Sweden), 18pp.

Landsberg, J.J.(1977). Effect of weather on plant development. In: Environmental Effects on Crop Physiology. (eds.) Landsberg, J.J. and Cutting, C.V., pp.289-307. London: Academic Press.

Legg, B.J., Day, W., Lawlor, D.W. and Parkinson, K.J. (1979). The effects of drought on barley growth: Models and measurements showing the relative importance of leaf area and photosynthetic rate. Journal of Agricultural Science **92:** 703-716.

Leslie, P.H. (1945). On the use of matrices in certain population mathematics. Biometrika **33:** 183-212.

Lewisi, E.G. (1942). On the generation and growth of a population. Sankhya **2:** 93-96.

Liu, W.T. and Kogan, F.N. (1996). Monitoring regional drought using the vegetation condition index. International Journal of Remote Sensing **17** (14): 2761-2782.

Mall, R.K. and Aggarwal, P.K. (2002). Climate change and rice yield in diverse agro-envrionments of India. I Evaluation of impact assessment models. Climate Change **52:** 315-330.

Mani, M.S. (1940). Biological notes on Indian parasitic chalcidoidea. Misc. Bull. Indian Coun. Agric. Res. **30:** 4-5.

Manibhushanrao, K. (1988). EPIBLA: A proposed computer modeling for epidemiology of blast disease of rice. Proceedings of the Annual workshop of Directorate of Rice Research, April 26-29, held in TNAU, Coimbatore.

Marshall, D.R. and Jain, S.K. (1969). Interference in pure and mixed population of *Avena fatua* and *A. barbata*. J. Ecol. **57:** 251-270.

Mass, S.J. and Arkin, G.F. (1980). TAMW: A Wheat Growth and Development Simulation Model. Research centre program and model documentation No. 80-3. TEMPLE: Blackland Research Centre, Texas Agricultural Experiment Station.

Mc Cown, R.L., Hammer, G.L., Hargreaves, J.N.G., Holzeworth, D.P. and Freebairn, D.M. (1996). APSIM: A novel software system for model development, model testing and simulation in agricultural systems research. Agricultural Systems **50:** 255-271.

Mc Kinion, J.M., Jones, J.W. and Hesketh, J.D. (1975). A system of growth equations for the continuous simulation of plant growth. Transactions of the American Society of Agricultural Engineers **18:** 975-984.

Mc Mennamy, J.A. and O'Toole, J.C. (1983). RICEMOD: A physiologically based rice growth and yield model. Research Paper Series No.87. Los Banos: IRRI.

Mcgilchrist, C.A.(1965). Analysis of Competition Experiments. Biometrics **21:** 975-985.

Mead, R. (1967). A mathematical model for estimates of interplant competitions, Biometrics **23:** 189-205.

Medhavy, T.T., Sharma, T., Dubey, R.P., Mahey, R.K. and Sharma, P.K. (1993). Development of wheat yield model for Punjab using remotedly sensed data and historical yield trends. In: ISRS Symposium, November 25-27, Guwahati.

Medhavy, T.T., Sharma, T., Dubey, R.P., Mahey, R.K. and Sharma, R.K. (1995). Development of wheat yield models for Punjab using remotedly sensed data and historical yield trends. Journal of Indian Society of Remote Sensing **23** (1): 23-30.

Meyer, G.E., Curry, R.B., Streeter, J.G. and Mederski, H.J. (1979). A Dynamic Simulation of Soybean Growth, Development and Seed Yield: I. Theory, structure and validation. Research bulletin of the Ohio agricultural research and development centre, No.1113.

Miller, D.R. (1975). Experiment in sensitivity analysis on an uncertain model. Simulation Today No.26. LaJolla: Simulation Councils, Inc. Cal.

Miller, D.R., Weidhaas, D.E. and Hall, R.C. (1973). Parameter sensitivity in insect population modeling. J. Theoret. Biol. **42:** 263-274.

Miyasaka, S.C., Ogoshi, R.M., Tusuji, G.Y. and Kodani, L.S. (2003). Site and planting date effects on taro growth. Comparison with aroid model production. Agron. J. **95**: 545-557.

Modawal, C.N. (1941). Biological note on *Chilomenes sexmaculata* Fabr. Indian J. Ent. **3**: 139-140.

Monteith, J.L. (1981). Does light limit production? In: CB. Johnson (ed.) (1981), Physiological Limiting Plant Productivity. pp.23-38. London: Butterworths.

Monteith, J.L., Huda, A.K.S. and Midya, D. (1989). RESCAP. A resource capture model for sorghum and pearl millet. In: S.M. Virmani, H.L.S. Tondon, and G. Algarswami (eds.), Modeling the growth and development of sorghum and pearl millet. Research Bulletin No.12 , pp.30-34, ICRISAT, Patancheru.

Murata, Y. (1975). Estimation and simulation of rice yield from climatic factors. Agricultural meteorology **15**: 117-131.

Murthy, C.S., Chari, S.T., Raju, P.V. and Johna, S. (1996). Improved ground sampling and crop yield estimation using satellite data. International Journal of Remote Sensing **17** (5): 945-956.

Nagarajan, S. (1983). Plant Disease Epidemiology, pp. 269. New Delhi; Oxford and IBH Publ.

Narayanan, E.S., Subbarao, B.R. and Kaur, R.B. (1956). Studies on the parasites of pea leaf-miner, *Phytomyza atricornis* Meigan, Proc. Indian Acad. Sci. B. **44**: 137-147.

Narayanan, E.S., Subbarao, B.R., Rao, M.R. and Sharma, A.K.-(1960). Biology and morphology of the immature stages of *Microtonus indicus* (Braconidae: Hymenoptera), a parasite of *Phyllotreta cruciferae* Goeze (Chrysomelidae: Coleoptera). Proc. Indian Acad. Sci. B **51**: 280-287.

Nasir, M.M. (1947). Biology of Chrysopa scelests Bank, Indian J. Ent **9**: 177-189.

National Remote Sensing Agency (NRSA) (1990). National Agricultural Drought Assessment and Monitoring System (NADAMS), Hyderabad.

Newkirk, K.M., Parker, J.C., Baker, J.C., Carson, E.W., Brumback, JR. T.B. and Balci, O. (1989). User's Guide to VT – Maize Version 1.0 (R), Virginia Water Resources Research Centre. Blacksburg; Virginia Polytechnic Institute and State University.

Ng. E. and Loomis, R.S. (1984). Simulation of growth and yield of potato crop. Simulation Monograph Series, Wageningen: PUDOC.

Norman, J.M. and Campbell, G. (1983). Application of a plant-environment model to problems in irrigation. Advances in Irrigation **2**: 155-188.

Ono, K. (1965). Principles, methods and organization of blast disease forecasting. In: Rice Blast Disease, pp.173-194. Baltimore: John Hopkins Press.

Pandey, P.C., Dadhwal, V.K., Sahai, B. and Kale, P.P. (1992). An optimal estimation technique for increasing the accuracy of crop forecast by combining remotedly sensed and conventional forecast results. International Journal of Remote Sensing **13** (14): 2735-2741.

Pant, M.M. (1979). Dependence of plant yield on density and planting pattern. Annals of Botany **44:** 513-516.

Patel, N.K., Singh, T.P., Navalgund, R.R. and Sahai, B. (1982). Spectral signature of moisture stressed wheat. Journal of Indian Society of Remote Sensing **10:** 27-34.

Patel, N.K., Ravi, N., Navalgund, R.R., Dash, R.N., Das, K.C. and Patnaik, S. (1991). Estimation of rice yield using IRS-IA digital data in coastal tract of Orissa. International Journal of Remote Sensing **12** (11): 2259-2266.

Patel, N.K. (1996). Personal communication as quoted in Dadhwal, V.K. and Ray, S.S. (2000). Crop assessment using remote sensing – Part-II: Crop condition and yield assessment. In: Proceedings of National Seminar on Remote Sensing and Agricultural Statistics: Rational Scope and AIMS. April 21-22, 1998. Supplement to Indian Journal of Agricultural Economics, **55** (2): 55-67, April-June 2000.

Penning De Vries, F.W.T. and Keulen, H. Van. (1982). La production actuelle et l'action de l'azole et du phosphore. In: Penning De Vries, F.W.T. and Djiteye (eds.) La productivite des paturages saheliens. Une etude des sols des vegetation et de l'exploitatioon de cette resource naturelle, pp. 196-226. Wageningen: PUDOC.

Penning De Vries, F.W.T., Jansen, D.M., Ten Berge, H.F.M. and Bakema, A. (1989). Simulation of Ecophysiological Processes of Growth in Several Annual Crops. Wageningen: PUDOC.

Pielou, E.C. (1960). A single mechanism to account for regular and aggregated populations. J. Ecol. **48**: 575-584.

Pielou, E.C. (1962). The use of plant-to-neighbor distances for the detection of competition. J. Ecol. **50**: 357-367.

Pitter, R.L. (1977). The effect of weather and technology on wheat yields in Oregon. Agricultural Meteorology **18**: 115-131.

Pokharna, S.S., Ray, S.S. and Nanavati, S.C. (1995). Production Estimate of Rapeseed/Mustard in Gujarat for Years 1993-94 and 1994-95, Scientific Note: RTSAM/SAC/CAPE-II/SN/51/95, Space Application Centre, ISRO, Ahmedabad.

Poole, R.W. (1974). An Introduction to Quantitiative Ecology, 532p. New York: Mc Graw-Hill.

Potdar, M.B. (1993). Sorghum yield modeling based on crop growth parameters determined from visible and near–IR channel NOAA – VHRR data. International Journal of Remote Sensing **14** (5): 585-905.

Potdar, M.B., Sudha, R., Ravi, N., Navalgund, R.R. and Dubey, R.C. (1995). Spectro-meteorological modeling of sorghum yield using single data IRS LISS –1 and rainfall distribution data. International Journal of Remote Sensing **10** (3): 467-485.

Rabbinge, R., Ankersmit, G.W. and Pak, G.A. (1979). Epidemiology and simulation of population development of *silobion avenae* in winter wheat. Netherlands Journal of Plant Pathology **85**: 197-200.

Rakshpal, R. (1949). Notes on the biology of *Bagrada cruciferarum* Kirk. Indian J. Ent. **11**: 11-16.

Ram, D. and Singh, P. (1995). Low temperature and high vapour pressure deficit risks to cool season crops in the sub-tropics. Annals of Arid Zone 34 (4): 267-271.

Rao, B.M., Thapliyal, P.K., Pal, P.K., Manikiam, B. and Dwivedi, A. (2001). Large scale soil moisture estimation using microwave radiometer data. Journal of Agrometeorology 3 (1, 2): 179-187.

Rao, K.H.V.D., Kumar, C.S.K. and Prasad, H. (2001). Irrigation water requirement and supply analysis in Dehradun region—An integrating remote sensing and GIS approach. Journal of the Indian Society of Remote Sensing 29 (1, 2): 59-67.

Rao, P.P., Nageshwara, M.V., Rao, K. and Ayyangar, R.S. (1985). Infrared leaf senescence of rice crop and its relationship with grain yield. Journal of Indian Society of Remote Sensing **10**: 1-18.

Rao, P.S., Rathore, L.S., Gillespie, T.J. and Kushwaha, H.S. (1999). Estimating potato crop wetness duration from agrometeorological data. Mausam **50**: 71-76.

Rao, P.S., Saseendran, S.A., Rathore, L.S. and Bahadur, J. (1996). Medium range weather forecasts in India during Monsoon, 1994. Meteorol. Appl. **3**: 317-324.

Rastogi, A., Kalra, N., Aggarwal, P.K., Sharma, S.K., Harit, R.C., Navalgund, R.R. and Dadhwal, V.K. (2000). Estimation of wheat leaf area index from satellite data using price model. Intl. J. Remote Sensing **21** (15): 2943-2949.

Rataul, H.S. (1959). Studies on the biology of cabbage butterfly. Indian J. Hort. **16**: 255-265.

Rathore, L.S., Mendiratta, N. and Singh, K.K. (1998). Soil moisture prediction under maize in sandy loam. Annals of Arid Zone **37** (1): 47-52.

Rathore, L.S., Singh, K.K. and Saseendran, S.A. (2001). Modeling the impact of climate change on rice production in India. Mausam **52**: 263-274.

Rathore, L.S., Srinivasan, G. and Singh, K.K. (1994). Soil moisture and evapotranspiration simulations for irrigated wheat using soil-plant-atmosphere-water (SPAW) model. Mausam **45** (1): 63-68.

Raut, S., Sarma, K.S.S. and Das, D.K. (2001). Evaluation of irrigation management in a canal command area based on agrometeorological and remote sensing. Journal of the Indian Society of Remote Sensing **29** (4): 225-228.

Ray, S.S., Pokharna, S.S. and Ajai (1999). Cotton yield estimation using agrometerological model and satellite derived spectral profile. International Journal of Remote Sensing (In Press).

Reddy, M.V., Sharma, S.B. and Nene, Y.L. (1990). Pigeonpea disease management. In: The pigeonpea, (eds.) Nene, Y.L., Hall, S.D. and Sheila, V.K. pp.310-316. CAB International U.K. and ICRISAT, Patancheru, A.P., India.

Ritchie, J.T. (1991). Specification of the ideal model for predicting crop yields. In: Climatic Risk in Crop Production: Models and Management for the Semiarid Tropics and Subtropics, (eds.) Muchow, R.C. and Bellamy, J.A. Chapter 6. pp.97-122. Wallingford; C.A.B. International.

Ritchie, J.T. and Algarswamy, G. (2003). Model concepts to express genetic differences in maize yield components. Agronomy J. **95**: 4-9.

Ritchie, J.T. et al. (1989b). CERES-Maize V2.0. As quoted in Ritchie, J.T. 1991: Specification of the ideal model for predicting crop yields. In: Climatic Risk in Crop Production: Models and Management for the Semiarid Tropics and Subtropics, (eds.) Muchow, R.C. and Bellamy, J.A. Chapter 6. pp.97-122. Wallingford; C.A.B. International.

Ritchie, J.T., Alocilja, E.C., Singh, U. and Uehara, G. (1986). IBSNAT and the CERES – Rice model. In: Proceedings of the Workshop on the Impact of Weather Parameters on Growth and Yield of Rice. April 1986, Los Banos: IRRI.

Ritchie, J.T., Godwin, D.C. and Otter-Nacke, S. (1985). CERES-Wheat. A Simulation Model of Wheat Growth and Development. Texas A & M University Press, College Station.

Ritchie, J.T., Johnson, B.S., Otter-Nacke, S. and Godwin, D.G. (1989a). Development of a Barley Yield Simulation Model. Final progress report, United States Department of Agriculture No. 86-CRSR.2-2867. East Lansing ; Michigan State University.

Ritchie, J.T. and Algarswamy, G. (1989). Simulation of Sorghum and Pearl Millet. Research Bulletin No.12. pp.24-26. ICRISAT, Patancheru.

Robertson, G.W. (1968). A bio-meteorological time scale for a cereal crop involving day and night temperature and photoperiod. International Journal of Biometeorology. **12**: 191-223.

Romig, R.W. and Calpouzos, L. (1970). The relationship between stem rust and loss of yield of spring wheat. Phytopathology **60**: 1376-1380.

Rosenthal, W.D., Vanderlip, R.L., Jackson, B.S. and Arkin, G.F. (1989). SORKAM: A Grain Sorghum Crop Growth Model. Computer Software Documentation Series MP 1669. College Station: Texas Agricultural Experiment Station.

Rout, G. and Senapati, B. (1968). Biology of mustard aphid, *Lipaphis erysimi*, in India. Ann. Ent. Soc. Am. **61**: 259-261.

Sakamoto, C.M.(1978). The Z-index as a variable for crop yield estimation. Agricultural Meteorology **19**: 305-313.

Sall, M.A. (1980). Epidemiology of grape powdery mildew: A model. Phytopathology, **70**: 338-342.

Samuel, C.K. (1942). Biological notes on two egg parasites of *Bagrada picta* Fabr. Pentatomidae. Indian J. Ent. **4**: 92-93.

Sankaran, V.M., Aggarwal, P.K. and Sinha, S.K. (2000). Improvement in wheat yields in northern India since 1965: Measured and simulated trends. Field Crops Res. **66**: 141-149.

Sarangi, R.K., Chauhan, P. and Nayak, S. (2001). Chlorophyll A concentration along West Coast of India using IRS-P3 MOS – B data. Journal of the Indian Society of Remote Sensing **29** (4): 197-202.

Saseendran, S.A., Rathore, L.S. and Datta, R.K. (1996). Distribution of monsoon rainfall in India during *El Nino* associated drought situations. Annals of Arid Zone **35** (1): 9-16.

Saseendran, S.A., Singh, K.K., Rathore, L.S., Rao, G.S.L.H.V.P., Mendiratta, N., Lakshminarayan, K. and Singh, S.V. (1998a). Evaluation of the CERES- Rice version 3.0 model for the climatic conditions of the state of Kerala, India. Meteorol. Appl. **5**: 385-392.

Saseendran, S.A., Hubbard, K.G., Singh, K.K., Mendiratta, N., Rathore, L.S. and Singh, S.V. (1998b). Optimum transplanting dates for rice in Kerala, India, determined using both CERES V.3.0 and Clim Prob. Agron. J. **90**: 185-190.

Sathe, P.V. and Jadhav, N. (2001). Retrieval of chlorophyll from the sea-leaving radiance in the Arabian sea. Journal of the Indian Society of Remote Sensing **29** (1, 2): 97-106.

Sehgal, M. (2003). Integrated pest management of rice in rice-wheat cropping systems: RWC-CIMMYT. 2003. Addressing Resource Conservation Issues in Rice-Wheat-System of South Asia: A Resource Book. Rice-Wheat consortium for Indo-Gangetic Plains - International Maize and Wheat Improvement Centre, New Delhi, India. 305p.

Sehgal, M., Jeswani, M.D. and Kalra, N. (2001a). Management of insect, disease and nematode pests of rice and wheat in Indo-Gangetic Plains. J. of Crop Production 4 (1): 167-226.

Sehgal, M., Jeswani, M.D. and Kalra, N. (2001b). Management of insect, disease and nematode pests of rice and wheat in the Indo-Gangetic plains. In: the Rice-Wheat cropping systems of South Asia: Efficient production management, (ed.) Kataki, P.K., Food Production Press, pp.167-226.

Seif, E. and Pederson, D.G. (1978). Effect of rainfall on the grain yield of spring wheat, with an application. Australian Journal of Agricultural Research **29**: 1107-1115.

Sethi, S.L. and Atwal, A.S. (1963). Influence of temperature and humidity on the development of different stages of lady-bird-beetle, *Coccinella septempunctata* L. (Coleoptera: Coccinellidae). Indian J. Agric. Sci. **34**: 166-171.

Shaner, G.E., Peart, R.M., Newman, J.E. and Stirm, W.L. (1971). EPIMAY: An Evaluation of a Plant Disease Display Model. Purdu Univ. Agric. Exptl. Stan. 14pp. West Lafayette, Indiana.

Sharma, A.K. and Subbarao, B.R. (1964). A further contribution to the knowledge of the taxonomy and biology of Aphididae (Ichneumonoidae: Hymenoptera) with particular reference to Indian forms. Indian J. Ent. **26**: 458-460.

Sharma, P.K., Chaurasia, R. and Mahey, R.K. (2000). Wheat production forecasts using remote sensing and other techniques—experience of Punjab State. Supplement to Indian Journal of Agricultural Economics, **55** (2): 68-80.

Sharma, T., Sudha, K.S., Ravi, N., Navalgund, R.R., Tomar, K.P., Chakravarti, N.V.K. and Das, D.K. (1993). Procedures for wheat yield prediction using Landsat MSS and IRS-1A data. International Journal of Remote Sensing **14**, (13): 2509-2519.

Sheehy, J.E., Cobby, J.M. and Ryle, G.J.A. (1980). The use of a model to investigate the influence of some environmental factors on the growth of perennial ryegrass. Annals of Botany, **46**: 343-365.

Sidhu, H.S. and Singh, S. (1964). Biology of the mustard aphid *Lipaphis erysimi* Kalt. in Punjab. Indian Oilseeds J. **8**: 348-359.

Singh, H. and Malik, V.S. (1993). Biology of painted bug, *Bagrada ruciferarum*. Indian J. Agric. Sci **63** (10): 672-674.

Singh, H., Kalra, V.K. and Rohilla, H.R. (1990). Effect of nitrogenous fertilizer on the development of mustard aphid, *Lipaphis erysimi* (Kalt.). J. Aphidol. **4** (1-2): 6-8.

Singh, K.K., Rathore, L.S., Attri, S.D. and Baxla, A.K. (2000). Water management for wheat through soil moisture simulation using SPAIN model. Annals of Agri-Bio Research **5** (2): 121-126.

Singh, P. (1980). Influence of host plant on the biology of the pea leaf-miner, *Phytomyza horticola* Goureau. M.Sc. Thesis, Punjab agric univ., Ludhiana, 90pp.

Singh, P. (1995a). Sensitivity of crop productivity to climate change. In: Impact of Modern Agriculture on Environment. (eds.) Arora, Behl, Tauro and Joshi. **3**: 1-7. New Delhi: Soc. Sust. Agri. and N.R.M. and Max Mueller Bhavan.

Singh, P. (1995b). Sensibility of crop productivity of morpho-physiological variation for characters in mungbean (*Vigna radiata* L.) under drought conditions. Proceedings, 2nd European Conference in Grain Legume, 9-13 July, 1995, pp.484. Copenhagan (Denmark).

Singh, P. and Ram, D. (1993). Modeling of renewable resources for agricultural research and development. Intern. J. Trop. Agric. **XI** (3): 187-197.

Singh, P. and Ram, D. (2000). Utilization of photoperiod and solar radiation resources in further enhancement of genetic crop yield potential in south, southeast and far-east Asia. Natnl. J. Pl. Improv. **2**: 5-10.

Singh, P. and Singh, I. (2000). Sensitivity of morpho-physiological variation for characters under global environmental change. Natnl. J. Pl. Improv. **2**: 87-88.

Singh, P. and Upadhyaya (2001). Biological interaction in tropical grassland ecosystem. In: Structure and Function in Agro-Ecosystem Design and Management. (eds.) M. Shiyomi and H. Koizumi, pp.113-143. London: CRC Press.

Singh, P., Morgan, J.M., Pal, R., Singh, D.P. and Sharma, H.C. (1982). Predicting wheat yield and its components in a semiarid region. Trans Isdt & Ucds, **7** (2): 9-14.

Singh, P., Pal, R., Singh, D.P. and Singh, V.P. (1989). Validation of SIMWHEAT model in a semi-arid region. Indian J. Ecol. **16** (1): 25-29.

Singh, P., Jarwal, S.D. and Ram, D. (1992). Potential crop biomass conversion, solar radiation and effective photoperiod to phenology at Hisar (India) and Reading (United Kingdom). In: Proceedings of 2nd World Renewable Energy Congress Reading, United Kingdom, September 13-18, 3: 1474-1478.

Singh, P., Boote, K.J., Rao, A.Y., Iruthayaraj, M.R., Sheikh, A.M., Hundal, S.S., Narang, R.S. and Singh, P. (1994). Evaluation of the groundnut model PNOTGRO for crop response to water availability, sowing dates, and seasons. Field Crops Research 39: 147-162.

Singh, P., Mehta, A.K., Ram, D. and Jarwal, J.D. (1996). Potential and water limited production of mungbean in south, southeast and far-east Asia. Haryana. Agric. Univer. J. Res. 26: 23-33.

Singh, R. and Malhotra, R.K. (1979). Some studies on biology of *Coccinella undecimpunctata memetresi* Muls, a predator of mustard aphid. Curr. Sci. 48 (2): 904-905.

Singh, R., Goyal, R.C., Saha, S..K. and Chhikara, R.S. (1992). Use of satellite spectral data in crop yield estimation surveys. International Journal of Remote Sensing 14: 2583-2592.

Singh, R.P. and Kogan, F. (2002). Monitoring vegetation condition from NOAA Operational Polar–Orbiting Satellites over Indian region. Journal of the Indian Society of Remote Sensing 30 (3): 117-118.

Singh, R.P., Dhadhwal, V.K. and Navalgund, R.R. (1999). Wheat crop inventory using high spectral resolution IRS – P3 MOS B Spectrometer data. Journal of the Indian Society of Remote Sensing 27 (3): 167-173.

Singh, S.P. and Verma, A.N. (1988). Antibiosis mechanism of resistance to stem borer, *Chilo partellus* (Swinhoe) in sorghum. Insect Sci. Applic. 9 (5): 579-582. (Great Britain).

Singh, S.P., Bishnoi, O.P., Niwas, R. and Singh, M. (2001). Relationship of crop yield with spectral indices. Journal of the Indian Society of Remote Sensing 29 (1, 2): 93-96.

Singh, S.V., Datta, R.K., Rathore, L.S. and Rao, P.S. (1999). Role of weather forecasts in dryland agriculture. In: Fifty years of Dryland Agricultural Research in India, H.P. Singh, Y.S. Ramkrishna, K.L. Sharma and B. Venkateswarlu (eds.), Central Research Institute for Dryland Agriculture, Hyderabad, India, pp.227-234.

Sridhar, V.N., Dadhwal, V.K., Chowdhary, K.N., Sharma, R., Bairagi, G.D. and Sharma, A.K. (1994). Wheat production forecasting for predominantly unirrigated region in Madhya Pradesh (India). International Journal of Remote Sensing 15: 1304-1316.

Srivastava, A.S. and Srivastava, J.L. (1961). Note on the life history of mustard aphid, *Lipaphis erysim* Kault). Proc. Natn. Acad. Sci. India (B), 31: 422-424.

Srivastava, A.S., Siddiqui, M.S. and Saxena, H.P. (1965). Note on the life and seasonal history of *Tanymecus indicus* Fst. (Coleoptera: Curculionidae), a serious pest of wheat crop. Labdev J. Sci. Technol. 3: 61-62.

Stapper, M. (1984). SIMTAG: A simulation model of wheat genotypes. Armidale: University of New England, Department of Agronomy and Soil Science.

Stapper, M. and Arkin, G.F. (1980). CORNF. A Dynamic Growths and Development Model for Maize (*Zea mays* L.). Program and Model Documentation No.80.2. College Station; Texas Agricultural Experiment Station.

Stewart, D.W., Cober, E.R. and Bernard, R.L. (2003). Modeling genetic effects on the photothermal response of soybean phenological development. Agron. J. 95: 65-70.

Streifer, W. (1974). Realistic models in population biology. Advances in Ecological Research, **8**: 199-266.

Sweeney, D.G., Hand, D.W., Slack, G. and Thornley, J.H.M. (1981). Modeling the growth of winter lettuce. In: Mathematics and Plant Physiology, (eds.) D.A. Rose and D.A. Charles Edwards, pp.217-229. London: Academic Press.

Teng, P.S. (1980). Exploratory and optimization computer experiments for designing management systems of barley leaf rust. Phytopathology, **71**: 260.

Teng, P.S. (1985). A comparison of simulation approaches to epidemic modeling. Ann. Rev. Phytopathol. **23**: 351-379.

Teng, P.S., Blackie, M.J. and Close, R.C. (1977). A simulation analysis of crop yield loss due to rust disease. Agr. Systems **2**: 189-198.

Teng, P.S., Blackie, M.J. and Close, R.C. (1978). Simulation of Barley Leaf Rust: Structure and Validation of BARSIM, I (with editors, Ag. Systems).

Thornley, J.H.M. (1983). Crop yield and planting density. Annals of Botany **52**: 257-259.

Tripathi, M.P., Panda, R.K., Pradhan, S. and Sudhakar, S. (2000). Runoff modeling of a small watershed using satellite data and GIS. Journal of the Indian Society of Remote Sensing **30** (1, 2): 39-52.

Tucker, C.J., Holden, B.N., Elgin JR., G.H. and Mc Murtrey, J.E. III (1980). Remote sensing of total dry matter accumulation in winter wheat. Remote Sens. Environ **11**: 267-277.

Van der Plank, J.E. (1963). Plant Diseases: Epidemic and Control, New York: Academic Press.

Verma, R.M. (1945). Prepupal and pupal changes in the mustard sawfly, *Athalia proxima* Klug. Indian J. Ent. **7**: 238.

Waggoner, P.E. (1974). Simulation of epidemic. In: Epidemics of Plant Disease: Mathematical Analysis and Modeling, (ed.) J. Kranz, pp.137-160. London: Chapman and Hall.

Waggoner, P.E. and Horsfall, J.G. (1969). EPIDEM: A simulator of plant disease written for computer. Conn. Agr. Exp. Sta. Bull. **698**. pp.80.

Waggoner, P.E., Horsfall, J.G. and Lukens, R.J. (1972). EPIMAY: A simulator of southern corn leaf blight. Conn. Agr. Exp. Sta. Bull. **729**. 87 pp.

Wann, M., Raper, C.D. and Lucas, H.L. (1978). A dynamic model for plant growth: A simulation of dry matter accumulation for tobacco, Phytosynthetica **12**: 121-136.

Weiss, A. (2003). Symposium papers: Introduction. Agronomy J. **95**:1-3.

Welch, S.M., Wilkerson, G., Whiting, K., Sun, N., Vagts, T., Buol, G. and Mavromatis, T. (2000a). Estimating genetic coefficients for a soybean growth model. ASAE paper. 00-3043, ASAE, St. Joseph, MI.

Welch, S.M., Zhang, J., Sun, N. and Yu Mak, T. (2000b). Efficient estimation of genetic coefficients for crop models [CD-ROM]. In: Bowen et al. (ed.) Proc. SAAD - 3 Symp., Universidad Nacional Agraria, La Molina (UNALM), Lima, Peru, 8-10 Nov. 1999, Int. Potato Cent, Lima, Peru.

Welch, S.M., Roe, J.L. and Dong, Z. (2003). A genetic neural network model of flowering time control in *Arabidopsis thaliana*. Agron. J. **95**: 71-81.

Welch, S.M., Jones, J.W., Brennan, M.W., Reeder, G. and Jacobson, B.M. (2002). PC yields: Model based decision support for soybean production. Agric. Syst. **74**: 79-98.

White, J.W. and Hoogenbooms, G. (2003). Gene-based approaches to crop simulation. Past experiences and future opportunities. Agron. J. **95**: 52-64.

Wilkerson, G.G., Jones, J.W., Boote, K.J. and Mishoe, J.W. (1985). SOYGRO V5.0 Soybean Crop Growth and Yield Model. Technical documentation. Gainesville: Department of Agricultural Engineering, University of Florida.

Wilkerson, G.G., Jones, J.W., Boote, K.J., Ingram, K.T. and Mishoe, J.W. (1983). Modeling soybean growth for crop management. Transactions of the American Society of Agricultural Engineers **26**: 63-73.

Willey, R.W. and Heath, S.B.(1969). Plant population and crop yield. Advances in Agronomy **21**: 281-321.

Williams, E.J. (1962). The analysis of competition experiments. Aust. J. Biol. Sci. **15**: 509-525.

Williams, J.R., Jones, C.A. and Dyke, P.T. (1984). A modeling approach to determining the relationship between erosion and soil productivity. Transactions of the American Society of Agricultural Engineers **27**: 129-144.

Williamson, M.H. (1972). The Analysis of Biological Population. London: Arnold.

Wit, C.T. DE and Goudriaan, J. (1978). Simulation of Ecological Processes. Simulation Monograph, 175p. Wageningen: PUDOC.

Wit, C.T. DE and Others. (1978). Simulation of assimilation, respiration and transpiration of crops, Wageningen: PUDOC.

Zadoks, J.C. (1984). EPIPRE: A computer-based scheme for pest and disease control in wheat. In: Cereal production, (ed.) E.J. Gallagher, Chapter 25, pp.215-225. London; Butterworths.

Zadoks, J.C., Chang, T.T. and Konzak, C.F.(1974). A Decimal Code for the Growth Stage of Cereals. EUCARPIA Bull. No. 7.

Index

Epilog

"I could come across only a small pebble
on the vast shore of knowledge".

Isac Newton

"Those, who do not know that they do not know, are ignorant.
But they, who know that they do not know, feel
themselves more ignorant."

Ishopnishad

"The more our knowledge increases the
more our ignorance unfolds."

John F. Kennedy

T - #0450 - 101024 - C0 - 234/156/29 - PB - 9781578084180 - Gloss Lamination